CLASSIFICATION OF PLANT COMMUNITIES

CLASSIFICATION OF PLANT COMMUNITIES

edited by

Robert H. Whittaker

Dr W. Junk bv Publishers The Hague, Boston 1978

ISBN 90 6193 566 0
© Dr. W. Junk b.v. - Publishers - The Hague 1978
Cover design Max Velthuijs

PREFACE

The natural communities of the world are diverse, and many schools of ecology have developed classifications of communities in partial independence of one another. There is consequently a vast and widely dispersed literature on the classification of plant and animal communities, comprising divergent approaches of different schools and representing a great experiment on the usefulness of different possibilities for classification. The editor sought in a review monograph of 1962 to summarize these schools and their history, and in 1973 published a treatise on 'Ordination and Classification of Communities' as volume 5 of the *Handbook of Vegetation Science*. We were fortunate, in preparing the latter work, to have a truly international panel of authors to discuss different major approaches to classification.

This second edition of the book of 1973 is intended to make the work more widely available in a less expensive form as companion volumes on ordination and on classification of plant communities. We have not sought to revise the chapters on classification but consider that these are articles of lasting value. This volume offers the reader an introduction to concepts about, and approaches to, classifying natural communities, a survey of the different techniques that have developed and might be of interest to him, and more detailed descriptions and scholarly reviews of the most important approaches. In the survey the reader may recognize how the diversity of communities and interesting questions about them have led to different approaches, each with its value and contribution to ecology as a science.

CONTENTS

12 Approaches to Classifying Vegetation, by ROBERT H. WHITTAKER. 1
13 The Physiognomic Approach, by JOHN S. BEARD 33
14 Dominance-Types, by ROBERT H. WHITTAKER. 65
15 The Finnish School and Forest Site-Types, by TOOMAS E. A. FREY 81
16 Synusial Approaches to Classification, by JAN J. BARKMAN 111
17 Russian Approaches to Classification, by VERA D. ALEKSANDROVA 167
18 North European Approaches to Classification, by HANS TRASS and NILS MALMER 201
19 Numerical Classification, by DAVID W. GOODALL. . . . 247
20 The Braun-Blanquet Approach, by VICTOR WESTHOFF and EDDY VAN DER MAAREL 287

Index 400

(A companion volume on Ordination of Plant Communities comprises articles 1 to 11.)

Authors and addresses

VERA D. ALEKSANDROVA
  Komarov Botanical Institute, Popova Str. 2, Leningrad 22, U.S.S.R.
JAN BARKMAN
  Laboratorium voor Plantensystematiek en -Geographie van de Landbouwhogeschool, Biologisch Station, Kampsweg 27, Wijster (Dr.), The Netherlands
JOHN S. BEARD
  6 Fraser Road, Applecross, Western Australia
TOOMAS E. A. FREY
  Institute of Zoology and Botany, 21 Vanemuise St., 202400 Tartu, Estonian S.S.R., U.S.S.R.
DAVID W. GOODALL
  CSIRO Division of Land Management, Private Bag, Wembley, W. A. 6014, Australia
EDDY VAN DER MAAREL
  Botanical Laboratory, Geo-Botany Division, University of Nijmegen, Toernooiveld, Nijmegen, The Netherlands
NILS MALMER
  Department of Plant Ecology, Lunds University, Östra Vallgatan 14, 223 61 Lund, Sweden
HANS TRASS
  Department of Plant Taxonomy and Ecology, Tartu State University, 40 Michurin St., Tartu, Estonian S.S.R., U.S.S.R.
VICTOR WESTHOFF
  Botanical Laboratory, Geo-Botany Division, University of Nijmegen, Toernooiveld, Nijmegen, The Netherlands
ROBERT H. WHITTAKER
  Ecology and Systematics, Cornell University, Ithaca, N.Y. 14853, U.S.A.

12   APPROACHES TO CLASSIFYING VEGETATION

Robert H. Whittaker

Contents

12.1	The Basis of Classification	3
12.1.1	Class Concepts	3
12.1.2	Natural vs. Artificial Units	4
12.1.3	Implications	6
12.2	History: The Traditions	6
12.2.1	The Physiognomic Tradition	7
12.2.2	The Northern Tradition	7
12.2.3	The Southern Tradition	8
12.2.4	The Russian Tradition	9
12.2.5	The British Tradition	9
12.2.6	The American Tradition	10
12.3	Major Approaches	11
12.3.1	Physiognomic Units	11
12.3.2	Environmental Units	11
12.3.3	Landscape Units	12
12.3.4	Biotic Areas	13
12.3.5	Zones and Series	14
12.3.6	Species Dominance	15
12.3.7	Vegetation Dynamics	15
12.3.8	Stratal Units	16
12.3.9	Sociations	16
12.3.10	Forest Site-Types	17
12.3.11	Numerical Classification	17
12.3.12	Floristic Units (of Braun-Blanquet)	18
12.4	Summary	19

12 APPROACHES TO CLASSIFYING VEGETATION

12.1 The Basis of Classification

12.1.1 CLASS-CONCEPTS

Plant communities are classified through a process of interaction between a phytosociologist and vegetation (WHITTAKER 1962). The phytosociologist moves through the landscape and observes that certain kinds of communities repeat themselves: similar combinations of species appear in similar biotopes at a number of points in the landscape. Similar communities, once observed, are grouped together in the phytosociologist's memory as an informal, preliminary conception of a community-type. The manner in which the phytosociologist groups communities, and frames his conception of a type, is likely to be influenced by what he thinks is most important and interesting about these communities. He may also change his conception of a type as further observations influence his interpretation of it. In time, however, the phytosociologist wishes to abstract from his conception a more formal definition of a community-type which can be communicated to others. This definition will describe a *class*, a grouping of individuals or observations by their shared characteristics. In this case the class unites a number of stands, or particular communities in the field (or samples representing these), that are alike in ways the phytosociologist regards as significant.

The formal definition is a *class-concept* which should, preferably, be simply and clearly stated. The phytosociologist will state the class-concept according to his interests and view of what is important for classification. If he is most interested in vegetation stucture, he may state a definition of a formation: Forests dominated by needle-leaved trees in subarctic and subalpine climates are members of community-type (formation) A. If his interests center on floristic composition, he may define an association: Plant communities in which several of the following character-species ... occur together belong to community-type (association) B. If he is concerned mainly with dominant species, he may define a dominance-type: Grasslands in which the two species a and b are more important than any others represent community-type (dominance-type) C. Such statements

3

of class-concepts are also intensional definitions of community-types. The particular communities in the field for which a defining statement becomes true, which therefore conform to the class-concept, constitute its extensional definition. A phytosociological table giving the species composition of a number of observed communities that fit the definition of Association B is thus an approach toward extensional definition of that association. The intensional definition is an abstraction narrowly limited to some most essential characteristics of the association. A community table as an approach to extensional definition is a much less selective abstraction that communicates many of the characteristics of the community-type as these are shared by, or differ between, the vegetation samples compiled in the table.

12.1.2 NATURAL VS. ARTIFICIAL UNITS

At an earlier period ecologists and phytosociologists were agitated by debate on whether associations and other community-types were real or abstract, artificial or natural units; some of the history of such debate has been reviewed elsewhere (WHITTAKER 1962). The first issue, of the community-type as a reality or abstraction, is the more easily resolved. A community-type, as a class-concept, is inescapably an abstraction. This abstraction is a reality as such in its psychological and cultural context. As such it is to be clearly distinguished from the plant communities in the field, that are also accepted as realities, as phenomena subject to observation and abstraction. The second issue, of the community-type as a natural or artificial unit may require somewhat more consideration, for it leads to the very basis of the problems of classifying vegetation (WHITTAKER 1956, 1962).

The question must be stated in terms of relationships between species and plant communities being classified before it can be answered. To consider the kinds of relationships that characterize a 'natural' unit, we may take an example from a different area of study — the 'good' species in the taxonomy of individual organisms. A large number of characteristics are shared by members of the species and distinguish those members from the members of other species. Individuals with characteristics intermediate between species A and other species (i.e. hybrids) are absent or rare. The species as a class is thus relatively homogeneous within itself, and discontinuous with other, comparable classes. Genetic exchange among its members by sexual reproduction, and genetic or other barriers to hybridization, maintain its internal homogeneity and external discontinuity. Because many of the characteristics of its members are correlated with one another, different taxonomists, chosing different characteristics for their class-concepts of the species, might come to the same extensional definitions — they might group the same individual specimens into the species by different characteristics. The species as a unit is then 'natural,' for the characteristics of its members have a relatively strong effect in determining the membership and boundary of the class in extension, in relative independence of the characteristics that particular taxonomists choose for emphasis in their intensional definitions.

With plant communities, it is not thus. We have commented (articles 3,5) on the two principles stated by RAMENSKY (1926) and GLEASON (1926). (i) Species are distributed 'individualistically,' each according to its own way of relating to environment, hence no two alike. Species do not fit naturally into groupings that correspond to community-types and are discontinuous with other such groupings. Ecological groups and character-species groups are primarily arbitrary groupings of species by similarity of distributional relationships; the limits of such a grouping and the number of species it is to comprise must be decided by the ecologist. Because individualistic species are combined with one another in varied combinations and proportions in communities, the phytosociologist must choose what combinations he is to recognize as associations or other community-types. (ii) Plant communities often (usually, if not affected by different disturbances or by environmental discontinuity) intergrade continuously. Boundaries of community-types are in consequence often arbitrary. The number of community-types into which a vegetational continuum is divided, and the locations of the boundaries of these types, are determined by the phytosociologist's choice of the characteristics he uses to define those types. In these respects plant communities are widely different from 'good' species. Community-types are not natural but arbitrary units in the sense that their extensional definitions are strongly influenced if not wholly determined by phytosociologists' choices of the characteristics by which communities are to be classified.

Classifications are further affected by the relative independence of these different characteristics (see article 3). Not only species but also groupings of species by growth-forms, life-forms, or strata show relative independence of one another, and may be differently combined into particular communities. Each growth-form has a different relation to environments, and a different area of those environments in which it is important or dominant, from any other growth-form (WHITTAKER & NIERING 1965). Over-all community characteristics such as total coverage, biomass, productivity, and species-diversity are differently related to environmental gradients. Plant communities are complex, many-faceted phenomena whose many characteristics are in some respect functionally interrelated, and yet are not simply correlated with one another.

12.1.3 IMPLICATIONS

The continuity of communities and the relative independence of their characteristics have major implications for principles in the classification of communities (WHITTAKER 1962). Among these implications:

i) There are many possible approaches to classifying communities, emphasizing different kinds of characteristics as bases of classification.

ii) The community-types, or units, that result from different approaches may not correspond in their boundaries and memberships. Units of one system may bear no simple relation to those of another. This fact makes interesting the study of comparative classifications — application of several methods to a given problem, so that the effects of these approaches may be compared (MILLER 1951, ELLENBERG 1967).

iii) No one approach to classification can claim exclusive merit. Different approaches, based on different characteristics of communities, may reveal different relationships of interest. Different approaches to classification are to be judged by their efficiency, utility, and productiveness for the study of different kinds of vegetation by ecologists or phytosociologists of different interests.

iv) There may thus be merit, as English-language ecologists have often felt (WHITTAKER 1962, article 14), in an open-minded and informal approach to classification, seeking for each vegetation area and research purpose a most useful classification.

v) There is also value, as many phytosociologists would stress, in a standardization of the approach to classification by many researchers (BRAUN-BLANQUET 1951, article 20). Such standardization favors the efficient collection, organization, and communication of information about plant communities and makes possible the effective relation of the research of each individual to that of many others.

12.2 History: the Traditions

Because of the varied possibilities for classifying vegetation many schools have developed, distinctive in their responses to the vegetation problems of different areas but complexly interrelated by sharing of ideas, forming a kind of braided stream of phytosociological history. The history of schools has been reviewed elsewhere (WHITTAKER 1962) and is to be treated in another part of the *Handbook*; it cannot be given in detail here. We shall outline only briefly some aspects of that history involving classification, in terms of the major traditions of ecology and phytosociology (WHITTAKER 1962, cf. ALEKSANDROVA 1969, SHIMWELL 1971).

12.2.1 THE PHYSIOGNOMIC TRADITION

The origins of scientific classification of vegetation appear in the work of early students of plant geography, notably HUMBOLDT (1807) and GRISEBACH (1838). From HUMBOLDT may be traced the idea of growth-forms as major types of plants characterizing communities and of associations of plants, from GRISEBACH the concept of the formation as a distinctive kind of plant community, characterized by its growth-forms. Growth-forms determine the visible structure, or physiognomy, of plant communities. Vegetation study in which physiognomy and its relations to climate are central may be considered part of the Physiognomic Tradition. From its origins in Europe the physiognomic approach spread to students on all continents and a wide range of work in plant ecology and phytosociology, geography, and climatology. The Physiognomic Tradition thus both was parental to the five regional traditions, and is still a major approach to vegetation in its own right (article 13).

12.2.2 THE NORTHERN TRADITION

Emphasis of physiognomy was a guiding influence in the early development of the Northern Tradition of phytosociology in the Scandinavian and Baltic area. The northern authors were dealing with vegetation mostly rather poor in species, but in which the layers or strata (e.g. the tree, shrub, herb, and moss strata of a forest) were conspicuous. As schools of more intensive vegetation research developed in the North, concern shifted from broad physiognomic relations to study of communities through their strata and the major or dominant species of these. Three directions characterized different major schools: (i) Vegetation was classified by the dominant species of strata (and the sociation as a unit) in the school of Uppsala (HULT 1881, SERNANDER 1898, FRIES 1913, DU RIETZ 1921, article 18.) (ii) Treatment of the strata themselves as units (unions) was developed by GAMS (1918) in Austria, LIPPMAA (1939) in Estonia, and others (article 16). (iii) The undergrowth strata of forests were used as the basis of classification and indication of site quality by the Finnish School of CAJANDER (1909, 1949, article 15). The Northern Tradition included several other schools, among them MARKUS' (1930) study of nature-complexes in Estonia, the Danish schools of RAUNKIAER (1934) and of BÖCHER (1933, 1954), Icelandic work relating narrowly defined formations to biotopes and environmental gradients (HANSEN 1930, STEINDÓRSSON 1954), and the Dutch grassland research of DE VRIES (1948). Vegetation research in the Northern Tradition has been both distinctive and diverse in its response to the properties of northern vegetation (article 18), but many recent authors replace the northern approaches by, or fit them into, those of the Southern Tradition's school of BRAUN-BLANQUET.

12.2.3 THE SOUTHERN TRADITION

In its earlier period the Southern Tradition was centered in the cities of Zürich and Montpellier under the leadership of SCHRÖTER (SCHRÖTER & KIRCHNER 1902), and FLAHAULT (1893, FLAHAULT & SCHRÖTER 1910). The largest school in the tradition, that of BRAUN-BLANQUET, is often termed the Zürich-Montpellier or Swiss-French School; but other schools in those cities have included that of BROCKMAN-JEROSCH & RÜBEL (1912, RÜBEL 1930) in Zürich, concerned with physiognomic relations of vegetation, SCHMID's (1941, 1952, 1961) approach in Zürich through regional floristic units, and the emphasis of succession and habitat relations of LÜDI (1921, 1948) in Zürich, and of KUHNHOLZ-LORDAT (1952) in Montpellier. The Southern Tradition in a broader sense includes other distinctive work such as that of MEUSEL (1939a, 1954) in Germany, concerned with landscape interpretation and chorological relations, AICHINGER's (1952, 1954) system for treating vegetational dynamics in Austria, GAUSSEN's (1933, 1951) approach to vegetation mapping in France, and the work of NEGRI (1927, 1954) and others in Italy and of VILLAR (1929) in Spain. The center of the greatest single school of phytosociology, that of BRAUN-BLANQUET (1913, 1921, 1951) has been at Montpellier, at the 'Station Internationale de Géobotanique Mediterranéene et Alpine'; but the 'Zentralstelle für Vegetationskartierung' at Stolzenau, Germany, became under TÜXEN (1937, 1968b, 1970) a second center. The school of BRAUN-BLANQUET has been characterized by concern with full floristic composition of vegetation, and by the creation for community taxonomy of a formal hierarchy of which the association is the basic unit (article 20). Approaches concerned with gradient relations and continuities of communities, related to American gradient analysis, have been developed by ELLENBERG (1950, 1952), DAGNELIE (1960), MAAREL (1969, MAAREL & LEERTOUWER 1967), and MOORE et al. (1970). The school has produced an enormous literature on the vegetation of Europe and has established beachheads of application by investigators on other continents.

12.2.4 THE RUSSIAN TRADITION

Early work on vegetation in Russia was physiognomic. As more intensive study of vegetation developed, the concept of the sociation was imported from the school of Uppsala; research was based on this unit (which the Russians term 'association') by the schools of SUKACHEV (1928, 1932) at Leningrad and of ALEKHIN (1926, 1932) and KATZ (1929, 1933) at Moscow. Much Russian work following that of KELLER (DIMO & KELLER 1907, KELLER 1925-6) arranged communities in sequences or ecological series along environmental gradients. Studies of forests by SUKACHEV (1928, 1932) and others, employing sociations as units arranged in ecological series, relates to the Finnish school though the units differ from CAJANDER's sitetypes. SUKACHEV (1954), like his predecessors MOROSOV (1928) and KRUEDENER (1926), conceives of forest communities as landscape units or biogeocoenoses, interrelating the various climatic, topographic, edaphic, biological, and communal factors of forests and their environments. Russian phytocoenology has included also the stratal or synusial approaches developed by KELLER (1932) and GROSSHEIM (1930) and RAMENSKY's (1926, 1930, article 5) emphasis of species individuality and vegetational continua. Other developments in Russia are discussed by ALEKSANDROVA (1969, article 17).

12.2.5 THE BRITISH TRADITION

Early work of Moss (1910) and TANSLEY (1911, 1920) used the formation as a unit but interpreted this in terms of successional processes, not simply as a regional climax unit. Unlike the American author CLEMENTS, TANSLEY (1920, 1939) and most other British ecologists accepted the 'polyclimax' view that there might be a number of stable or climax communities in a given area in addition to the prevailing or climatic climax. TANSLEY (1939) consequently recognizes formations in British vegetation that would be treated as successional by the School of CLEMENTS. Formations were divided into broadly defined 'associations' characterized by their dominant species (TANSLEY 1939, BEARD 1955). British Commonwealth authors have applied the ideas of TANSLEY and CLEMENTS in other parts of the world; but Australian ecology developed indigenous approaches to classification of some interest (WOOD 1939, CROCKER & WOOD 1947, BEADLE 1948, BEADLE & COSTIN 1952). Much Commonwealth ecology has been physiognomic; to the use of formations BEARD (1955, article 13) has added the concept of formation series, an ecological series of formations along a major environmental gradient. Recent developments in British ecology include the work of POORE (1955-6, 1962) approaching vegetation through *noda* conceived as points of reference in vegetation that often intergrades continuously, GIMINGHAM's (1961, 1969) treatment of heath vegetation as a network of variation, the numerical classification through 'association-analysis' and related techniques of WILLIAMS & LAMBERT (1959, LAMBERT & DALE 1964, article 19), and interest in indirect ordination through principal components analysis (GOODALL 1954, AUSTIN & ORLOCI 1966, GREIG-SMITH et al. 1967, YARRANTON 1967, GITTINS 1969, article 10).

12.2.6 THE AMERICAN TRADITION

Concern with vegetation development characterized the early work of COWLES (1899, 1901), CLEMENTS (1905, 1916) and COOPER (1913). In the hands of CLEMENTS (1916, 1936, WEAVER & CLEMENTS 1938, article 14) emphasis of succession and climax became the basis of a distinctive school and system of vegetation study. The basic units were formations and associations as climax communities adapted to the climates of geographic regions (article 14). COUPLAND (1961) applied CLEMENTS' classification to grasslands, and BRAUN (1938, 1947, 1950, 1956) studied the eastern forests with some of the concepts of CLEMENTS employed in a different perspective. Other approaches include the physiognomic systems of DANSEREAU (1951, 1957) and KÜCHLER (1949), article 13, the use of stratal units by CAIN (1936) and BILLINGS (1945) and of the sociation by DAUBENMIRE (1952) and HANSON (1953), use of concepts of BRAUN-BLANQUET by CONARD (1935, 1952), DANSEREAU (1943, 1946), BECKING (1957), and JANSSEN (1967), and application of the Finnish site-types by HEIMBERGER (1934). GLEASON (1926, 1939) dissented from the school of CLEMENTS to assert (like RAMENSKY) the principles of species individuality and vegetational continuity. Research in gradient analysis, approaching vegetation through gradient relations of environment, species populations, and communities was developed by WHITTAKER (1951, 1956, 1960, 1967, articles 2 and 3) and the Wisconsin school (CURTIS & McINTOSH 1951, CURTIS 1959, BRAY & CURTIS 1957, McINTOSH 1967, article 8).

12.3 Major Approaches

Though the schools of ecology and phytosociology are many, the basic major approaches to classification are fewer. We shall distinguish twelve for a brief review of different possibilities in classification, with reference to a few publications representing each. Some of these are given more extended discussion in the articles that follow:

12.3.1 PHYSIOGNOMIC UNITS

The first classifications of vegetation, developed in classic works on plant geography, were based on the physiognomy, or structure, of vegetation. Physiognomy is defined by the structural types of plants, or growth-forms (e.g. grasses, or broadleaved deciduous trees) that dominate or are most conspicuous in communities. A community-type on a given continent defined by growth-form dominance (and major features of environment) is a *formation*. A grouping of similar formations occurring in similar climates of the different continents is a *formation-type*. These concepts have been the basis of extensive investigations in biogeography. Of the rich literature in the area we may mention only a few classic or recent works (GRISEBACH 1872, WARMING 1909, RÜBEL 1930, SCHIMPER & FABER 1935, KÜCHLER 1949, BEARD 1955, DANSEREAU 1957, SCHMITHÜSEN 1959, FOSBERG 1961, and ELLENBERG & MUELLER-DOMBOIS 1967). Ecologists concerned with animals as well as plants have often used as corresponding units the *biome* and *biome-type* (CLEMENTS & SHELFORD 1939, ALLEE et al., 1949, ODUM 1953, KENDEIGH 1954, 1961, TISCHLER 1955). The physiognomic approach is discussed in the following article (13) by BEARD.

12.3.2 ENVIRONMENTAL UNITS

Physiognomic units classify vegetation by structure as an expression of environment. It is quite possible to turn this approach around, and to classify the environments themselves, with the classification guided by the intention that the environmental units will also be characterized by difference of structure. Some systems for the classification of climates (KÖPPEN 1923, THORNTHWAITE 1948) seek to define climatic units that will correspond to major formation-types. A number of authors (e.g. MERRIAM 1894, EMBERGER 1936, 1942) have sought to define altitudinal zones primarily on the basis of climate; the system of HOLDRIDGE (1947, 1967, TOSI 1960) employs a more complex classification of zones by both temperature and moisture conditions and seeks to assign formation-types of the world to these. Some authors have designed classifications including *biotope-types* or *habitat-types* based on characteristics of local environments as well as climates; among these may be mentioned VILLAR (1929), DANSEREAU (1957: 130), PEARSE (1926), and ELTON & MILLER (1954).

12.3.3 LANDSCAPE UNITS

Because of the many-faceted character of communities and the functional interrelation of their various characteristics, it is natural to think that all characteristics of communities and environments should be considered together in classifying them. The ideal becomes difficult in practice. Confronted with an almost unmanageable number of characteristics, the phytosociologist must choose among them that which is to be emphasized in a given classification. There is no very clear basis for the choice, and different choices will produce different classifications. Most plant geographers and phytosociologists have preferred to specify one kind of community characteristic as a primary basis for a clearer and more consistent classification. The many-factor approach has, however, been applied to units on two levels. (i) Landscapes. All characteristics of geographic areas – climate, geological substrate and topography, flora and fauna, kinds of natural communities, and cultural effects of man – may be considered together to define geographic or landscape units (PASSARGE 1921, 1927, 1929, REGEL 1939, TÜXEN 1968a). When landscapes as wholes are treated as units or grouped into classes by considering their many characteristics, the results are *landscape-types* (REGEL 1949, TROLL 1950). (ii) Microlandscapes. Smaller, homogeneous parts of the earth's surface may be treated as ecosystems, considering the community and its environment together, and classified into microlandscape-types. MARKUS designated a local landscape unit as a *nature-complex*, and classified these (MARKUS 1926, 1929, 1930, REGEL 1939). In Russia the local landscape unit has been designated a *biogeocoenose*, and the concept has had extensive influence on the classification of forests as biogeocenose-types

(Morosov 1928, Kruedener 1926, Paczoski 1930, Sukachev 1944, 1954, 1960) and on other aspects of ecology. Among other West-European approaches considering environment and community together may be mentioned Negri (1927), Gams (1927), Hansen (1930), Scharfetter (1932), Sørensen (1937), Däniker (1939a,b), Gradmann (1941), Lüdi (1948), Kuhnholtz-Lordat (1952), Etter (1954), Ehrendorfer (1954), and Sjörs (1955).

12.3.4 Biotic Areas

In some approaches the geographic areas of species are emphasized as a basis of dividing of the earth's surface into biotic units. Schmid (1922, 1941, 1942, 1955, 1956b, 1961, 1963) has developed a distinctive approach to vegetation classification and interpretation. Geographic or regional units, for the recognition of which areas of species distributions are emphasized, are designated *Vegetationsgürtel*. From the flora of a regional unit, its *Artengarnitur*, species are variously distributed in different combinations into intergrading, local *Biocoenosen* that are not classifiable in any real sense, but among which abstract types or models may be recognized (Schmid 1940, 1942, 1950, 1952, Ehrendorfer 1954). Among the species themselves structural types or growth-forms and ecological response types (representation-forms) are distinguished (Schmid 1950, 1952, 1956a, 1963, Heuer 1949, Schwartz 1955, Saxer 1967); these many-sided ecological types are preferred to the traditional life-forms. The spectrum of growth-forms contributes to the characterization of the biocenose-type and permits its comparison with other biocoenoses. Major phytocoenoses of normal climates and soils in a vegetation-girdle are termed *regional* phytocoenoses, as distinguished from *local* phytocoenoses of distinctive biotopes (Schmid 1961). Although definition of the vegetation-girdles is in principle floristic, they correspond in general to regional vegetation types that may be recognized by physiognomy and dominance. Schmid's system has been applied to both intensive analysis of forest communities (Schmid 1936b, Heuer 1949, Schwartz 1955, Saxer 1955, 1967), and to broad geographic interpretation of European vegetation in terms of vegetation-girdles and their historic relations (Schmid 1949, 1961). The work of Meusel (1939a, 1939b, 1940, 1943, 1951, 1954) is distinct from that of Schmid but related in its concern with many-sided landscape relationships and with species distributions. Meusel (Meusel et al. 1965, cf. Böcher 1954) develops the concept of

13

areal-types or distributional classes of species as an approach to vegetation interpretation and classification. In the United States DICE (1938, 1943, 1952, GOLDMAN & MOORE 1946, KENDEIGH 1961) recognizes as *biotic provinces* geographic units characterized by climate, physiography, soils, and major natural communities that develop in relation to these. Biotic provinces are on this basis landscape units, in the definition of which characteristic communities are emphasized. Their principal use, however, has been as faunal or biogeographic units appropriate for interpretation of species distributions and the evolutionary differentiation of species and subspecies.

12.3.5 ZONES AND SERIES

Some approaches to classification are based on dividing a vegetational gradient into segments, or *zones*. MERRIAM (1890, 1898, 1899, MUESEBECK & KROMBEIN 1952, KENDEIGH 1954) recognized altitudinal zones in mountains of the United States and sought to relate the zones to temperature in transcontinental belts. In many areas the zones were defined by dominant plant species, but they were also biotic units in the sense of the preceding paragraph, units that were thought to be definable by the distributions of animal species. Many authors have treated as zones (which together form successional series) the bands of vegetation surrounding water bodies. Treatment of vegetation in terms of zones has had a longer history in Europe (BRAUN-BLANQUET 1964). More recently it has been applied to relate the vegetations of tropical mountains to one another (TROLL 1961, HEDBERG 1951) and to classify communities of the marine littoral (DOTY 1957, STEPHENSON & STEPHENSON 1961, SOUTHWARD 1958, DAHL 1952-3). BEARD (1944, 1955, article 13) has related formations in the American Tropics to one another as *formation-series* – sequences of formations along major environmental gradients. Finnish (CAJANDER 1903, CAJANDER & ILVESSALO 1921, KUJALA 1945, article 15) and Russian (KELLER 1925-6, SUKACHEV 1928, 1932, article 17) work has related smaller vegetation units to one another in sequences along environmental gradients; such a sequence is an *ecological series*. All these approaches have in common a combination of classification (dividing community gradients into units, segments, or zones) and ordination (relating these units to one another along environmental axes). In many cases the units are dominance-types; but in other treatments they have been defined by climate, or as formations, associations, or sociations (see below). Ecological series can be used in any circumstances in which relations of community-types to environmental gradients can be recognized. Use of *zones* usually implies, however, that one environmental gradient has so great an effect on communities that other environmental variables may be subordinated to it. The communities are then treated as belts or zones in relation to that complex-gradient. Zones as dominance-types seem more clearly defined in communities of more severe environments, such as those of mountain ranges in dry climates, and the seacoast and shores of inland water bodies.

12.3.6 SPECIES DOMINANCE

It is natural to classify plant communities by what is, after physiognomy, their most obvious characteristic: dominant species. Classification by dominance-types has a number of limitations, but is often convenient and adequate to the purpose of a given study. Discussions of classification by dominance include DU RIETZ (1932), GAMS (1933), CROCKER & WOOD (1947), BEADLE & COSTIN (1952), TANSLEY & CHIPP (1926), TANSLEY (1939), BEARD (1955), POORE (1955-6), MUNZ & KECK (1949-50), and WHITTAKER (1956, 1960). Many authors have combined dominance-types for lower-level with formations for higher-level classification. This approach, widespread among English-language ecologists particularly, is further discussed in a review by the author (WHITTAKER 1962) and in articles 13, 14, and 17 below. Many authors have used the term 'association' for dominance-types, but there may be advantage in applying 'association' only to the floristically conceived units of the School of BRAUN-BLANQUET.

12.3.7 VEGETATION DYNAMICS

A successional sequence is an 'ecological series' in time (or considered in movement along an environmental gradient in the present, as in the case of a pond margin). Successions in different habitats tend to converge toward climax communities that are more similar than the early stages of the successions. It is consequently possible to group the diverse successional communities of an area around a kind of climax toward which they lead, as a basis of classification. This combination of successional relationships with definition of the climax community-types as formations or dominance-types has been applied in English-language ecology (Moss 1910, TANSLEY

1911, 1920, 1939, ADAMSON & OSBORN 1924, STAMP 1925, PHILLIPS 1930, WOOD 1937, ROSAYRO 1950, COUPLAND 1961). The approach of most British ecologists to classification is informal, but a distinctive approach to formal classification developed in the United States under the leadership of CLEMENTS (1916, 1928, 1936, WEAVER & CLEMENTS 1929, CLEMENTS & SHELFORD 1939). CLEMENTS considered that there was a single, climatic climax community for a given area, toward which all successions converged. Climax communities were conceived as formations, and geographic subdivisions of formations were recognized as 'associations.' Successional or developmental types were termed 'associes.' A quite different system for intensive study of local vegetation by relationships of narrowly defined vegetation-development -types (*Vegetationsentwicklungstypen*) was developed in the Austrian mountains by AICHINGER (1952, 1967, WENDELBERGER 1951).

12.3.8 STRATAL UNITS

It is posssible to classify fractions of plant communities, and in particular to use different units of classification for the different strata or life-forms. Plants of the tree, shrub, herb, and thallophyte strata of a given community may in this case belong to different units, and there may be no simple distributional correspondence of these units when they are observed over a wider range of environments. This synusial treatment of vegetation has a long history from formulation of the approach by GAMS (1918, 1927) through the Estonian school of LIPPMAA (1933, 1935, 1939) to the present. The units of such classification are often termed *unions*. Synusial approaches through stratal or life-form units have had most extensive use for communities of epiphytic (BARKMAN 1958, HOSOKAWA 1954, WILMANNS 1962), and aquatic (e.g. VAARAMA 1938, JØRGENSEN 1948, KORNAŚ & MEDWECKA-KORNAŚ 1950, HÄYRÉN 1956, HARTOG & SEGAL 1964, SCHMITZ 1965) plants. The approach is reviewed in more detail by BARKMAN in article 16.

12.3.9 SOCIATIONS

It is possible also to classify communities by combinations of unions (or of dominant species of strata) as an approach related to the preceding one. By this approach two plant communities, one combining unions A, B, and C, and the other unions A, D, and C belong to different community-types. The classification of vegetation by combinations of stratal dominants developed in Sweden as a characteristic of the Uppsala school (FRIES 1913, DU RIETZ 1921, 1932, 1936) and has been applied to other northern and alpine vegetation (BOLLETER 1921, GAMS 1927, NORDHAGEN 1928, 1937, 1943, BÖCHER 1933, 1954, HANSON 1953, DAUBENMIRE 1952, 1954), and in Russia (article 17). Several terms have been used, but *sociation* is now accepted for such units. The most extensive use of sociations has been in Scandinavia, where much of the recent work employing these units fits them into the system of Braun-Blanquet (NORDHAGEN 1937, 1954, DAHL 1957). Article 18 by TRASS and MALMER discusses Scandanavian phytosociology further.

12.3.10 FOREST SITE-TYPES

A somewhat different approach with a stratal emphasis was developed in Finland by CAJANDER (1909, 1949, CAJANDER & ILVESSALO 1921). CAJANDER considered that the undergrowth of northern forests was relatively independent of the species of the canopy dominant. Forest site-types could be defined by composition of the undergrowth and used to indicate environment and its potential for growth of canopy trees. The site-types were grouped into classes by moisture conditions, and were arranged into ecological series in relation to major environmental gradients. Site-types have been widely used in northern Europe and have had more limited application in North America (HEIMBURGER 1934, KUJALA 1945). In many of the applications outside Finland the site-type concept departs from that of CAJANDER to converge with that of the sociation (ARNBORG 1953, LINDQUIST 1954, SUKATSCHEW 1932, GAMS 1933) or the association of BRAUN-BLANQUET. The site-type approach is considered in article 15 by FREY.

12.3.11 NUMERICAL CLASSIFICATION

Numerical classifications are those based directly on measurements of relative similarity of either the distributions of species or the compositions of samples. TUOMIKOSKI (1942), DAGNELIE (1960, 1962), and HEGG (1965) have considered groups of correlated species as a basis of classification. GOODALL (1953) used absence of species correlations as a criterion for homogeneity of lower-level vegetation units; WILLIAMS & LAMBERT (1959, 1961, LAMBERT & DALE 1964, LAMBERT & WILLIAMS 1962, GREIG-SMITH 1964) used correlations among species to classify samples in a dichotomous hierarchy. Measures of similarity of plant community samples (KULCZYŃSKI 1928) were used by MOTYKA (1947, MOTYKA et al. 1950) and MATUSZKIEWICZ (1948, 1950, 1952) to arrange the samples in an ordered matrix and classify the vegetation into units; SØRENSEN (1948), GUINOCHET (1954), ELLENBERG (1956), DAHL (1957), BARKMAN (1958) and others have used sample similarities as an aid in classification. More formal techniques of classifying samples by quantitative comparisons are described by EDWARDS & CAVALLI-SFORZA (1965), MAC NAUGHTON-SMITH et al. (1964), and ORLOCI (1967). POORE (1955-6, 1962) and GIMINGHAM (1961) used similarity measurements along with dominant and constant species to derive vegetation units termed *noda*. Vegetation is conceived as a multi-dimensional pattern or network of variation, in which a nodum is an abstract reference point. The term *nodum* might well be used for community-types on any level that are derived from numerical techniques. Such community-types may or may not correspond to any other unit of classification described here. Numerical techniques cannot solve the fundamental problems of classifying continuously intergrading communities (see also ASHBY 1936, 1948, DAHL 1957, GREIG-SMITH 1964, WHITTAKER 1962, 1967) but can provide results of interest. Numerical approaches are developed further in article 19 by GOODALL, see also Ordination, article 6 by McINTOSH and recent developments in article 11.

12.3.12 FLORISTIC UNITS

It is possible, finally, to base classification on the full floristic composition of communities, without reliance on numerical techniques. In the school of BRAUN-BLANQUET (1913, 1921, 1932, 1951, 1964, ELLENBERG 1956, BECKING 1957, TÜXEN 1968b, 1970) vegetation samples (relevés) are grouped into community-types by similarity of composition, especially by representation of character-species. Character-species are species whose distributions are centered in, or largely limited to, a given community-type. The basic unit of the system is the *association*. Associations are grouped into higher units termed alliances, alliances into orders, and orders into classes to produce a formal hierarchy of community classification. A community-type or unit, on any level, in this system may be termed a *syntaxon* (BARKMAN et al. 1958). The higher units also may be defined by character-species. Associations may be divided into subassociations, and these into variants and facies as lower-level units. Subassociations and variants may be defined by differential-species – species that are (whatever their over-all distributions) largely limited to one or the other of two lower units being compared. The system of BRAUN-BLANQUET is the most widely applied and most effectively standardized of all approaches to classification, and has been adapted to diverse kinds of communities. The range of units from narrow to broad has made the system effective for such varied purposes as intensive study of local communities for environmental indication and land management on the one hand, and classification into higher units of the vegetation of all western Europe on the other. The system has limitations that have been commented on (POORE 1955-6, WHITTAKER 1962), but so do all approaches to vegetation classification. The final article (20) by WESTHOFF and MAAREL, among those following, will review this most successful of approaches to formal classification of plant communities.

12.4 SUMMARY

Plant communities are complex phenomena which can be variously classified. Because of the individualistic distributions of species and the continuity of communities, there is no single, natural unit of classification. Different choices of ways of defining community-types imply different classifications of the same vegetation.

Many schools have developed different systems of classification adapted to different kinds of landscapes and research interests. The earliest means of classification was by vegetation structure or physiognomy, but divergent later approaches are briefly described in terms of five regional traditions, each including several schools – the Northern or Scandinavian and Baltic, the Southern European, the Russian, the British, and the American.

Concepts of classification are shared by some of the schools, and the number of major ways of classifying communities is smaller than the number of schools. Twelve basic approaches to classification (and the units they employ) are distinguished: (1) physiognomic or structural (formation, formation-type), (2) environmental (biotope-type, etc.), (3) many-factor or landscape (landscape-type, microlandscape-type or biogeocoenose-type), (4) biotic areas (vegetation-girdle, biotic province), (5) segments of community-gradients (life-zone, ecological series), (6) dominant species (dominance-type), (7) vegetation dynamics (American formation and association as regional vegetation types, associes and vegetation-development-type), (8) stratal or life-form divisions (union), (9) stratal combinations (sociation), (10) forest undergrowth types (site-type), (11) numerical comparisons (nodum), and (12) floristic units of BRAUN-BLANQUET (association and other syntaxa). Of these numbers 1, 6, 8, 9, 10, 11, and 12 are further discussed in the articles that follow.

REFERENCES

ADAMSON, R. S. & T. G. B. OSBORN, – 1922 – On the ecology of the Ooldea District. *Trans. R. Soc. S. Aust.* 46: 539—564.
AICHINGER, E., – 1952 – Fichtenwälder und Fichtenforste als Waldentwicklungstypen. Ein forstwirtschaftlicher Beitrag zur Beurteilung der Fichtenwälder and Fichtenforste. *Angew. PflSoziol.*, Wien 7: 1—179.
AICHINGER, E., – 1954 – Statische und dynamische Betrachtung in der pflanzensoziologischen Forschung. *Veröff. geobot. Inst. Rübel,* Zürich 29: 9—28.
AICHINGER, E., – 1967 – Die Waldentwicklungstypen im Raume von Kirchleerau. (Engl. and French summs.) *Veröff. geobot. Inst. ETH, Stftg. Rübel,* Zürich 39: 187—270, 283, 293.
ALECHIN, W. W., – 1926 – Was ist eine Pflanzengesellschaft? Ihr Wesen und ihr Wert als Ausdruck des sozialen Lebens der Pflanzen. *Beih. Repert. Spec. nov. Regni veg.,* Beih. 37: 1—50.
ALECHIN, W. W., – 1932 – Die vegetationsanalytischen Methoden der Moskauer Steppenforscher. *Handb. biol. ArbMeth.* 11, 6: 335—373.
ALEKSANDROVA, VERA, D., – 1969 – Classification of Vegetation. Principles of Classification and Classification Systems of Various Phytocoenological Schools. (Russian). Nauka, Leningrad. 275 pp.
ALLEE, W. C., A. E. EMERSON, O. PARK, T. PARK & K. P. SCHMIDT, – 1949 – Principles of Animal Ecology. Saunders, Philadelphia. 837 pp.
ARNBORG, T., – 1953 – Det nordsvenska skogstypsschemat. 3rd ed. Svenska Skogsvardsför. Förlag, Stockholm. 20 pp.
ASHBY, E., – 1936 – Statistical ecology. *Bot. Rev.* 2: 221—235.
ASHBY, E., – 1948 – Statistical ecology. II. —A reassessment. *Bot. Rev.* 14: 222—234.
AUSTIN, M. P. & L. ORLOCI, – 1966 – Geometric models in ecology. II. An evaluation of some ordination techniques. *J. Ecol.* 54: 217—227.
BARKMAN, J. J., – 1958 – Phytosociology and Ecology of Cryptogamic Epiphytes, Including a Taxonomic Survey and Description of their Vegetation Units in Europe. Van Gorcum, Assen. 628 pp.
BARKMAN, J. J., H. DOING KRAFT, C. G. VAN LEEUWEN, & V. WESTHOFF, – 1958 – Enige opmerkingen over de terminologie in de vegetatiekunde. *Corresp Bl. Florist. Veget. Onderz. Neder.* 8: 87—93.
BEADLE, N. C. W., – 1948 – The Vegetation and Pastures of Western New South Wales, with Special Reference to Soil Erosion. Tennant, Govt. Printer, Sidney. 281 pp.
BEADLE, N. C. W. & A. B. COSTIN, – 1952 – Ecological classification and nomenclature, with a note on pasture classification by C. W. E. MOORE. *Proc. Linn. Soc. N. S. W.* 77: 61—82.
BEARD, J. S., – 1944 – Climax vegetation in tropical America. *Ecology* 25: 127—158.
BEARD, J. S., – 1955 – The classification of tropical American vegetation-types. *Ecology* 36: 89—100.
BECKING, R. W., – 1957 – The Zürich-Montpellier school of phytosociology. *Bot. Rev.* 23: 411—488.
BILLINGS, W. D., – 1945 – The plant associations of the Carson Desert Region, western Nevada. *Butler Univ. bot. Studs.* 7: 89—123.
BÖCHER, T. W., – 1933 – Studies on the vegetation of the east coast of Greenland between Scoresby Sound and Angmagssalik (Christian IX.s Land). *Meddr. Grønland* 104 (4): 1—132.
BÖCHER, T. W., – 1954 – Oceanic and continental vegetational complexes in Southwest Greenland. *Meddr. Grønland* 148 (1): 1—336.

BOLLETER, R., – 1921 – Vegetationsstudien aus dem Weisstannental. *Jb. St. Gall. naturw. Ges.* 57 (2): 1—140.
BRAUN, E. LUCY, – 1938 – Deciduous forest climaxes. *Ecology* 19: 515—522.
BRAUN, E. LUCY, – 1947 – Development of the deciduous forests of eastern North America. *Ecol. Monogr.* 17: 211—219.
BRAUN, E. LUCY, – 1950 – Deciduous Forests of Eastern North America. Blakiston, Philadelphia. 596 pp.
BRAUN, E. LUCY, – 1956 – The development of association and climax concepts: their use in interpretation of the deciduous forest. *Am. J. Bot.* 43: 906—911.
BRAUN-BLANQUET, J., – 1913 – Die Vegetationsverhältnisse der Schneestufe in den Rätisch-Lepontischen Alpen: Ein Bild des Pflanzenlebens an seinen äussersten Grenzen. *Neue Denkschr. Schweiz. naturf. Ges.* 48: 1—347.
BRAUN-BLANQUET, J., – 1921 – Prinzipien einer Systematik der Pflanzengesellschaften auf floristischer Grundlage. *Jb. St. Gall. naturw. Ges.* 57 (2): 305—351.
BRAUN-BLANQUET, J., – 1932 – Plant Sociology, the Study of Plant Communities. Transl. by G. D. FULLER and H. S. CONARD. Mc Graw-Hill, New York. 439 pp.
BRAUN-BLANQUET, J., – 1951 – Pflanzensoziologie: Grundzüge der Vegetationskunde. 2nd ed. 631 pp. 3rd ed. 1964. 865 pp. Springer, Wien.
BRAY, J. R. & J. T. CURTIS, – 1957 – An ordination of the upland forest communities of southern Wisconsin. *Ecol. Monogr.* 27: 325—349.
BROCKMAN-JEROSCH, H. & E. RÜBEL, – 1912 – Die Einteilung der Pflanzengesellschaften nach ökologisch-physiognomischen Gesichtspunkten. Engelmann, Leipzig. 72 pp.
CAIN, S. A., – 1936 – Synusiae as a basis for plant sociological field work. *Am. Midl. Nat.* 17: 665—672.
CAJANDER, A. K., – 1903 – (1906) – Beiträge zur Kenntniss der Vegetation der Alluvionen des nördlichen Eurasiens. I. Die Alluvionen des unteren Lena-Thales. *Acta Soc. Sci. fenn.* 32 (1): 1—182.
CAJANDER, A. K., – 1909 – Ueber Waldtypen. *Acta for. fenn.* 1 (1): 1—175.
CAJANDER, A. K., – 1949 – Forest types and their significance. *Acta for. fenn.* 56 (4): 1—71.
CAJANDER, A. K. & Y. ILVESSALO, – 1921 – Über Waldtypen II. *Acta for. fenn.* 20 (1): 1—77.
CLEMENTS, F. E., – 1905 – Research Methods in Ecology. Univ. Publ. Co., Lincoln, Nebr. 334 pp.
CLEMENTS, F. E., – 1916 – Plant succession: an analysis of the development of vegetation. *Publs Carnegie Instn,* Washington 242: 1—512.
CLEMENTS, F. E., – 1928 – Plant Succession and Indicators. Wilson, New York. 453 pp.
CLEMENTS, F. E., – 1936 – Nature and structure of the climax. *J. Ecol.* 24: 252—284.
CLEMENTS, F. E. & V. E. SHELFORD, – 1939 – Bio-ecology. Wiley, New York. 425 pp.
CONARD, H. S., – 1935 – The plant associations of central Long Island; a study in descriptive plant sociology. *Am. Midl. Nat.* 16: 433—516.
CONARD, H. S., – 1952 – The vegetation of Iowa: an approach toward a phytosociologic account. *Stud. nat. Hist. Iowa Univ.* 19 (4): 1—166.
COOPER, W. S., – 1913 – The climax forest of Isle Royale, Lake Superior, and its development. *Bot. Gaz.* 55: 1—44, 115—140, 189—235.
COUPLAND, R. T., – 1961 – A reconsideration of grassland classification in the northern Great Plains of North America. *J. Ecol.* 49: 135—167.

Cowles, H. C., – 1899 – The ecological relations of the vegetation on the sand dunes of Lake Michigan, I. Geographical relations of the dune floras. *Bot. Gaz.* 27: 95—117, 167—202, 281—308, 361—391.

Cowles, H. C., – 1901 – The physiographic ecology of Chicago and vicinity; a study of the origin, development, and classification of plant societies. *Bot. Gaz.* 31: 73—108, 144—182.

Crocker, R. L. & J. G. Wood, – 1947 – Some historical influences on the development of the South Australian vegetation communities and their bearing on concepts and classification in ecology. *Trans. R. Soc. S. Aust.* 71: 91—136.

Curtis, J. T., – 1959 – The Vegetation of Wisconsin; An Ordination of Plant Communities. Univ. Wisconsin, Madison. 657 pp.

Curtis, J. T. & R. P. Mc Intosh, – 1951 – An upland forest continuum in the prairie-forest border region of Wisconsin. *Ecology* 32: 476—496.

Dagnelie, P., – 1960 – Contribution à l'étude des communautés végétales par l'analyse factorielle. (Engl. summ.) *Bull. Serv. Carte phytogéogr.*, Paris, Sér. B, 5: 7—71, 93—195.

Dagnelie, P., – 1962 – L'étude des communautés végétales par l'analyse des liaisons entre les espèces et les variables écologiques. Inst. Agron. de l'État, Gembloux. 135 pp.

Dahl, E., – 1952-3 – Some aspects of the ecology and zonation of the fauna on sandy beaches. *Oikos* 4: 1—27.

Dahl, E., – 1957 – Rondane: mountain vegetation in South Norway and its relation to the environment. *Skr. norske Vidensk-Akad.* Mat-naturv. Kl. 1956 (3): 1—374.

Däniker, A. U., – 1939a – Die Pflanzengesellschaft, ihre Struktur und ihr Standort. *Ber. schweiz. bot. Ges.* 49: 522—540.

Däniker, A. U. – 1939b – Die Biozönose als Einheit der Vegetation. *Verhandl. schweiz. naturf. Ges.* 1939: 65—67.

Dansereau, P., – 1943 – L'érablière laurentienne. I. Valeur d'indice des espèces. *Can. J. Res.*, Sect. C, Bot. 21: 66—93.

Dansereau, P., – 1946 – L'érablière laurentienne. II. Les successions et leurs indicateurs. *Can. J. Res.*, Sect. C, Bot. 24: 235—291.

Dansereau, P., – 1951 – Description and recording of vegetation upon a structural basis. *Ecology* 32: 172—229.

Dansereau, P., – 1957 – Biogeography: An Ecological Perspective. Ronald, New York. 394 pp.

Daubenmire, R., – 1952 – Forest vegetation of northern Idaho and adjacent Washington, and its bearing on concepts of vegetation classification. *Ecol. Monogr.* 22: 301—330.

Daubenmire, R., – 1954 – Vegetation classification. *Veröff. geobot. Inst. Rübel*, Zürich 29: 29—34.

Dice, L. R., – 1938 – The Canadian biotic province with special reference to the mammals. *Ecology* 19: 503—514.

Dice, L. R., – 1943 – The Biotic Provinces of North America. Univ. Michigan, Ann Arbor. 78 pp.

Dice, L. R., – 1952 – Natural Communities. Univ. Michigan, Ann Arbor. 547 pp.

Dimo, N. A. & B. A. Keller, – 1907 – Im Gebiet der Halbwüste. Bodenkundliche und botanische Untersuchungen in Süden des Zarizyner Kreises im Saratowschen Gouvernement. (In Russian) Saratow.

Doty, M. S., – 1957 – Rocky intertidal surfaces. In 'Treatise on Marine Ecology and Paleoecology,' ed. J. W. Hedgpeth, vol. 1, Ecology. *Mem. geol. Soc. Am.* 67 (1):535—585.

Du Rietz, G. E., – 1921 – Zur methodologischen Grundlage der modernen Pflanzensoziologie. Holzhausen, Wien. 267 pp.
Du Rietz, G. E., – 1932 (1930) – Vegetationsforschung auf soziationsanalytischer Grundlage. *Handb. biol. ArbMeth.* 11, 5: 293—480.
Du Rietz, G. E., – 1936 – Classification and nomenclature of vegetation units 1930—1935. *Svensk bot. Tidskr.* 30: 580—589.
Edwards, A. W. F. & L. L. Cavalli-Sforza, – 1965 – A method for cluster analysis. *Biometrics* 21: 362—375.
Ehrendorfer, F., – 1954 – Gedankungen zur Frage der Struktur und Anordnung der Lebensgemeinschaften. *Angew. PflSoziol.*, Wien, Festschr. Aichinger 1: 151—167.
Ellenberg, H., – 1950 – Landwirtschaftliche Pflanzensoziologie. I. Unkrautgemeinschaften als Zeiger für Klima und Boden. Ulmer, Stuttgart. 141 pp.
Ellenberg, H., – 1952 – Landwirtschaftliche Pflanzensoziologie. II. Wiesen und Weiden und ihre standörtliche Bewertung. Ulmer, Stuttgart. 143 pp.
Ellenberg, H., – 1956 – Aufgaben und Methoden der Vegetationskunde. In 'Einführung in die Phytologie' by H. Walter, Vol. IV, Pt. 1. Ulmer, Stuttgart 136 pp.
Ellenberg, H. ed., – 1967 – Vegetations- und Bodenkundliche Methoden der forstlichen Standortskartierung: Ergebnisse eines internationalen Methodenvergleichs im Schweizer Mittelland. *Veröff. geobot. Inst. ETH, Stiftg Rübel*, Zürich 39: 1—296.
Ellenberg, H. & D. Mueller-Dombois, – 1967 – Tentative physiognomic-ecological classification of plant formations of the earth. *Ber. geobot. Inst. ETH, Stiftg. Rübel*, Zürich 1965/6, 37: 21—55.
Elton, C. S. & R. S. Miller, – 1954 – The ecological survey of animal communities; with a practical system of classifying habitats by structural characters. *J. Ecol.* 42: 460—496.
Emberger, L., – 1936 – Remarques critiques sur les étages de végétation dans les montagnes marocaines. *Ber. schweiz. bot. Ges.* 46: 614—631.
Emberger, L., – 1942 – Un projet d'une classification des climats du point de vue phytogéographique. *Bull. Soc. Hist. nat. Toulouse* 77: 97—124.
Etter, H., – 1954 – Grundsätzliche Betrachtungen zur Beschreibung und Kennzeichnung der Biochore. (French summ.) *Vegetatio* 5/6: 500—510.
Flahault, C., – 1893 – Les zones botaniques dans le Bas-Languedoc et les pays voisins. *Bull. Soc. bot. Fr.* 40 (Sess. extraord.): xxxvi—lxii.
Flahault, C. & C. Schröter, – 1910 – Rapport sur la nomenclature phytogéographique. *Actes 3me Int. bot. Congr.*, Bruxelles 1910, 1: 131—164.
Fosberg, F. R., – 1961 – A classification of vegetation for general purposes. *Trop. Ecol.* 2: 1—28. Also in 'Guide to the check sheet for IBP areas,' by G. F. Peterken. *IBP Handbook* 4: 73—120. Blackwell, Oxford and Edinburgh.
Fries, T. C. E., – 1913 – Botanische Untersuchungen im nördlichsten Schweden: Ein Beitrag zur Kenntnis der alpinen und subalpinen Vegetation in Torne Lappmark. Vetensk. och prakt. unders. i Lappland, anordn af Luossavaara-Kiirunavaara Aktiebolag. Flora och Fauna 2: 1—361. Almqvist & Wiksells, Stockholm.
Gams, H., – 1918 – Prinzipienfragen der Vegetationsforschung: Ein Beitrag zur Begriffsklärung und Methodik der Biocoenologie. *Vjschr. naturf. Ges. Zürich* 63: 293—493.
Gams, H., – 1927 – Von den Follatères zur Dent de Morcles: Vegetationsmonographie aus dem Wallis. *Beitr. geobot. Landesaufn. Schweiz* 15: 1—760.
Gams, H., – 1933 – Die Stellung der Waldtypen im Vegetationssystem. *Forstarchiv* 9: 53—59.

GAUSSEN, H., – 1933 – Géographie des plantes. Colin, Paris. 222 pp. 2nd ed., 1954, 223 pp.
GAUSSEN, H., – 1951 – Le dynamisme des biocénoses végétales. *Année biol.*, 3e Sér., 27: 89—101.
GIMINGHAM, C. H., – 1961 – North European heath communities; a 'network of variation' *J. Ecol.* 49: 655—694.
GIMINGHAM, C. H., – 1969 – The interpretation of variation in north-European dwarf-shrub heath communities. (Germ. summ.) *Vegetatio* 17: 89—108.
GITTINS, R., – 1969 – The application of ordination techniques. In 'Ecological Aspects of the Mineral Nutrition of Plants,' ed. I. H. RORISON, *Symp. Brit. Ecol. Soc.* 1968, 9: 37—66.
GLEASON, H. A., – 1926 – The individualistic concept of the plant association. *Bull. Torrey bot. Club* 53: 7—26.
GLEASON, H. A., – 1939 – The individualistic concept of the plant association. *Am. Midl. Nat.* 21: 92—110.
GOLDMAN, E. A. & R. T. MOORE, – 1946 – The biotic provinces of Mexico. *J. Mammal.* 26: 347—360.
GOODALL, D. W., – 1953 – Objective methods for the classification of vegetation. I. The use of positive interspecific correlation. *Aust. J. Bot.* 1: 39—63.
GOODALL, D. W., – 1954 – Objective methods for the classification of vegetation. III. An essay in the use of factor analysis. *Aust. J. Bot.* 2: 304—324.
GRADMANN, R., – 1941 – Methodische Grundfragen und Richtungen der Pflanzensoziologie. *Beih. Repert. Spec. nov. Regni Veg.* 131: 1—41.
GREIG-SMITH, P., – 1964 – Quantitative Plant Ecology. 2nd ed. Butterworths, London. 256 pp.
GREIG-SMITH, P., M. P. AUSTIN, & T. C. WHITMORE, – 1967 – The application of quantitative methods to vegetation survey. I. Association-analysis and principal component ordination of rain forest. *J. Ecol.* 55: 483—503.
GRISEBACH, A., – 1838 – Ueber den Einfluss des Climas auf die Begränzung der natürlichen Floren. *Linnaea* 12: 159—200.
GRISEBACH, A., – 1872 – Die Vegetation der Erde nach ihrer klimatischen Anordnung. Ein Abriss der vergleichenden Geographie der Pflanzen. Engelmann, Leipzig. 2 vols., 603 and 709 pp.
GROSSHEIM, A. A., – 1930 – Zur Frage nach dem Zustandekommen der Pflanzendecke. *Beitr. Biol. Pfl.* 18: 225—286.
GUINOCHET, M., – 1954 – Sur les fondements statistiques de la phytosociologie et quelques unes de leurs conséquences. *Veröff. geobot. Inst. Rübel*, Zürich 29: 41—67.
HANSEN, H. MØLHOLM, – 1930 – Studies on the vegetation of Iceland. In 'The Botany of Iceland,' ed. J. L. A. KOLDERUP-ROSENVINGE & E. WARMING, vol. 3, pt. 1, no. 10: 1—186. Frimodt, Copenhagen.
HANSON, H. C., – 1953 – Vegetation types in northwestern Alaska and comparisons with communities in other Arctic regions. *Ecology* 34: 111—140.
HARTOG, C. DEN, & S. SEGAL, – 1964 – A new classification of the water-plant communities. *Acta bot. neerl.* 13: 367—393.
HÄYRÉN, E., – 1956 – Über die Algenvegetation des sandigen Geolitorals am Meere in Schweden und in Finnland. *Svensk bot. Tidskr.* 50: 257—269.
HEDBERG, O., – 1951 – Vegetation belts of the East African mountains. *Svensk. bot. Tidskr.* 45: 141—202.
HEGG, O., – 1965 – Untersuchungen zur Pflanzensoziologie und Ökologie im Naturschutzgebiet Hohgant (Berner Voralpen), mit einem Beitrag zur Methodik der floristisch-statistischen Erfassung pflanzensoziologischer Zusammenhänge. (French and Engl. summs.) *Beitr. geobot. Landesaufn. Schweiz* 46: 1—188.

HEIMBURGER, C. C., – 1934 – Forest-type studies in the Adirondack region. *Mem. Cornell Univ. (N.Y.) agric. Exp. Stn.* 165: 1—122.

HEUER, ILSE, – 1949 – Vergleichende Untersuchungen an der Föhrenbeständen des Pfynwaldes (Wallis); Versuch einer biocoenologischen Analyse. *Beitr. geobot. Landesaufn. Schweiz* 28: 1—185.

HOLDRIDGE, L. R., – 1947 – Determination of world plant formations from simple climatic data. *Science, N. Y.* 105: 367—368.

HOLDRIDGE, L. R., – 1967 – Life Zone Ecology. Rev. ed. Tropical Science Center, San Jose. 206 pp.

HOSOKAWA, T., – 1954 – On the *Campnosperma* forests of Kusaie in Micronesia, with special reference to the community units of epiphytes. *Vegetatio* 5/6: 351—360.

HULT, R., – 1881 – Försök till analytisk behandling af växtformationerna. *Meddn. Soc. Fauna Flora fenn.* 8: 1—155.

HUMBOLDT, A. VON, – 1807 – Ideen zu einer Geographie der Pflanzen nebst einem Naturgemälde der Tropenländer. Cotta, Tübingen. 182 pp.

JANSSEN, C. R., – 1967 – A floristic study of forests and bog vegetation, northwestern Minnesota. *Ecology* 48: 751—765.

JØRGENSEN, E. G., – 1948 – Diatom communities in some Danish lakes and ponds. *Biol. Skr.*, K. Danske Vidensk. Selsk. 5 (2): 1—140.

KATZ, N. J., – 1929 – Die Zwillingsassoziationen und die homologen Reihen in der Phytosoziologie. *Ber. dt. bot. Ges.* 47: 154—164.

KATZ, N. J., – 1933 – Die Grundprobleme und die neue Richtung der Phytosoziologie. *Beitr. Biol. Pfl.* 21: 133—166.

KELLER, B. A., – 1925-6 – Die Vegetation auf den Salzböden der russischen Halbwüsten und Wüsten. *Z. Bot.* 18: 113—137.

KELLER, B. A., – 1932 – Die Methoden zur Erforschung der Ökologie der Steppen- und Wüstenpflanzen. *Handb. Biol. ArbMeth.* 11, 6: 1—128.

KENDEIGH, S. C., – 1954 – History and evaluation of various concepts of plant and animal communities in North America. *Ecology* 35: 152—171.

KENDEIGH, S. C., – 1961 – Animal Ecology. Prentice-Hall, Englewood Cliffs. 468 pp.

KÖPPEN, W., – 1923 – Die Klimate der Erde: Grundriss der Klimakunde. de Gruyter, Berlin & Leipzig. 369 pp.

KORNAŚ, J. & A. MEDWECKA-KORNAŚ, – 1950 – Associations végétales sous-marines dans le Golfe de Gdańsk (Baltique polonaise). (Engl. summ.) *Vegetatio* 2: 120—127.

KRUEDENER, A. VON, – 1926 – Waldtypen als kleinste natürliche Landschaftseinheiten bzw. Mikrolandschaftstypen. *Petermanns geogr. Mitt.* 72: 150—158.

KÜCHLER, A. W., – 1949 – A physiognomic classification of vegetation. *Ann. Ass. Am. Geogr.* 39: 201—210.

KUHNHOLTZ-LORDAT, G., – 1952 – Le tapis végétale dans ses rapports avec les phénomènes actuels de surface en Basse-Provènce (de Cassis à Bandol). *Encyclopéd. Beogéogr. et Écol.* 9: 1—208. Lechevalier, Paris.

KUJALA, V., – 1945 – Waldvegetationsuntersuchungen in Kanada mit besonderer Berücksichtigung der Anbaumöglichkeiten kanadischer Holzarten auf natürlichen Waldböden in Finnland. *Ann. Acad. Scient. fenn.* (*Suomal. Tiedeakat. Toim.*) Ser. A, 4 Biol. 7: 1—434.

KULCZYŃSKI, S., – 1928 – Die Pflanzenassoziationen der Pieninen. *Bull. int. Acad. pol. Sci. Lett.*, Cl. Sci. Math. Nat., Sér. B 1927 (Suppl. 2) : 57—203.

LAMBERT, JEAN M. & M. B. DALE, – 1964 – The use of statistics in phytosociology. *Adv. ecol. Res.* 2: 59—99.

LAMBERT, JEAN M. & W. T. WILLIAMS, – 1962 – Multivariate methods in plant ecology. IV. Nodal analysis. *J. Ecol.* 50: 775—802.

LINDQUIST, B., – 1954 – Ein Waldtypenschema für die skandinavischen Buchenwälder. *Angew. PflSoziol.* Wien, Festschr. Aichinger 2: 965—970.
LIPPMAA, T., – 1933 – Aperçu général sur la végétation autochtone du Lautaret (Hautes - Alpes) avec des remarques critiques sur quelques notions phytosociologiques. (Eston. summ.) *Acta Inst. Horti. bot. tartu.* 3 (3): 1—108.
LIPPMAA, T., – 1935 – Une analyse des forêts de l'île estonienne d'Abruka (Abro) sur la base des associations unistrates. *Acta Inst. Horti bot. tartu.* 4 (no. 1/2, art. 5): 1—97.
LIPPMAA, T., – 1939 – The unistratal concept of plant communities (the unions). *Am. Midl. Nat.* 21: 111—145.
LÜDI, W., – 1921 – Die Pflanzengesellschaften des Lauterbrunnentales und ihre Sukzession: Versuch zur Gliederung der Vegetation eines Alpentales nach genetisch-dynamischen Gesichtspunkten. *Beitr. geobot. Landesaufn. Schweiz* 9: 1—364.
LÜDI, W., – 1948 – Die Pflanzengesellschaften der Schinigeplatte bei Interlaken und ihre Beziehungen zur Umwelt. Eine vergleichend ökologische Untersuchung. *Veröff. geobot. Inst. Rübel*, Zürich, 23: 1—400.
MAAREL, E. VAN DER, – 1969 – On the use of ordination models in phytosociology. (Germ. summ.) *Vegetatio* 19: 21—46.
MAAREL, E. VAN DER & J. LEERTOUWER, – 1967 – Variation in vegetation and species diversity along a local environmental gradient. *Acta bot. neerl.* 16: 211—221.
MACNAUGHTON-SMITH, P., W. T. WILLIAMS, M. B. DALE, & L. G. MOCKETT, — 1964 – Dissimilarity analysis: a new technique of hierarchial sub-division. *Nature, Lond.* 202: 1034—1035.
MARKUS, E., – 1926 – Verschiebung der Naturkomplexe in Europa. *Geogr. Z.* 32: 516—541.
MARKUS, E., – 1929 – Die Grenzverschiebung des Waldes und des Moores in Alatskivi. *Acta Comment., Univ. tartu.*, Ser. A, 14 (3): 1—157.
MARKUS, E., – 1930 – Naturkomplexe von Alatskivi. *Acta Comment., Univ. tartu.*, Ser. A, 18 (8): 1—13.
MATUSZKIEWICZ, W., – 1948 – Roślinność lasów okolik Lwowa. (Engl. summ.: The vegetation of the forests of the environs of Lvov). *Annls. Univ. Mariae Curie-Skłodowska*, Lublin, Sect C, 3: 119—193.
MATUSZKIEWICZ, W., – 1950 – Badania fitosocjologiczne nad lasami bukowymi w Sudetach. (Russ. & Engl. summs.: Phytosociological researches on the beech--forests in the Sudetts-Mtns). *Annls. Univ. Mariae Curie-Skłodowska*, Lublin, Sect. C, Suppl. 5: 1—196.
MATUSZKIEWICZ, W., – 1952 – Zespoły leśne Białowskiego Parky Narodowego. (Russ, & Germ. summs.: Die Waldassoziationen von Białowieza-Nationalpark). *Annls. Univ. Mariae Curie-Skłodowska*, Lublin, Sect. C, Suppl. 6: 1—218.
MCINTOSH, R. P., – 1967 – The continuum concept of vegetation. *Bot. Rev.* 33: 130—187. Responses, *ibid.* 34: 253—332, 1968.
MERRIAM, C. H., – 1890 – Results of a biological survey of the San Francisco Mountain region and desert of the Little Colorado, Arizona. *N. Am. Fauna* 3: 1—136.
MERRIAM, C. H., – 1894 – Laws of temperature control of the geographic distribution of terrestrial animals and plants. *Natn. geogr. Mag.* 6: 229—238.
MERRIAM, C. H., – 1898 – Life zones and crop zones of the United States. *Bull. Bur. biol. Surv. U. S. Dep. Agric.* 10: 1—79.
MERRIAM, C. H., – 1899 – Results of a biological survey of Mount Shasta, California. *N. Am. Fauna* 16: 1—179.

MEUSEL, H., – 1939a – Die Vegetationsverhältnisse der Gipsberge im Kyffhäuser und im südlichen Harzvorland. Ein Beitrag zur Steppenheidefrage. *Hercynia* 2: 1—372.
MEUSEL, H., – 1939b – Pflanzensoziologische Systematik. *Lotos* 4: 393—401.
MEUSEL, H., – 1940 – Die Grasheiden Mitteleuropas: Versuch einer vergleichenden-pflanzensoziologischen Gliederung. *Bot. Arch.*, Leipzig 41: 357—519.
MEUSEL, H., – 1943 – Über die Grundlagen der Vegetationsgliederung. *Forschn. Fortschr.* 19: 34—36.
MEUSEL, H., – 1951 – Über Pflanzengemeinschaften: Probleme der Vegetationskunde, behandelt en einigen Pflanzenvereinen der Heimat. *Urania*, Jena 14: 95—106, 178—188.
MEUSEL, H., – 1954 – Über die umfassende Aufgabe der Pflanzengeographie. *Veröff. geobot. Inst. Rübel*, Zürich 29: 68—80.
MEUSEL H., E. JÄGER & E. WEINERT, – 1965 – Vergleichende Chorologie der zentraleuropäischen Flora. Fischer, Jena. 583+258 pp.
MILLER, A. H., – 1951 – An analysis of the distribution of the birds of California. *Univ. Calif. Publs. Zool.* 50: 531—644.
MOORE, J. J., P. FITZSIMONS, E. LAMBE, & J. WHITE, – 1970 – A comparison of some phytosociological techniques (Germ. summ.) *Vegetatio* 20: 1—20.
MOROSOW, G. F., – 1928 – Die Lehre vom Walde. Neumann, Neudamm. 375 pp.
MOSS, C. E., – 1910 – The fundamental units of vegetation. Historical development of the concepts of the plant association and plant formation. *New Phytol.* 9: 18—53.
MOTYKA, J., – 1947 – O zadaniach i metodach badán géobotanicznych. (French summ.: Sur les buts et les méthodes des recherches geobotaniques). *Annls. Ûniv. Mariae Curie-Skłodowska*, Lublin, Sect. C, Suppl. 1: 1—168.
MOTYKA, J., B. DOBRZAŃSKI, & S. ZAWADZKI, – 1950 – Wstępne badania nad łąkami południowo-wschodniej Lubelszczyzny. (Russ. & Engl. summs.: Preliminary studies on meadows in the south-east of the province of Lublin). *Annls. Univ. Mariae Curie-Skłodowska*, Lublin, Sect. E, 5: 367—447.
MUESEBECK, C. F. W. & K. V. KROMBEIN, – 1952 – Life zone map. *Syst. Zool.* 1: 24—25.
MUNZ, P. A. & D. D. KECK, – 1949-50 – California plant communities. *Aliso* 2: 87—105, 199—202. Also in 'A California Flora,' by P. A. MUNZ, 1959, pp. 10—20. Univ. California, Berkeley & Los Angeles.
NEGRI, G., – 1927 – Recenti contributi alla concezione sinecologica dei consorzi vegetali. *Nuovo G. bot. ital.*, N. S. 34: 872—885. (Abstr. in *Atti Soc. ital. Progr. Sci.* 16: 578, 1928).
NEGRI, G., – 1954 – Interpretazione individualistica del paesaggio vegetale. *Nuovo G. bot. ital.*, N. S. 61: 579—694.
NORDHAGEN, R., – 1928 – Die Vegetation und Flora des Sylenegebietes. I. Die Vegetation. *Skr. norske Vidensk-Akad.*, Math-naturv. Kl. 1927 (1): 1—612.
NORDHAGEN, R., – 1937 – Versuch einer neuen Einteilung der subalpinen-alpinen Vegetation Norwegens. *Bergens Mus. Árbok*, Naturv. rekke 1936 (7):1—88.
NORDHAGEN, R., – 1943 – Sikilsdalen og Norges Fjellbeiter: En plantesosiologisk monografi. *Bergens Mus. Skr.* 22: 1—607.
NORDHAGEN, R., – 1954 – Vegetation units in the mountain areas of Scandinavia. *Veröff. geobot. Inst. Rübel*, Zürich 29: 81—95.
ODUM, E. P., – 1953 – Fundamentals of Ecology. Saunders, Philadelphia. 384 pp. 2nd ed., 1959, 546 pp. 3rd ed., 1971, 574 pp.
ORLOCI, L., – 1967 – An agglomerative method for classification of plant communities. *J. Ecol.* 55: 193—206.
PACZOSKI, J. K., – 1930 – Lasy Białowieży. (Germ. summ.: Die Waldtypen von

Białowieza). *Państwowa Rada Ochrony Przyrody*, Monogr. Nauk 1: 1—575. Poznań.
PASSARGE, S., – 1921 – Vergleichende Landschaftskunde. I. Aufgaben und Methoden der vergleichenden Landschaftskunde. Reimer/Vohsen, Berlin. 71 pp.
PASSARGE, S., – 1927 – Aufgaben und Methoden der Landschaftskunde, erläutert an den Elementen der nordwestdeutschen Landschaft. *Geogr. Anz.* 28: 44—49.
PASSARGE, S., – 1929 – Botanische und geographische Pflanzenvereine. *Naturwissenschaften* 17: 565—566.
PEARSE, A. S., – 1926 – Animal Ecology. McGraw-Hill, New York. 2nd ed., 1939, 642 pp.
PHILLIPS, J. F. V., – 1930 – Some important vegetation communities in the Central Province of Tanganyika Territory (formerly German East Africa): A preliminary account. *J. Ecol.* 18: 193—234.
POORE, M. E. D., – 1955a – The use of phytosociological methods in ecological investigations. I. The Braun-Blanquet system. *J. Ecol.* 43: 226—244.
POORE, M. E. D., – 1955b – The use of phytosociological methods in ecological investigations. II. Practical issues involved in an attempt to apply the Braun-Blanquet system. *J. Ecol.* 43: 245—269.
POORE, M. E. D., – 1955c – The use of phytosociological methods in ecological investigations. III. Practical application. *J. Ecol.* 43: 606—631.
POOORE, M. E. D., – 1956 – The use of phytosociological methods in ecological investigations. IV. General discussion of phytosociological problems. *J. Ecol.* 44: 28—50.
POORE, M. E. D., – 1962 – The method of successive approximation in descriptive ecology. *Adv. ecol. Res.* 1: 35—68.
RAMENSKY, L. G., – 1926 – Die Grundgesetzmässigkeiten im Aufbau der Vegetationsdecke. (Abstr. from *Věstn. opytn. Děla*, Voronezh 1924: 37—73). *Bot. Zbl.* N. F. 7: 453—455.
RAMENSKY, L. G., – 1930 – Zur Methodik der vergleichenden Bearbeitung und Ordnung von Pflanzenlisten und anderen Objekten, die durch mehrere, verschiedenartig wirkende Faktoren bestimmt werden. *Beitr. Biol. Pfl.* 18: 269—304.
RAUNKIAER, C., – 1934 – The Life Forms of Plants and Statistical Plant Geography. Clarendon, Oxford. 632 pp.
REGEL, C., – 1939 – Komplexe, Landschaft, Vegetationsprovinz. *Verh. schweiz. naturf. Ges.* 1939: 68—70.
REGEL, C., – 1949 – Landschaft und Pflanzenverein, mit besonderer Berücksichtigung russischer Forschungen. (French and Ital. summs.) *Geographica helv.* 4: 243—254.
ROSAYRO, R. A. DE, – 1950 – Ecological conceptions and vegetational types with special reference to Ceylon. *Trop. Agric., Ceylon* 106: 108—121.
RÜBEL, E., – 1930 – Pflanzengesellschaften der Erde. Huber, Bern-Berlin. 464 pp.
SAXER, A., – 1955 – Die Fagus-Abies und Piceagürtelarten in der Kontaktzone der Tannen und Fichtenwälder der Schweiz. *Beitr. geobot. Landesaufn. Schweiz.* 36: 1—198.
SAXER, A., – 1967 – Eine Waldkartierung im aargauischen Suhrental nach der Methode von E. Schmid. (Engl. and French summs.) *Veröff. Geobot. Inst. ETH, Stftg. Rübel*, Zürich 39: 149—185, 282, 292—3.
SCHARFETTER, R., – 1932 (1928) – Die kartographische Darstellung der Pflanzengesellschaften. *Handb. Biol. ArbMeth.* 11, 5: 77—164.
SCHIMPER, A. F. W. & F. C. VON FABER, – 1935 – Pflanzengeographie auf physiologischer Grundlage. 3rd ed. Fischer, Jena. 2 vols, 588 & 1612 pp.
SCHMID, E., – 1922 – Biozönologie und Soziologie. *Naturw. Wschr.*, N. F. 21: 518—523.

SCHMID, E., – 1936 – Die Reliktföhrenwälder der Alpen. *Beitr. geobot. Landesaufn. Schweiz* 21: 1—190.
SCHMID, E., – 1940 – Die Vegetationskartierung der Schweiz im Masstab 1: 200,000. *Ber. geobot. Inst. Rübel*, Zürich 1939: 76—85.
SCHMID, E., – 1941 – Vegetationsgürtel und Biocoenose. *Ber. schweiz. bot. Ges.* 51: 461—474.
SCHMID, E., – 1942 – Über einige Grundbegriffe der Biocoenologie. *Ber. geobot. Inst. Rübel*, Zürich 1941: 12—26.
SCHMID, E., – 1949 – Prinzipien der natürlichen Gliederung der Vegetation des Mediterrangebietes (mit einer halb schematischen Karte). *Ber. schweiz. bot. Ges.* 59: 169—200.
SCHMID, E., – 1950 – Zur Vegetationsanalyse numidischer Eichenwälder. *Ber. geobot. Inst. Rübel*, Zürich 1949: 23—39.
SCHMID, E., – 1952 – Natürliche Vegetationsgliederung am Beispiel des Spanischen Rif. *Ber. geobot. Inst. Rübel*, Zürich 1951: 55—79.
SCHMID, E., – 1955 – Der Ganzheitsbegriff in der Biocoenologie und in der Landschaftskunde. *Geographica helv.* 10: 153—162.
SCHMID, E., – 1956a – Die Wuchsformen der Dikotyledonen. *Ber. geobot. Inst. Rübel*, Zürich 1955: 38—50.
SCHMID, E., – 1956b – Die Vegetationsgürtel der Iberisch-Berberischen Gebirge. *Veröff. geobot. Inst. Rübel*, Zürich 31: 124—163.
SCHMID, E., – 1961 – Erläuterung zur Vegetationskarte der Schweiz (French & Engl. summs.) *Beitr. geobot. Landesaufn. Schweiz* 39: 1—52.
SCHMID, E., – 1963 – Die Erfassung der Vegetationseinheiten mit floristischen und epimorphologischen Analysen. *Ber. schweiz. bot. Ges.* 73: 276—324.
SCHMITHÜSEN, J., – 1959 – Allgemeine Vegetationsgeographie. In, 'Lehrbuch der allgemeinen Geographie,' ed. E. OBST, vol. 4. de Gruyter, Berlin. 261 pp.
SCHMITZ, W., – 1965 – Die Soziologie aquatischer Mikrophyten. (Engl. summ.) In, 'Biosoziologie,' ed. R. TÜXEN, *Ber. Symp. int. Vereinig. Vegetationskunde*, Stolzenau/Weser 1960, 4: 120—139.
SCHRÖTER, C. & O. KIRCHNER, – 1902 – Die Vegetation des Bodensees. 2. Teil von C. SCHRÖTER. *Bodensee-Forschungen* 9 (2): 1—86. Komm.-Verlag Schr. Ver. Gesch. Bodensees und seiner Umgebung von J. T. Stettner, Lindau.
SCHWARZ, U., – 1955 – Die natürlichen Fichtenwälder des Juras. *Beitr. geobot. Landesaufn. Schweiz* 35: 1—143.
SERNANDER, R., – 1898 – Studier öfver vegetationen i mellersta Skandinaviens fjälltrakter. 1. Om tundraformationer i svenska fjälltrakter. *Ofvers. K. svenska Vetensk. Akad. Förhandl.* 6: 325—367.
SHIMWELL, D. W., – 1971 – Description and Classification of Vegetation. Sidgwick & Jackson, London. 322 pp.
SJÖRS, H., – 1955 – Remarks on ecosystems. *Svensk. bot. Tidskr.* 49: 155—169.
SØRENSEN, T., – 1937 – Remarks on the flora and vegetation of Northeast Greenland 74° 30' — 79° N. lat. *Meddr. Grønland* 101(4): 108—140.
SØRENSEN, T., – 1948 – A method of establishing groups of equal amplitude in plant sociology based on similarity of species content and its application to analysis of the vegetation on Danish commons. *Biol. Skr.*, K. danske Vidensk. Selsk. 5 (4): 1—34.
SOUTHWARD, A. J., – 1958 – The zonation of plants and animals on rocky sea shores. *Biol. Rev.* 33: 137—177.
STAMP, L. D., – 1925 – The vegetation of Burma from an ecological standpoint. *Res. Monogr. Univ. Rangoon* 1: 1—58.
STEINDÓRSSON, S., – 1954 – The coastline vegetation at Gásar in Eyjafjörður in the North of Iceland. *Nytt. Mag. Bot.* 3: 203—212.

STEPHENSON, T. A. & ANNE STEPHENSON, – 1961 – Life between tide-marks in North America. IVA. Vancouver Island. I. *J. Ecol.* 49: 1—29.
SUKACHEV, V. N., – 1928 – Principles of classification of the spruce communities of European Russia. *J. Ecol.* 16: 1—18.
SUKATSCHEW, W., – 1932 – Die Untersuchung der Waldtypen des osteuropäischen Flachlandes. *Handb. biol. ArbMeth.* 11, 6: 191—250.
SUKATSCHEW, W., – 1944 – On principles of genetic classification in biocenology. (Russ. with Engl. summ.) *Zhur. Obschch. Biol.* 5: 213—227.
SUKATSCHEW, W., – 1954 – Die Grundlagen der Waldtypen. *Angew. PflSoziol.*, Wien, Festschr. Aichinger 2: 956—964.
SUKATSCHEW, W., – 1960 – The correlation between the concept of 'forest ecosystem' and 'forest biogeocenose' and their importance for the classification of forests. *Silva fenn.* 105: 94—97.
TANSLEY, A. G., ed., – 1911 – Types of British Vegetation. Cambridge Univ., Cambridge. 416 pp.
TANSLEY, A. G., – 1920 – The classification of vegetation and the concept of development. *J. Ecol.* 8: 118—149.
TANSLEY, A. G., – 1939 – The British Islands and their Vegetation. Cambridge Univ., Cambridge. 930 pp.
TANSLEY, A. G. & T. F. CHIPP, eds., – 1926 – Aims and Methods in the Study of Vegetation. Brit. Emp. Veget. Comm. & Crown Agents for Colonies, London. 383 pp.
THORNTHWAITE, C. W., – 1948 – An approach toward a rational classification of climate. *Geogrl. Rev.* 38: 55—94.
TISCHLER, W., – 1955 – Synökologie der Landtiere. Fischer, Stuttgart. 414 pp.
TOSI, J. A., Jr., – 1960 – Zonas de vida natural en el Perú. Memoria explicativa sobre el mappa ecológico del Perú. (Engl., French, Germ. summs.) *Bol. Técn. Inst. interamer. cienc. agríc.* OEA Zona Andina, 5: 1—271.
TROLL, C., – 1950 – Die geographische Landschaft und ihre Erforschung. *Studium gen.* 3: 163—181.
TROLL, C., – 1961 – Klima und Pflanzenkleid der Erde in dreidimensionaler Sicht. *Naturwissenschaften* 48: 332—348.
TUOMIKOSKI, R., – 1942 – Untersuchungen über die Vegetation der Bruchmoore in Ostfinnland. I. Zur Methodik der pflanzensoziologischen Systematik.(Finn. summ.) *Ann. Bot., Soc. zool. -bot. fenn. Vanamo (Suomal. eläin-ja kasvit. Seur. van. kasvit. Julk.)* 17(1):1-203.
TÜXEN, R., – 1937 – Die Pflanzengesellschaften Nordwestdeutschlands. *Jber. naturh. Ges. Hannover* 1929/30—1935/36, 81—87: 1—170. *Mitt. flor.-soz. Arb Gemein Neidersachsen* 3: 1—170.
TÜXEN, R., ed., — 1968a — Pflanzensoziologie und Landschaftsökologie. *Ber. Symp. int. Vereinig. Vegetationskunde*, Stolzenau/Weser 1963, 7 : 1—426.
TÜXEN R., ed., — 1968b — Pflanzensoziologische Systematik. *Ber. Symp. int. Vereinig. Vegetationskunde*, Stolzenau/Weser 1964, 8 : 1—347.
TÜXEN, R., – 1970 – Pflanzensoziologie als synthetische Wissenschaft. In 'Vegetatiekunde als synthetische wetenschap.' ed. H. J. VENEMA, H. DOING, and I. S. ZONNEVELD. *Misc. Pap. Landbouwhogeschool Wageningen*, 5: 141—159.
VAARAMA, A., – 1938 – Wasservegetationstudien am Grosssee Kallavesi. (Finn. summ.) *Ann. Bot., Soc. zool. -bot. fenn. Vanamo (Suomal. eläin-ja kasvit. Seur. van. kasvit. Julk.)* 13 (1): 1—318.
VILLAR, E. H. DEL, – 1929 – Geobotánica. Editorial Labor, Barcelona-Buenos Aires. 339 pp.
VRIES, D. M. DE, – 1948 – Method and survey of the characterization of Dutch grasslands. (Esperanto summ.) *Vegetatio* 1: 51—57.

WARMING, E., – 1909 – Oecology of Plants: An Introduction to the Study of Plant-Communities. Oxford Univ., Oxford. 422 pp.
WEAVER, J. E. & F. E. CLEMENTS, – 1929 – Plant Ecology. McGraw-Hill, New York. 520 pp. 2nd ed., 1938, 601 pp.
WENDELBERGER, G., – 1951 – Das vegetationskundliche System Erwin Aichingers und seine Stellung im pflanzensoziologischen Lehrgebäude Braun-Blanquets. *Angew. PflSoziol.*, Wien 1: 69—92.
WHITTAKER, R. H., – 1951 – A criticism of the plant association and climatic climax concepts. *NW. Sci.* 25: 17—31.
WHITTAKER, R. H., – 1956 – Vegetation of the Great Smoky Mountains. *Ecol. Monogr.* 26: 1—80.
WHITTAKER, R. H., – 1960 – Vegetation of the Siskiyou Mountains, Oregon and California. *Ecol. Monogr.* 30: 279—338.
WHITTAKER, R. H., – 1962 – Classification of natural communities. *Bot. Rev.* 28: 1—239.
WHITTAKER, R. H., – 1967 – Gradient analysis of vegetation. *Biol. Rev.* 42: 207—264.
WHITTAKER, R. H. & W. A. NIERING, – 1965 – Vegetation of the Santa Catalina Mountains, Arizona. (II) A gradient analysis of the south slope. *Ecology* 46: 429—452.
WILLIAMS, W. J. & JEAN M. LAMBERT, – 1959 – Multivariate methods in plant ecology. I. Association-analysis in plant communities. *J. Ecol.* 47: 83—101.
WILLIAMS, W. J. & JEAN M. LAMBERT, – 1961 – Multivariate methods in plant ecology. III. Inverse association-analysis. *J. Ecol.* 49: 717—729.
WILMANNS, O., – 1962 – Rindenbewohnende Epiphytengemeinschaften in Südwestdeutschland. *Beitr. naturk. Forsch. Südwestdeutschl.* 21 (2): 87—164.
WOOD, J. G., – 1937 – The Vegetation of South Australia. Trigg, Govt. Printer, Adelaide. 164 pp.
WOOD, J. G., – 1939 – Ecological concepts and nomenclature. *Trans. R. Soc. S. Aust.* 63: 215—223.
YARRANTON, G. A., – 1967 – Principal components analysis of data from saxicolous bryophyte vegetation at Steps Bridge, Devon. III. Correlation of variation in vegetation with environmental variables. *Can. J. Bot.* 45: 249—258.

13 THE PHYSIOGNOMIC APPROACH

John S. Beard

Contents

13.1	Introduction	35
13.2	History and Major Concepts	36
13.3	Physiognomic Systems	38
13.3.1	Formation-Types	38
13.3.2	Climatic Correlation	41
13.3.3	Descriptive Systems	46
13.3.4	Formation-Series	48
13.4	Physiognomy in British Commonwealth Ecology	51
13.4.1	Forest Types and Associations	51
13.4.2	Profile Diagrams	52
13.4.3	Formations and Formation-Types	54
13.4.4	Formation-Series	56
13.5	Floristic and Physiognomic Units	58
13.6	Conclusion	59
13.7	Summary	60

13 THE PHYSIOGNOMIC APPROACH

13.1 Introduction

Vegetation possesses two principal properties, floristic composition and physiognomy. Different workers have tended to be more interested in the one than the other, but in general in any locality it has been the physiognomy which has received first attention with a later progression to more detailed floristic studies. This happened for example in Europe, where in the later stage interest in physiognomy declined among phytosociologists but remained active among biogeographers working on vegetation on a world scale. At the same time there are countries, in the tropics especially, where local vegetation studies are still in an early stage and where physiognomy is consequently the primary concern. There are thus two developments to be discussed further – the physiognomic approach in biogeography, and physiognomic approaches to more local areas, particularly in British Commonwealth countries.

13.2 History and Major Concepts

Many of the great early biogeographers were primarily interested in the floristics of vegetation, e.g. DE CANDOLLE (1855), DRUDE (1890), and SCHIMPER (1898). GRISEBACH (1838, 1872) was the first great physiognomist and it was he who at a very early date (1838) introduced the term *formation* in its physiognomic connotation: 'I give the name phytogeographical formation to a group of plants, such as a meadow or a forest, that has a fixed physiognomic character.' The term formation was used in this sense by some later authors of the 19th Century but others used it for floristic units, that is as a synonym for association.

WARMING was the next great physiognomist and his papers appearing from 1895 attracted great attention culminating in the publication of his book '*Oecology of Plants*' (1909) specially written for English translation. For WARMING (p. 145) 'an association is a community of definite floristic composition within a formation', and (p. 140), 'a formation is an expression of certain defined conditions of life and is not concerned with floristic differences.' Whatever other influences later affected thought in the physiognomic field, these two propositions of WARMING's have remained basic. WARMING's formations were mainly characterised by dominant growth-form, and he was relatively little concerned with further details of vegetation structure, as these were studied later by the British school.

The other great Danish physiognomist, RAUNKIAER, whose collected papers from 1904 also found their way into a special English edition (1934), influenced the main stream of thought rather less than WARMING. RAUNKIAER's system is based purely on *life-forms*, a classification of plant forms according to the position and character of their perennating buds. His system is more floristic than physiognomic; the *biological spectra* (percentages of the species in a flora that belong to the various life-forms) facilitate comparisons of the floras of different geographic areas and communities but do not effectively characterize vegetation structure. Biological spectra have been employed by many workers, but this approach is of little value in, for instance, the recognition of forest types in the tropics. Just one element of RAUNKIAER's work, the *leaf-size classes*, was picked out as of practical value by the British physiognomic school and incorporated as a standard criterion of classification. The RAUNKIAER approach is further discussed by BRAUN-BLANQUET (1951, 1964), CAIN (1950, CAIN & OLIVEIRA 1959) and DANSEREAU (1957) and in Part II of this *Handbook*.

From the work of GRISEBACH, WARMING and others the plant *formation* has emerged as a key concept, equivalent to the *biome* of animal ecologists. The formation can be more broadly or more narrowly defined but the essential concept is a classification of vegetation by its physiognomy or structure, particularly growth-form dominance. By structure we understand here the organisation of the height and degree of density of these layers. Environment must usually also be part of the definition: A formation is a major kind of plant community on a given continent, characterized by physiognomy and a range of environments to which that physiognomy is a response. Thus, tropical grassland (savanna), temperate grassland (steppe), alpine meadow, and saltmarsh are regarded as different formations though all are dominated by the same growth-form (grasses and grass-like plants). Table I (WHITTAKER 1970) gives a simple outline of major growth-forms; for more detailed treatments see DU RIETZ (1931), ELLENBERG (1956), and SCHMITHÜSEN (1968). Many authors, especially RICHARDS, TANSLEY and WATT who will be mentioned below, have used the term 'life-form' in this sense but, in order to avoid confusion, the latter should be restricted to RAUNKIAER's life-forms which are not at all the same thing.

TABLE I

Major plant growth-forms on land

Trees, larger woody plants, mostly well above three meters tall
    Needle-leaved (mainly conifers – pine, spruce, larch, redwood, and so on)
    Broad-leaved evergreen (many tropical and subtropical trees, mostly with medium -sized leaves)
    Evergreen-sclerophyll (with smaller, tough, evergreen leaves)
    Broad-leaved deciduous (leaves shed in the Temperate Zone winter, or in the tropical dry season)
    Thorn-trees (armed with spines, in many cases with compound, deciduous leaves)
    Rosette trees (unbranched, with a crown of large leaves – palms and tree-ferns)
Lianas (woody climbers or vines)
Shrubs, smaller woody plants, mostly below three meters in height
    Needle-leaved
    Broad-leaved evergreen
    Broad-leaved deciduous
    Evergreen-sclerophyll
    Rosette shrubs (yucca, agave, aloe, palmetto, and so on)
    Stem succulents (cacti, certain euphorbias, and so on)
    Thorn-shrubs
    Semishrubs (suffrutescent, that is, with the upper parts of the stems and branches dying back in unfavorable seasons)
    Subshrubs or dwarf-shrubs (low shrubs spreading near the ground surface, less than 25 cm high)
Epiphytes (plants growing wholly above the ground surface, on other plants)
Herbs, plants without perennial aboveground woody stems
    Ferns
    Graminoids (grasses, sedges, and other grasslike plants)
    Forbs (herbs other than ferns and graminoids)
Thallophytes
    Lichens
    Mosses
    Liverworts

The matter of scale also materially influences formation concepts. The essential definition of formation offers much play for individual choice of how broadly or narrowly formations are to be defined. Definition by only the major growth-form of the uppermost stratum (or the stratum of greatest coverage) produces relatively few broadly defined formations; distinction by different combinations of growth-forms of the same or different strata produces a larger number of more narrowly defined formations. For treatment of continental or world vegetation few, broad formations can be conveniently dealt with, whereas detailed local studies may require recognition of subdivisions and minor types as formations or subformations.

Within this framework it is now generally recognised that physiognomy expresses biotope conditions and that similar biotopes in different parts of the world will exhibit and can be recognised by vegetation of similar physiognomy. Local influences are sometimes pronounced because different growth-forms are available in different geographic areas. For example northern coniferous forests (taiga) and the Australian mallee (of giant shrubs) have physiognomies all their own, while Southern Africa, Australia and middle America have all produced growth-forms adapted to desert conditions, some strikingly convergent but others quite different. Broadly, however, there is a convergence in physiognomy in corresponding major environments on a worldwide scale, and world-wide systems of vegetation classification must inevitably be physiognomic in character. The evolutionary convergence of formations of similar environments on different continents has emerged as a major principle in biogeography. These generalised, world-scale units that group together similar formations from different continents are best termed *formation-types* (or *biome-types*).

13.3 Physiognomic Systems

We need to discuss (before going on to British applications to local areas) four major ways of approaching vegetation through physiognomy. These are: (i) Outline classifications of formation-types of the world, (ii) Interrelation of formations or formation-types with types of climates, (iii) Direct descriptive systems that need not rely on formations or formation-types, and (iv) Treatment in terms of sequences of formations along environmental gradients (formation-series) or of physiognomic continua along environmental gradients (ecoclines). Much of the difficulty of physiognomic research results from the fact that many formations intergrade continuously. We are dealing with merging phenomena which are not inherently suitable for classification, though they can be profitably classified. Given this difficulty, our four approaches entail: (i) Derivation of a classification by typification – recognition of major types of communities by physiognomy, without attempting to provide in the system for intergradations and minor physiognomic types. (ii) Classification that gives joint emphasis to climate and physiognomy, and must again limit itself to broad relationships. (iii) Use of descriptive characterization of physiognomy without formal classification. (iv) Concern with relations to environmental gradients and intergradations, with or without classification.

13.3.1 FORMATION-TYPES

There are a number of systems of formation-types or biome-types in the literature; but the systems that have had widest influence in

TABLE II

The formation-types of SCHIMPER and VON FABER (1935)

1. *Tropical Rain Forest* occupies regions of high and constant rainfall and temperature. The forest is many-layered, leaves are mainly evergreen, large, entire; trees tall and buttressed; epiphytes and lianas very common. The flora is very rich. Amazonia, Congo Basin, Malaysia.
2. *Subtropical rainforest* is found in humid subtropical regions with some seasonal variation in temperature and rainfall. Luxuriance in structure and composition is reduced. Brasil, African Highlands, Southeast Asia.
3. *Monsoon Forest*: Tropical and subtropical with a moderate winter dry season. Forest is tall, many-layered, with predominance of deciduous species in the canopy. Central America, India, Southeast Asia.
4. *Temperate Rain Forest* expresses high and constant rainfall in cooler regions. Forest is moderately tall, dense, few-layered, leaves are evergreen, small, or coriaceous. Much moss and lichen. A variant, montane rain forest or cloud forest, is found on tropical mountains. Tasmania, New Zealand, Chile.
5. *Summer-green Deciduous Forest* occupies regions with a pronounced seasonal change of temperature, a cold winter with snow and a mild to warm wet summer. Trees are tall, structure simple, leaves broad, fine and deciduous. Eastern North America, Europe, China.
6. *Needle-leaf Forest* is characteristic of cold regions with long winters and high rainfall. Trees are coniferous, needle or scale-leaved, and may be very large in size. Western North America, Northern Europa, Siberia.
7. *Evergreen Hardwood Forest* characterises regions of 'mediterranean' climate with a dry summer and wet, mild winter. Trees are small (except in Australia) and leaves sclerophyllous. Australia, California, Mediterrranean.
8. *Savanna Woodland* appears under a summer rainfall with a long dry season, i.e., more extreme than monsoon forest. Trees are small, evergreen, in open formation, with a ground layer of tropical bunch-grasses. Brazilian and African plateaux, North Australia.
9. *Thorn Forest and Scrub*. Tropical, dry climates. Trees are small, often thorny and deciduous. The ground layer includes many succulents, annuals and grasses. Brazil, Africa, India.
10. *Savanna* is a moist tropical grassland, with or without trees, and may owe its origin to fire or to adverse soil conditions or both. Pantropical.
11. *Steppe and Semidesert* occur in dry climates with winter-rainfall, i.e., more extreme than evergreen hardwood forest. Open shrublands with annual herbs and grasses, or dry grasslands. (Some tropical dry grasslands have been classed as steppe by other authors). North America, Australia, Russia, Argentina.
12. *Heath*. Like the tropical savanna the heath in temperate regions is governed by fire or adverse soil conditions or both. It is a formation of ericoid shrubs with scattered larger shrubs and small trees. Worldwide, locally.
13. *Dry Desert*. Warm regions of very low rainfall with open vegetation and special plant forms evolved in different parts of the world, e.g. succulent Cactaceae in North America, succulent Liliaceae, Aizoaceae, Euphorbia and Welwitschia in Southern Africa, hummock grasses in Australia.
14. *Tundra and Cold Woodland*. This is the semi-desert of cold regions where there is a short summer growing season. Lichens are especially abundant under sedges and grasses (tundra) or under stunted trees. On rocky areas mosses may be dominant. Northern hemisphere in high latitudes.
15. *Cold desert*. Edge of icecaps, glaciers and permanent snowfields. Vegetation sparse, mainly herbaceous.

TABLE III

RÜBEL's formation-types, with Latin, German, and English equivalents

Lignosa - Woody Formations

Pluviilignosa
1. Pluviisilvae - Regenwälder - rain forest
2. Pluviifruticeta - Regengebüsche - rain scrub

Laurilignosa
3. Laurisilvae - Lorbeerwälder - laurel-leaved forest
4. Laurifruticeta - Lorbeergebüsche - laurel-leaved scrub

Durilignosa
5. Durisilvae - Hartlaubwälder - sclerophyll forest
6. Durifruticeta - Hartlaubgebüsche - sclerophyll scrub

Ericilignosa
7. Ericifruticeta - Echte Heiden - heaths

Aestilignosa
8. Aestisilvae - Sommerwälder - summergreen deciduous forest
9. Aestifruticeta - Sommergebüsche - summergreen scrub

Hiemilignosa
10. Hiemisilvae - Regengrüne Wälder - raingreen (monsoon) forest
11. Hiemifruticeta - Regengrüne Gebüsche - raingreen scrub

Aciculilignosa
12. Aciculisilvae - Nadelwälder - needle-leaved or coniferous forest
13. Aciculifruticeta - Nadelgebüsche - needle-leaved scrub

Herbosa - Herbaceous Formations

Terriherbosa
14. Duriherbosa - Hartwiesen, Steppenwiesen - hardgrass prairie and steppe
15. Sempervirentiherbosa - Immergrüne Wiesen - evergreen grassland or meadow
16. Altherbosa - Hochstaudenwiesen - tall herbage or forbland

Aquiherbosa
17. Emersiherbosa - Sumpfwiesen - marsh
18. Submersiherbosa - Submerse Wasserwiesen - submerged aquatics
19. Sphagniherbosa - Hochmoor - peat bog

Deserta - Deserts
20. Siccideserta - Trockeneinöden - dry desert
21. Frigorideserta - Kälteneinöden - cold desert
22. Litorideserta - Strandsteppen - strand (saltspray) steppe
23. Mobilideserta - Wandereinöden - vegetation of dunes, and unstable soils
24. Rupideserta - Felsfluren - vegetation of screes, talus, etc.
25. Saxideserta - Stein- und Holzfluren - vegetation of rocks, tree trunks, etc.

Special Groups
26. Aquerrantia - Phytoplankton - free aquatic organisms, mostly microscopic
27. Solerrantia - Phytedaphon - terrestrial micro-organisms, soil bacteria and fungi
28. Aërerrantia - Phytaëron - aerial microorganisms

phytogeography are those of RÜBEL (1930, 1936) and SCHIMPER (SCHIMPER & FABER 1935). The two systems are summarized as Tables II and III. More recent systems (too detailed to reproduce here) have been proposed by FOSBERG (1961), ELLENBERG & MUELLER-DOMBOIS (1967), and SCHMITHÜSEN (1968). A major treatment of world vegetation by WALTER (1962, 1968) uses no formal classification. The SCHIMPER system seems in many respects the best. RÜBEL's system is more cumbersome and somewhat more artificial. His schematic pairing of forest and shrubland types results in overemphasis of shrublands in the classification, and underemphasis of woodlands (communities of small trees, usually in open spacing and with well-developed undergrowth). RÜBEL refers to his primary types as formation-classes, rather than using the more generally accepted term formation-type. Vegetation maps on a continental or subcontinental scale are almost inescapably based on physiognomy. For applications of formations in vegetation mapping, see KÜCHLER (1967).

Table IV is an effort to coordinate with one another these two classic systems and two current systems (SCHMITHÜSEN 1968, WHITTAKER 1970). The 'subclasses' of SCHMITHÜSEN's classification most nearly correspond to the formation-types of others, rather than his numerous and narrowly defined formation-types. It is possible to find common features as well as divergences among the classifications of different authors. Coordination of different systems is not, however, simple, and some of the types entered in Table IV do not correspond as neatly as their entry in the table might make it appear. Even types given the same name by three or four of the authors (e.g. heath, savanna, tundra) represent three or four different, though overlapping, concepts. Such difficulties reemphasize the subjective choices of the limits and numbers of formations recognized. There is thus a degree of artistry in the classification, or typification, of vegetation physiognomy, though the product of this artistry is valuable to our understanding of world vegetation.

13.3.2 CLIMATIC CORRELATION

Working with the SCHIMPER system, DANSEREAU (1957) sought to relate its units to climate. The formation-types are first rearranged into four major *biochores* – forest, savanna, grassland, and desert. Not everyone will concur with this, especially as DANSEREAU uses 'savanna' in a sense peculiar to himself, but the viewpoint is an interesting one. Within these biochores the formation-types may now be arranged into a triangle representing the range of major

TABLE IV

A comparison of four physiognomic classifications

Schimper & Faber 1935	Rübel 1930	Schmithüsen 1968	Whittaker 1970
1. Tropical rainforest 2. Subtropical rainforest 4. Temperate rainforest	1. Rainforest	IA. Evergreen forest	1. Tropical rainforest 3. Temperate giant rainforest 4. Montane rainforest 8. Elfin woodland
	3. Laurel-leaved forest		

6. Needleleaf forest	12. Needle-leaved forest		7. Taiga
			6. Temp. evergreen forest--
			   needleleaf
			   sclerophyll

7. Evergreen hardwood forest	5. Sclerophyll forest		
3. Monsoon forest 5. Summergreen deciduous forest	10. Raingreen forest 8. Summergreen forest	IB. Deciduous forest	2. Tropical seasonal forest 5. Temperate deciduous forest
8. Savanna woodland		IIA. Evergreen woodland	11. Temperate woodland
		IIB. Deciduous woodland	
9. Thorn forest and scrub	11. Raingreen scrub	IIC. Xeromorphic woodland IIIC. Xeromorphic shrubland	9. Thorn woodland 10. Thorn scrub

	9. Summergreen scrub	IIIB. Deciduous shrubland	12. Temperate shrublands--
			    deciduous
			    heath
			    sclerophyll
			    subalpine-needleleaf
			       -broadleaf

12. Heath	7. Heath	VIC. Heath (in part)	
	6. Sclerophyllous scrub 13. Needle-leaved scrub 4. Laurel-leaved scrub	IIIA. Evergreen shrubland	
10. Savanna 11. Steppe and semidesert	14. Hardgrass prairie	IVA. Savanna IVB. Steppe (in part)	13. Savanna 14. Temperate grassland 19. Cool-temperate desert scrub
13. Dry desert	20. Dry desert	VIIA. Desert	17. Tropical desert 18. Warm-temperate desert
14. Tundra 15. Cold desert	15. Meadows 21. Cold desert	IVC. Meadows VID. Tundra	15. Alpine grasslands 16. Tundra 20. Arctic-alpine desert
(edaphic)	19. Moss moor 2. Rain scrub 17. Marsh	VIE. Moss moor	21. Bog 24. Mangrove swamp 25. Saltmarsh

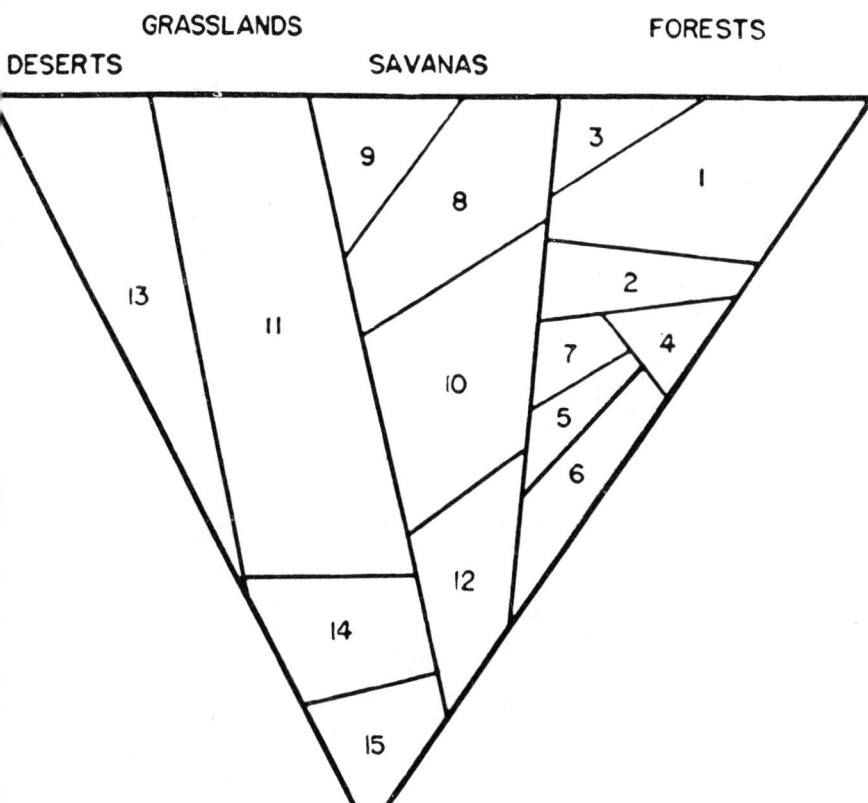

Fig. 1. A bioclimatic diagram arranged according to increasing temperatures (bottom up) and increasing rainfall (left to right) (DANSEREAU 1951). Formation-types are represented by numbers corresponding to those of SCHIMPER & VON FABER (1935), Table II. The upper-left angle is for extreme of heat and drought, the upper-right for the extreme of heat and humidity, the lower angle for extreme cold with low precipitation.

variations in the world's climates (Fig. 1). Fig. 2 represents a somewhat comparable diagram by WHITTAKER (1970). There are various ways in which these treatments, that start with a system of formation-types and then seek to relate these to climate, are imperfect. The formation-types overlap in relation to climate in a complex way. Effects of maritime vs. continental climates and of soil-and-fire influences are, in particular, difficult to deal with adequately. The alternative approach is to start from the other side of the relationship, to seek to define not formation-types as such but major climatic regions that may be characterized both by climate and by physiognomy adapted to that climate. The most widely used systems

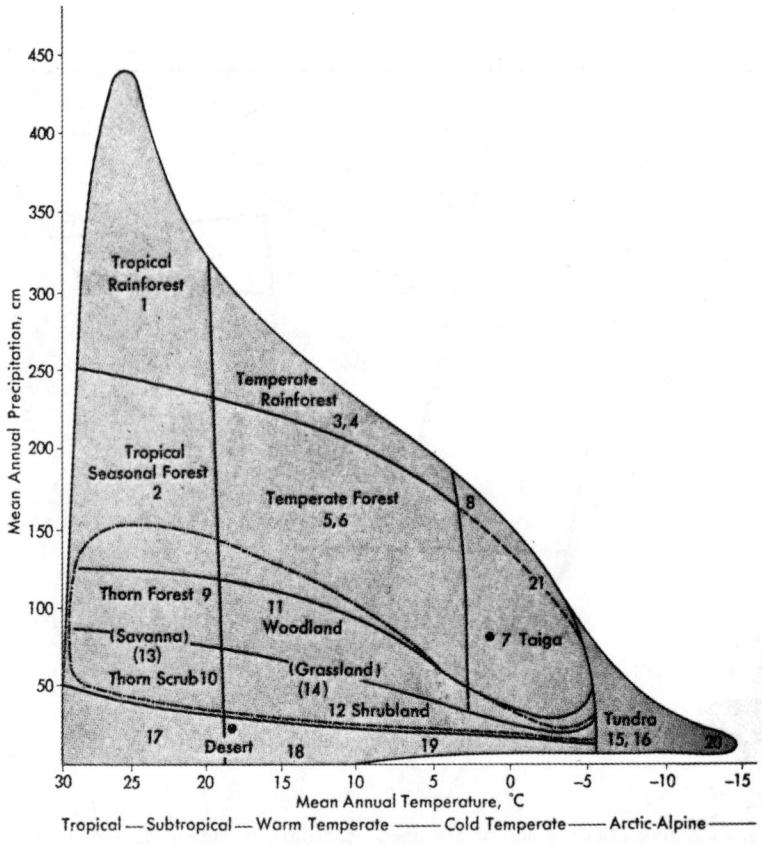

Fig. 2. A pattern of world formation-types in relation to climatic humidity and temperature (WHITTAKER 1970). The numbers refer to the formation-types of WHITTAKER (1970) as given in Table IV. The boundaries are necessarily approximate. In climates between forest and desert, maritime versus continental climate, soil effects, and fire effects can shift the balance between woodland, shrubland, and grassland types. The dot-and-dash line encloses a wide range of environments in which either grassland, or one of the types dominated by woody plants, may form the prevailing vegetation in different areas.

of this sort are those of KÖPPEN (1900, 1923), and THORNTHWAITE (1931, 1948) (see also STEFANOFF 1930, RUMNEY 1968). TROLL (1948, 1961, 1968) and STOCKER (1963) have dealt with the asymmetry of the Northern and Southern Hemispheres, and effects of altitude versus latitude, in the relations of climate and vegetation.

A recent system relating physiognomy to climate is that of HOLDRIDGE (1947, 1967), classifying vegetation of the world into a system of compartments defined by temperature and moisture (Fig. 3). The

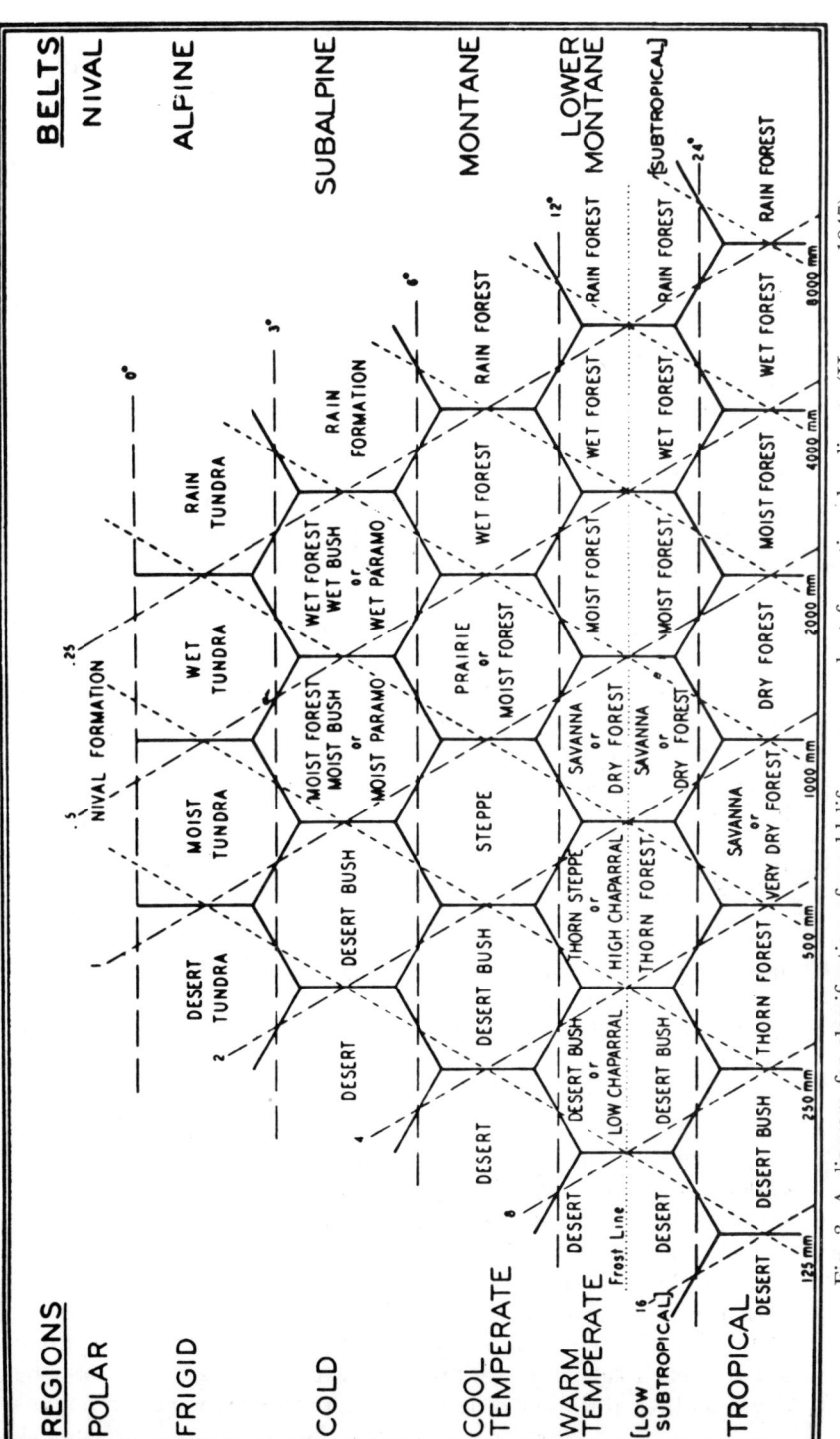

Fig. 3. A diagram for classification of world life zones or plant formations by climate (HOLDRIDGE 1947).

system suffers from its over-schematic character, which juxtaposes in compartments some quite disparate types of vegetation, while also requiring separation into different compartments of other, more similar, types. But all approaches have their difficulties, and HOLDRIDGE's has been gaining favor as an interpretation of tropical vegetation in particular.

13.3.3 DESCRIPTIVE SYSTEMS

An entirely different line of development in the physiognomic treatment of vegetation was opened by KÜCHLER (1947). Classification in its usual sense (here involving grouping entities around abstract type-concepts) was rejected. KÜCHLER proposed instead a system of descriptive notation using four series of symbols — predominant growth-form, height, density and special features — which can be combined so as to show the important features of plant cover. This represented essentially a geographer's approach to the characterisation of plant cover for mapping purposes and is most useful for over-all description of regional vegetation rather than for detailed work on plant communities. KÜCHLER has applied this system, amplified in 1949, in vegetation mapping (1948, 1950). As with RAUNKIAER's biological spectra the simplicity of the system has commended itself to later workers, and despite criticisms which can be levelled at it a recent paper has shown that it could be adapted to use in Australia (ROSS COCHRANE, 1963). In point of fact, as DANSEREAU has shown (1951), KÜCHLER's system for all its practical value to a geographer, is open to considerable criticism by ecologists. Major divisions in the system are of unequal significance and tend to ignore features not found in the North Temperate Zone. DANSEREAU set out to reform the system, making it more logical and comprehensive and appropriate to the needs of descriptive ecologists. He did so successfully but in the process transformed it from its original role as a mapping notation into an elegant ecological characterization which is too elaborate for use in mapping and may very well prove too complicated for widespread use. It is difficult to make any physiognomic classification fully inclusive without also making it cumbersome.

DANSEREAU (1951, 1957) proposed six categories of criteria in place of KÜCHLER's four, viz. growth-form (which he terms life-form), size, function, leaf size and shape, leaf texture, and coverage (Table V). The characters to be distinguished, varying from three to six in each category, are provided with two sets of symbols, the one alphabetical and the other pictorial, the former to be compound-

TABLE V

Six categories of criteria to be applied to a structural description of vegetation
(DANSEREAU 1957).

	1. FORM		4. LEAF SHAPE AND SIZE
T	trees	n	needle or spine
F	shrubs	g	graminoid
H	herbs	a	medium or small
M	bryoids	h	broad
E	epiphytes	v	compound
L	lianas	q	thalloid

2. SIZE

t tall (T: minimum 25m)
   (F: 2—8m)
   (H: minimum 2m)

5. LEAF TEXTURE

m medium (T: 10—25m)
   (F, H: 0.5—2m)   f filmy
   (M: minimum 10cm)   z membranous
                          x sclerophyll
l low (T: 8—10m)   k succulent or fungoid
   (F,H: maximum 50cm)
   (M: maximum 10cm)

	3. FUNCTION		6. COVERAGE
d	deciduous	b	barren or very sparse
s	semideciduous	i	discontinuous
e	evergreen	p	in tufts or groups
j	evergreen-succulent;	c	continuous
	or evergreen-leafless		

ed into formulae and the latter for the construction of diagrams. Note that this is no longer a system of major structural types of plants in the sense of Table I; it is a set of intersecting criteria by which plants (and communities) can be characterized without formal classification. To characterize the community a descriptive formula consisting of six letters (one from each of the prime categories) is made up for each synusia or stratum. Structurally very complex communities will qualify for a whole run of six-letter formulae, written with full-stops between them. This does in fact achieve a quite complete description of the community, but the result can be elaborate. The pictorial symbols are used to construct diagrams something like a stylised version of the RICHARDS profile diagram (see below), in which the symbols themselves suggest the shape and size of trees, shrubs, and other plants, and are spaced according to percentage cover (Fig. 4). These diagrams, which should become known as DANSEREAU Diagrams (Danserograms), have real value for comparative purposes. Because measured profile diagrams involve a certain amount of subjective interpretation and reliance

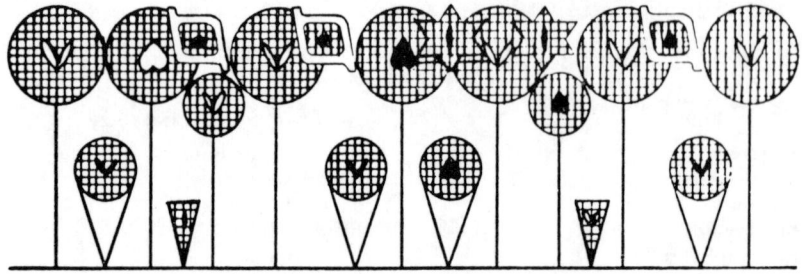

LECYTHIDETUM AMAZONENSE

Tteh(v)z(x)c. Ltehxi. Etegxp. Tmeh(v)z(x)b.
Ftev(h)xi. Hmegx(vz)b.

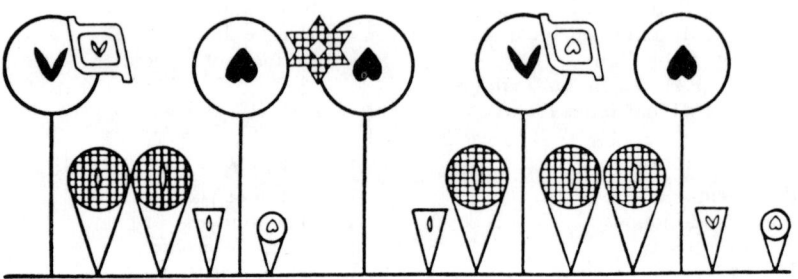

TECTONETUM GRANDIS

Ttdh(v)xi. Ltdv(h)zi. Eteazb. Ftegzp. Fldhzb.
Hmdg(v)zb.

Fig. 4. DANSEREAU diagrams for a tropical rainforest, above, and a more open monsoon (tropical deciduous) forest, below (DANSEREAU 1951). The upper figure may be compared with the profile diagram, Fig. 6.

on artistic skill, they may be difficult to compare, whereas the conventional symbols are always interpretable and readily comparable. A system specially adapted to the nature of Australian vegetation and based on the best features of KÜCHLER and DANSEREAU is now being applied in that country (BEARD 1969, BEARD & WEBB 1972).

13.3.4 FORMATION-SERIES

The fourth approach relates vegetation physiognomy to environmental gradients. BEARD (1944) working in the British tradition to be described below, developed a system of formations for tropical America. Unlike the systems of SCHIMPER and WARMING, which begin with a pre-conceived hierarchy of formations into which examples under study must be fitted, BEARD's system begins with typification of a number of actual communities and then proceeds to consider how they are related. BEARD's formations are more narrowly defined units than those of Tables II and III. The formations are related to one another and environment in terms of sequences along environmental gradients; such a sequence is a *formation-series*. Five formation-series were recognized, two considered to be climatically controlled and three as edaphic. The former are the seasonal series (lowland communities controlled by decreasing rainfall from rainforest through deciduous forests and thorn woodlands to desert) and the montane series (controlled by increasing altitude from lowland to montane rainforest to elfin woodland and the alpine paramo and puna). The edaphic formations were arranged into swamp, seasonal swamp, and dry evergreen series, the last being essentially on shallow soil or rock pavements. These series may be thought of as five directions of divergence from the most highly developed and focal community of most favorable environments, the tropical rainforest. In each case the component formations of the series show progressive reduction of structure and modification of growth-forms outwards toward a theoretical pessimum devoid of vegetation (Fig. 5).

Eleven years later additional information was available for the American tropics, the principal further contribution being that of FANSHAWE (1952). This information enabled BEARD (1955) to revise and expand his formation series into a framework capable in principle of embracing all types of vegetation likely to occur in tropical America. BEARD (1955) pointed out also that vegetation is in fact a continuum, and that the defined formations are only conveniently recognized stages in a continuous gradient. Many communities must be expected to show characteristics intermediate between two formations or even, less commonly, between two formation-series. In a study of temperate mountains WHITTAKER & NIERING (1965) showed the continuity of formations by plotting growth-form coverages in relation to elevation and topography. A formation-series with decreasing elevation, from subalpine forest (taiga) through other forest and woodland types to grassland and desert, was suggested (WHITTAKER & NIERING 1968). WHITTAKER (1970) has also extended BEARD's approach by relating more broadly defined formations to four major climatic gradients for the world. Three formation-series, here termed ecoclines again radiate (upward in mountains, northward, and toward drought) from the tropical rainforest; the fourth is from deciduous forest to desert in

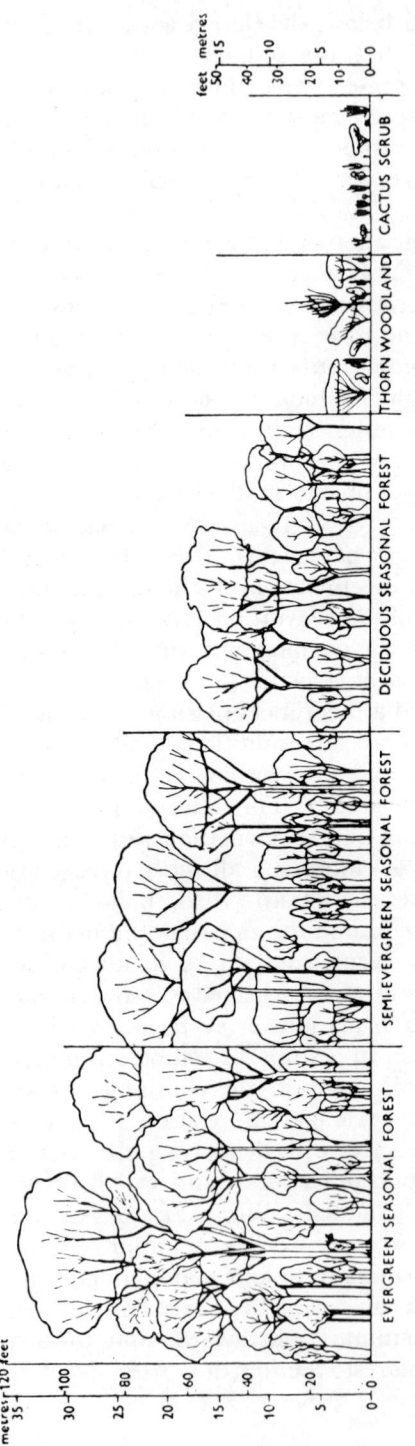

Fig. 5. A formation-series for communities on Trinidad affected by increasing drought from left to right (BEARD 1944). Community structure is represented by profile diagrams in the manner of DAVIS & RICHARDS (Fig. 6). This, the seasonal formation-series, diverges from the tropical rainforest of Fig. 6 along a climatic gradient of increasingly restricted and seasonal precipitation at low elevations in tropical America.

the North Temperate Zone. BEARD's (1944, 1955) most important advance may have been in showing that such higher-level ordination of vegetation is possible, relating formations as clearly characterized physiognomic types to five or more major axes of environmental variation. The system thus provides means of both relating physiognomy to environment, and comparing this relationship for different areas and continents.

13.4 Physiognomy in British Commonwealth Ecology

13.4.1 FOREST TYPES AND ASSOCIATIONS

Physiognomic approaches to local vegetation have been most actively employed in British ecology. In the first half of the present century administrators of British Colonial territories, which lay mainly in the tropics, were busy developing resources. Forest services were set up, forest officers proceeded to investigate the forest estates in their charge and to carry out inventories. Taxonomic investigations were made of species present, and ecological investigations were made of forest communities. The word 'forest-type' came into general use, and with the complexity of forest composition in the tropics it was natural that forest types should be based on physiognomy rather than species composition.
In 1924 a British Empire Vegetation Committee was appointed by an Imperial Botanical Conference in London with the object of promoting the study of the plant cover in British countries. While in the end only one of the ambitiously planned regional monographs ever came to fruition (ADAMSON 1938) there was a speedy production of a handbook for ecological workers, '*Aims and Methods in the Study of Vegetation,*' by TANSLEY & CHIPP (1926). The utility of this work, in the event, was diminished by its strongly Clementsian outlook. It represented in fact a Clementsian deviation in British ecology instead of a continuation and expansion of the physiognomic approach which would have been of most value to the forestry workers who were the people most concerned. Classification, to TANSLEY and CHIPP, was centered upon the association as 'the largest unit which consists of a definite assemblage of species (usually with definite dominants) and a definite habitat.' Smaller components of the association could be recognised, under the Clementsian terminology (faciation, location, society and clan, see 14.2). There was, however, no higher classification. The association was considered to possess a characteristic growth-form and structure as well as

floristic composition, but the word 'formation' was not mentioned. TANSLEY and CHIPP differed from CLEMENTS only in insisting (following SCHIMPER) upon a distinction between climatic and edaphic climax communities.

The principal weaknesses of this approach were that it proposed to define the association partly by habitat, which in practice was very difficult to assess accurately, and that it lacked physiognomic standards by which different associations could be compared. Field workers dealing with forest types were left to propose their own terminology and to speculate on similarities and differences between their own and other workers' communities. CHAMPION's (1936) classic work on the forest types of India and Burma is an example of the individualistic approach which became general. TANSLEY and CHIPP did not found a school of ecological thought.

13.4.2 PROFILE DIAGRAMS

The necessary steps to rationalise classification on a physiognomic basis were not taken until DAVIS & RICHARDS (1933-4) adopted the device of the *profile diagram*, a drawing of an accurately measured sample strip of forest (Fig. 6). A sample was to be selected by the field worker as typical of the formation which it is desired to represent and of mature structure, and the sample was to be measured by a definite procedure (quoted from RICHARDS 1952): —'A narrow rectangular strip of forest is marked out with cords, the right angles being obtained with the help of a prismatic compass. In Rain Forest the length of the strip should not usually be less than 200 ft. (61 m,); 25 ft. (7.6 m.) has proved a satisfactory width. All small undergrowth and trees less than an arbitrarily chosen lower limit of height are cleared away. The positions of the remaining trees are then mapped and their diameters noted. The total height, height to first (large) branch, lower limit of crown and width of crown of each tree are then measured. Often it is only possible to obtain these measurements by felling all the trees on the strip, and the trees must be felled in a carefully selected order, so that the heavier trees do not crush the smaller in falling, but it is sometimes possible to make measurements of sufficient accuracy from the ground by means of an Abney level. Felling has the advantage that herbarium material can be collected for the identification of species.'

The field measurements are plotted to produce an accurately drawn diagram which will show the structure of the community in a manner which is impossible by photography in dense forest. For this reason the profile diagram became popular in connexion with

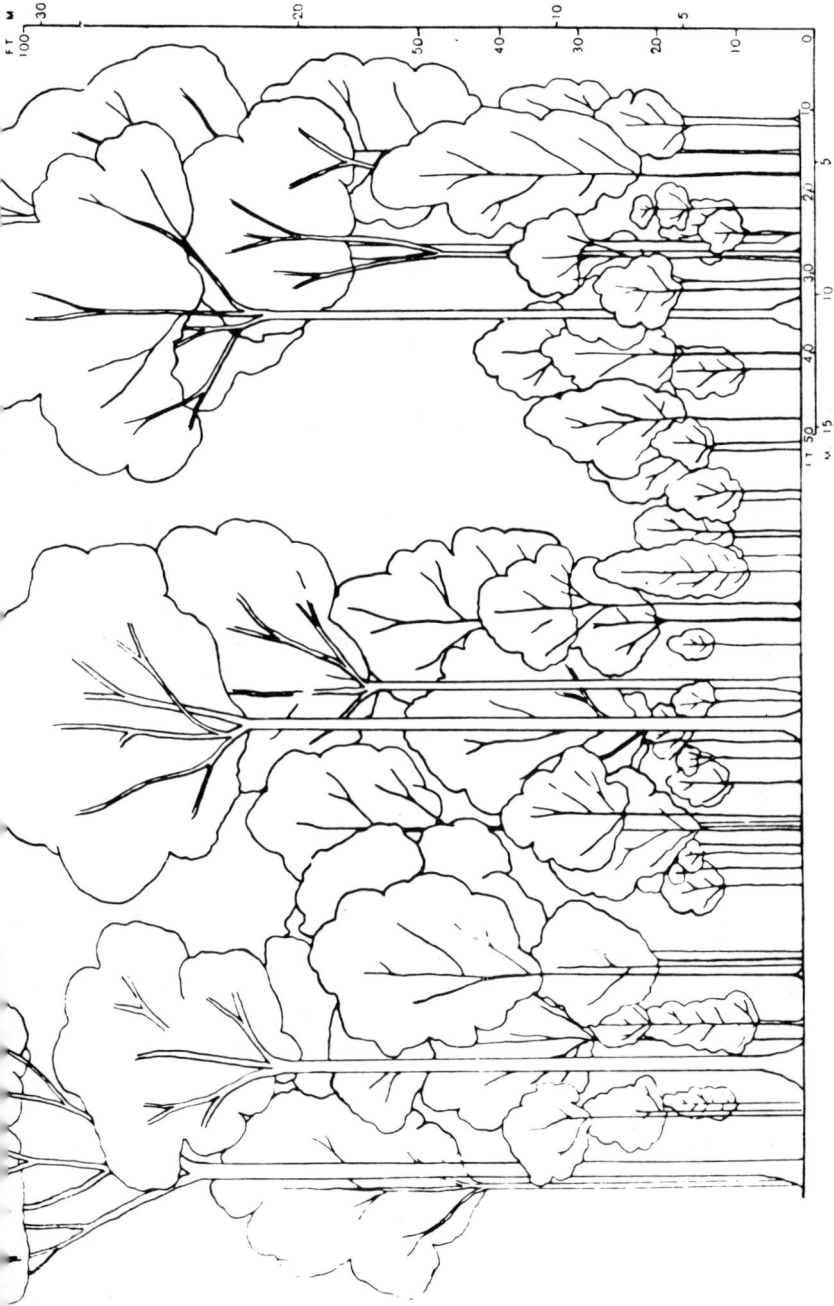

Fig. 6. A profile diagram for primary mixed tropical rainforest, Moriballi Creek, British Guiana, the first application of the profile technique as developed by Davis & Richards (1933-4). Only trees over 15 ft (4.6 m) are shown.

studies in tropical forests. After the initial publications by DAVIS & RICHARDS (1933-4) of the profile diagram in relation to studies of rain forest in British Guiana, the same methods were used in Sawarak (RICHARDS 1936) and Nigeria (1939), affording for the first time material for effective quantitative and qualitative comparisons of structure between examples from different parts of the world of tropical forests thought to belong to the same formation-type. A close correspondence was found between the examples of British Guiana and Sarawak, but the Nigerian forest showed differences. The practical utility of the profile diagram for comparative studies was amply demonstrated.

13.4.3 FORMATIONS AND FORMATION-TYPES

A paper by BURTT-DAVY (1938), who had been a member of the original British Empire Vegetation Committee, was read to the British Ecological Society drawing attention to the lack of progress and the general confusion in classification and terminology. After a review of the works of SCHIMPER (1903), WARMING (1909), CHIPP (1926) and CHAMPION (1936), he outlined a set of criteria for comprehensive description of vegetation and proposed a hierarchic physiognomic classification for tropical vegetation with appropriate standardised terminology. The British Ecological Society appointed a committee of three to examine the matter, and their report appeared as RICHARDS, TANSLEY & WATT (1939). These authors decided to 'call attention to certain basic principles' which are quoted as follows:

'There is an urgent demand for the stabilisation of the nomenclature of tropical forest vegetation, in order to avoid the confusion and misunderstanding that result from different (forest) officers calling the same thing by different names and possibly different things by the same name.... The authors hold that such a scheme must be inductive, based on the structure, life-form and flora of the communities themselves, not on mixed criteria which include necessarily vague characterisations of habitat....

'An actual sample of forest to be studied in the field is best regarded in the first place as a particular example (sometimes called a 'stand') of an *association* (dominated by two, several, or more species) or of a *consociation* (dominated by a single species), and characterised by the total *flora*, i.e., the whole collection of species present, though some (sometimes called *characteristic*) have a greater diagnostic or 'indicator' value than others.

'Certain associations and consociations are found to agree in the life-forms (i.e. growth-forms) of their component species, especially of their dominants, though the species may be totally different. This agreement in life-form indicates similarity of *essential habitat*, especially similarity of regional climate, and the associations showing such agreement may be put together in a larger unit – the formation.

'Formations of very well-marked life-form type are so strikingly and characteristically similar and are so obviously determined by similar climates, in whatever part of the world they may occur, that they have long been recognised as together forming distinct entities, which we may call *formation types*.'

RICHARDS, TANSLEY and WATT went on to recommend practical methods of recording structure and growth-form in order to typify formations. The RICHARDS profile diagram was brought in as an integral feature for the expression of structure and at the same time the growth-forms of component species were to be recorded in a quantitative manner. Percentages of the stand, or numbers of individuals per unit area, would be used to express occurrence of special characters such as plank buttresses, stilt roots, pneumatophores, thorny trunks or branches, cauliflory, succulence, peculiarities of bark; or of any special growth-forms present such as palms or cycads, bamboos, rattans and pandanus. Leaf characters are most important, and percentages of species and individuals in each stratum would express occurrence of leaves that are evergreen or deciduous, simple or compound, or of certain size-classes as proposed by RAUNKIAER (1934).

It has been shown that the Clementsian approach to vegetation led to certain difficulties in British practice. CLEMENTS (1916) drew attention to the phenomenon of plant succession which leads to the development of a mature and integrated community. Physiognomic studies, to be of comparative value, must in most cases deal with mature vegetation. A community subject to fire may have a structure directly related to the period since the last burning, and an immature example may be valueless except as a means of showing differences from a known mature example. It is possible that the failure of RICHARD's (1939) example of Nigerian rain forest to agree with the examples from British Guiana and Sarawak was due to disturbance of the Nigerian forest. In general, physiognomic treatments are scarcely relevant to questions of succession, but the British led by TANSLEY (1920) found it necessary to take a stand upon recognition of more than one kind of climax. To CLEMENTS there was but one climax in an area, determined by its climate, all other communities being regarded as in a state of succession toward that climax. TANSLEY objected that young, rapidly evolving communities which were demonstrably in a state of succession should be distinguished from others (e.g. fen, swamp, or saltmarsh) effectively controlled by local soil or water conditions. The latter would be considered by CLEMENTS as theoretically in succession, but in practice they are stable: they are self-maintaining or climax in their own particular habitats. The soils of such communities could not possibly evolve into soils which would support the climatic climax, and thus they should be regarded as edaphic climaxes. Later BEARD (1953) went further in suggesting that some landforms and soils could become senile and over-mature, developing adverse features which led to a devolution from the climatic climax. This was directly opposed to CLEMENT's dictum that the climax can be destroyed but cannot retrogress. Whatever the difficulties which surround the theory of the climax, and they are many (WHITTAKER 1953, and this *Handbook*, Part VIII), a nuclear idea remains important that there is a regionally characteristic vegetation, and it is this which forms the basis for the biogeographic treatment of formations in relation to climate.

13.4.4 FORMATION-SERIES

The methods of RICHARDS, TANSLEY and WATT for physiognomic typification were further developed by BEARD (1944) for comparison and classification of plant formations found in tropical America, only a few of which could be considered rain forest in a narrow sense, thus extending RICHARDS' work on the rain-forest to tropical vegetation in general. As described, BEARD's work led to recognition of a larger number of formations than in some other systems and to the concept of the rain forest as a kind of central or highest type of vegetation possible in the tropical environments, from which five 'formation-series' radiate in response to environmental gradients and interrelate the other formations to one another and the rain forest. The members of each series were united in sharing an essential habitat feature in some degree, either more or less favourable, which gives the formations some impress of common physiognomic characteristics. Thus deciduousness of leaves is a characteristic of the seasonal series, whereas a progressive simplification of structure can be traced through all the members of any series.

The basic principles of his work were expressed by BEARD (1955) in this way: 'The properties of the community express the habitat: the consistent dominance which defines the association expresses a constant local habitat within the regional flora and the consistent physiognomy which defines the formation expresses a constant essential habitat within the tropical zone. Types of habitat which are in general similar permit us further to group formations together into formation-series. There is thus a grouping at three levels: a floristic grouping — the association; a physiognomic grouping — the formation; and a habitat-grouping — the formation-series.'

To conclude this historical review, it has been shown that RICHARDS, TANSLEY and WATT, followed by BEARD, had aimed at the elimination of confusion in classification and terminology in the English-speaking world of tropical ecologists by introducing quantitative methods to define their physiognomic units. Confusion however still reigned in the international sphere, and an attempt to cope with this was produced in 1956 by the Scientific Council for Africa South of the Sahara in a curiously anonymous note appearing in the Publications of the Council under the title '*Phytogeography*.' This publication was the outcome of a conference designed to obtain some unanimity in classification and terminology of African vegetation, as the work done up to that time in different colonial territories by ecologists of different nationalities had produced a kaleidoscopic confusion of approach, method, concepts and nomenclature. The result was a framework similar to BEARD's for tropical America, which could be expected to embrace any plant community of tropical Africa. Twenty-three formations were typified, which is close to BEARD's number of twenty-eight, but there is a considerable difference in emphasis for climatic reasons. Most of tropical America receives rainfall in excess of 1000 mm per annum, whereas most of tropical Africa receives less than this. Thus most of BEARD's formations were moist forests and woodlands and most of the African formations are dry woodlands, savanna and steppe.

In the African publication, each formation is typified with a concise description and a standard name in both English and French, and a standard example taken from previously published work. Profile diagrams are also provided; though these appear to be diagrammatic rather than taken from actual measured samples, they are graphic and effective. The formations are organised into series similar to those of BEARD but the treatment stops short of a fully logical system because of a prior division into two main classes — Closed Forest Formations, and Mixed Forest-Grassland and Grassland Formations — an unnecessary reversion to the days of SCHIMPER and WARMING. Fully logical lowland-climate, montane-climate and edaphic series would each include a full range of types from forest to desert.

This African classification and terminology was followed by KEAY (1959) in the publication of the A.E.T.F.A.T. vegetation map of Africa south of the tropic of Cancer. In a recent paper BEARD

(1967) has discussed its application to tropical Australia. Physiognomic study and classification along the above lines have not yet been much applied outside the tropics. Notable exceptions have been the work of WEBB (1959) in Queensland and ROBBINS (1962) in New Zealand, both working in rain forest. ROBBINS made particular use of profile diagrams whereas WEBB laid the emphasis on growth-forms.

13.5 Floristic and Physiognomic Units

In the past, approaches to vegetation have tended to be more or less sharply divided between the floristic, which may ignore the structure and growth-form of vegetation, and the physiognomic, which need not take account of floristic composition. The former is undoubtedly the more suited to detailed studies of local areas and the latter to broadly-based studies on an international scale. It would seem that a desirable goal for the future should be the marrying of the two approaches into a single system of classification wherein floristic units are combined by their structure into physiognomic units.

Most students of physiognomy have been willing to recognise that their formations are divisible into floristically distinct units. All the British authors so far quoted accepted without question the existence of recognisable community-types characterised by consistent dominants and general composition, which they termed *associations*. At the same time it is true that the very existence of sociality (in the sense of species composition that is consistent between different stands) in tropical forest has been denied by other workers, mainly non-British, with experience in Africa and Malaysia. Thus VAN STEENIS (1958) describes the procedure for belt transects by forest surveyors in Indonesia from which 'it appears, however, that the plots are bewilderingly divergent and ... the picture does not suggest a constant specific composition.' On the other hand BEARD (1949) found from such enumerations in the British West Indian islands that consistent dominance and sociality were regular features of the vegetation, so much so that examples of the same forest type in different islands showed strikingly similar composition. It is evident therefore that sociality in tropical forest is different in different parts of the tropics, and that in areas with very rich floras the pattern tends to be obscured by sheer numbers of species. It has taken the recent work of ASHTON (1964) in North Borneo to demonstrate that modern statistical methods applied to this problem are capable of showing that sociality in a certain sense none the less exists.

The term *association* has been used throughout by British workers for their floristic units, but WHITTAKER (1962) has criticised this on the ground that the usage is not the same as that of the BRAUN-BLANQUET school which claims an exclusive right to the term. WHITTAKER (1962, article 14) suggested that the 'associations' of RICHARDS, TANSLEY and WATT, and BEARD, should be termed *dominance-types*. It is certainly true that the characterisation of the basic floristic unit of vegetation in a certain manner by diagnostic species has become identified with followers of BRAUN-BLANQUET. One may take the view, however, that the word 'association' means simply the basic floristic unit, and there is no reason to discard it if it is being defined in a different way. RICHARDS, TANSLEY and WATT in their full definition of the association quoted above make it clear that the association is known by its dominants but is characterised by its total flora, in which 'some species (sometimes called *characteristic*) have a greater diagnostic or 'indicator' value than others.' In selecting dominants to typify the association we are taking the easiest and most direct way of finding diagnostic species. It seems to the present writer that there is no real conflict. Can we not continue to accept WARMING's definition: 'An association is a community of definite floristic composition within a formation.'?

Different considerations apply to higher-level units. Floristic composition, the basis of the phytosociological approach, is only one of several important attributes of vegetation which merit attention and can be employed as a useful basis of classification. At the present time phytosociologists of the BRAUN-BLANQUET school use floristic groupings which become of less and less practical value toward the higher levels of the hierarchy. The association is the basic unit, alliances have been found useful, but orders and classes are another matter. Higher floristic groupings are applicable only to the geographical area of a particular flora, and even in the widest terms could do no more than extend across a single hemisphere. What valid comparison in floristic terms is there between vegetation units in the northern and southern hemispheres, or even between different parts of the equatorial zone? It does not seem that one can formulate meaningful world systems of vegetation classification except on a physiognomic basis.

13.6 Conclusion

From this review I would draw these conclusions: (i) The earliest approach to vegetation—through community structure or physiognomy—is still a vital and valuable one. It remains the principal basis of treating vegetation on a broad scale in relation to climate, and an important basis of vegetation research in more limited areas (particularly in the early stages of such work). (ii) Classification by physiognomy is subject to the difficulties (subjectivity in choice of community-types and their boundaries, and the conflict between reasonable classification that is incomplete and complete classification that is unmanageable) that afflict all classifications of natural communities. Nevertheless, physiognomy has proved to be the best means of gaining agreement on classification, and of comparing communities, among investigators on different continents. (iii) Physiognomy is adaptable to such different purposes as world-wide generalization about formation-series, and description of community structure without formal classification. The physiognomic approach most effectively relates salient structural characteristics of communities, which must be underlain by functional characteristics, to the qualities of environment to which these characteristics are adapted. (iv) There is advantage, for expressing these relationships, in both typification (in terms of formations and formation-types) and ordination (formation-series). Thus the major perspectives of classification and gradient analysis may well be combined not only in intensive local research, but also in study of the broader relationships of vegetation structure to environment.

13.7 SUMMARY

The physiognomic approach to vegetation is based on the study of community structure, as determined by the occurrence of major types of plants (growth-forms). Early sources of the approach led to the central concepts of formation (a major community-type as defined by physiognomy, with consideration also of broad environmental relations, on a given continent), and formation-type (a grouping of formations of convergent structure in response to similar climates on more than one continent). Four major physiognomic approaches are briefly reviewed: (i) Systems of formation-types for the land vegetation of the world, (ii) Treatments seeking joint classification of climate and community physiognomy expressing climate, (iii) Systems for the physiognomic description and characterization of vegetation without formal classification, and (iv) Relation of formations to one another in formation-series along environmental gradients. Emphasis of physiognomy in the study of local vegetation has been productive in British Commonwealth ecology, and has led to effective systems of formations and formation-series for tropical areas particularly. Formations can be reasonably subdivided into floristic units (dominance-types or associations). Floristic treatment of vegetation on a local level can thus be combined with physiognomic treatment, which is almost indispensible for broader biogeographic interpretation of plant communities.

REFERENCES

ADAMSON, R. S., – 1938 – The Vegetation of South Africa. Brit. Emp. Veget. Comm., London. 235 pp.
ASHTON, P., – 1964 – Ecological studies in the mixed dipterocarp forests of Brunei State. *Oxf. For. Mem.* 25: 1—75.
BEARD, J. S., –1944 – Climax vegetation in tropical America. *Ecology* 25: 127—158.
BEARD, J. S., – 1949 – The natural vegetation of the Windward and Leeward Islands. *Oxf. For. Mem.* 21: 1—192.
BEARD, J. S., – 1953 – The savanna vegetation of northern tropical America. *Ecol. Monogr.* 23: 149—215.
BEARD, J. S., – 1955 – The classification of tropical American vegetation-types. *Ecology* 36: 89—100.
BEARD, J. S., – 1967 – Some vegetation types of tropical Australia in relation to those of Africa and America. *J. Ecol.* 55: 271—290.
BEARD, J. S., – 1969 – The vegetation of the Boorabbin and Lake Johnston areas, Western Australia. *Proc. Linn. Soc. N. S. W.* 93: 239—269.
BEARD, J. S. & M. J. WEBB, – 1972 – The vegetation survey of Western Australia, its aims, objects and methods. Part I of Explan. Memoir to vegetation map 'Great Sandy Desert,' 1: 1,000,000 Series. Univ. West. Aust. Press, Perth. (in press.).
BRAUN-BLANQUET, J., – 1951 – Pflanzensoziologie: Grundzüge der Vegetationskunde. 2nd ed. Springer, Wien. 631 pp. 3rd ed. 1964, 865 pp.
BURTT DAVY, J., – 1938 – The classification of tropical woody vegetation-types. *Inst. Pap. Commonw. (Imp.) For. Inst.*, Oxford, 13: 1—85.
CAIN, S. A., – 1950 – Life-forms and phytoclimate. *Bot. Rev.* 16: 1—32.
CAIN, S. A. & G. M. DE OLIVEIRA CASTRO, – 1959 – Manual of Vegetation Analysis. Harper, New York. 325 pp.
CANDOLLE, A. DE, – 1855 – Géographie botanique raisonée. Paris.
CHANPION, H. G., – 1936 – A preliminary survey of the forest types of India and Burma. *Indian Forest Rec.*, New Ser. Silvic. 1 (1): 1—286.
CHIPP, T. F., – 1926 – The Gold Coast forest: A study in synecology. *Oxf. For. Mem.* 7: 1—94.
CLEMENTS, F. E., – 1916 – Plant succession: an analysis of the development of vegetation. *Publs. Carnegie Instn*, Washington 242: 1—512.
DANSEREAU, P., – 1951 – Description and recording of vegetation upon a structural basis. *Ecology* 32: 172—229.
DANSEREAU, P., – 1957 – Biogeography: An Ecological Perspective. Ronald, New York. 394 pp.
DAVIS, T. A. W. & P. W. RICHARDS, – 1933-4 – The vegetation of Moraballi Creek, British Guiana; an ecological study of a limited area of Tropical Rain Forest. *J. Ecol.* 21: 350—384, 22: 106—155.
DRUDE, O., – 1890 – Handbuch der Pflanzengeographie. Englehorn, Stuttgart. 582 pp.
DU RIETZ, G. E., – 1931 – Life forms of terrestrial flowering plants. *Acta phytogeogr. suec.* 3: 1—95.

ELLENBERG, H., – 1956 – Aufgaben und Methoden der Vegetationskunde. In 'Einführung in die Phytologie,' by H. WALTER Vol. 4, pt. 1. Ulmer, Stuttgart. 136 pp.

ELLENBERG, H. & D. MUELLER-DOMBOIS, – 1967 – Tentative physiognomic-ecological classification of plant formations of the earth. *Ber. geobot. Inst. ETH, Stiftg. Rübel,* Zürich 1965-6, 37: 21—55.

FANSHAWE, D. B., – 1952 – The vegetation of British Guiana: a preliminary review. *Inst. Pap. Commonw. (Imp.) For. Inst.,* Oxford, 29: 3—96.

FOSBERG, F. R., – 1961 – A classification of vegetation for general purposes. *Trop. Ecol.* 2: 1—28. Also in 'Guide to the check sheet for ecological IBP areas,' by G. F. PETERKIN. IBP *Handbook* 4: 73—120 Blackwell, Oxford & Edinburgh.

GRISEBACH, A., – 1838 – Ueber den Einfluss des Climas auf die Begränzung der natürlichen Floren. *Linnaea* 12: 159—200.

GRISEBACH, A., – 1872 – Die Vegetation der Erde nach ihrer klimatischen Anordnung. Englemann, Leipzig. 2 vols., 603 & 709 pp.

HOLDRIDGE, L. R., – 1947 – Determination of world plant formations from simple climatic data. *Science, N. Y.* 105: 367—368.

HOLDRIDGE, L. R., – 1967 – Life Zone Ecology. Rev. ed. Tropical Science Center, San José, Costa Rica. 206 pp.

KEAY, R. W., – 1959 – Vegetation Map of Africa South of the Tropic of Cancer. (Engl. & French) Oxford Univ. Press. 24 pp.

KÖPPEN, W., – 1900 – Versuch einer Klassifikation der Klimate, vorzugsweise nach ihren Beziehungen zur Pflanzenwelt. *Geogr. Z.* 6: 593—611.

KÖPPEN, W., – 1923 – Die Klimate der Erde: Grundriss der Klimakunde. de Gruyter, Berlin & Leipzig. 369 pp.

KÜCHLER, A. W., – 1947 – A geographic system of vegetation. *Geogrl. Rev., N. Y.* 37: 233—240.

KÜCHLER, A. W., – 1948 – A new vegetation map of Manchuria. *Ecology* 29: 513—516.

KUCHLER, A. W., – 1949 – A physiognomic classification of vegetation. *Ann. Ass. Am. Geogr.* 39: 201—210.

KÜCHLER, A. W., – 1950 – Die physiognomische Kartierung der Vegetation. *Petermanns geogr. Mitt.* 94: 1—6.

KÜCHLER, A. W., – 1967 – Vegetation Mapping. Ronald, New York. 472 pp.

RAUNKIAER, C., – 1934 – The Life Forms of Plants and Statistical Plant Geography. Clarendon, Oxford. 632 pp.

RICHARDS, P. W., – 1936 – Ecological observations on the rain forest of Mount Dulit, Sarawak. *J. Ecol.* 24: 1—37, 340—360.

RICHARDS, P. W., – 1939 – Ecological studies on the rain forest of southern Nigeria. I. The structure and floristic composition of the primary forest. *J. Ecol.* 27: 1—61.

RICHARDS, P. W., – 1952 – The Tropical Rain Forest: An Ecological Study. Cambridge Univ. Press. 450 pp.

RICHARDS, P. W., A. G. TANSLEY & A. S. WATT, 1939 – The recording of structure, lifeform and flora of tropical forest communities as a basis for their classification. *Inst. Pap. Commonw. (Imp.) For. Inst.,* Oxford. 19: 1—19; also *J. Ecol.* 28: 224—239, 1940.

ROBBINS, R. G., – 1962 – The podocarp-broadleaf forests of New Zealand. *Trans. R. Soc. N. Z. Bot.* 1: 33—75.

ROSS COCHRANE, G., – 1963 – A physiognomic vegetation map of Australia. *J. Ecol.* 51: 639—656.

RÜBEL, E., – 1930 – Pflanzengesellschaften der Erde. Huber, Bern & Berlin. 464 pp.

RÜBEL, E., – 1936 – Plant communities of the world. In 'Essays in Geobotany in Honor of William Albert Setchel,' ed. T. H. GOODSPEED, pp. 263—290. Univ. Calif. Press, Berkeley.
RUMNEY, G. R., – 1968 – Climatology and the World's Climates. Macmillan, New York & London. 656 pp.
SCHIMPER, A. F. W., – 1898 – Pflanzen-Geographie auf physiologischer Grundlage. Fisher, Jena. 876 pp.
SCHIMPER, A. F. W., – 1903 – Plant-geography upon a Physiological Basis. Clarendon, Oxford. 839 pp.
SCHIMPER, A. F. W. & F. C. VON FABER, – 1935 – Pflanzengeographie auf physiologischer Grundlage. 3rd ed. Fisher, Jena. 2 vols., 588 & 1612 pp.
SCHMITHÜSEN, J., – 1968 – Allgemeine Vegetationsgeographie. 3rd ed. de Gruyter, Berlin. 463 pp.
Scientific Council for Africa South of the Sahara, – 1956 – Phytogeography. (French & Engl.) Publ. Cons. scient. Afr. S. Sahara 53: 1—33.
STEENIS, C. G. G. J. VAN, – 1958 – Basic principles of rain forest sociology. (French summ.) In 'Study of Tropical Vegetation', pp. 159—165. Proc. Kandy Symp. 1956, Unesco, Paris.
STEFANOFF, B., – 1930 – Versuch zur Darstellung einer parallel Klassifikation der Klimate und der Vegetationstypen. (Bulgarian & German) Sb. bŭlg. Akad. Nauk. 26 (3): 1—122. (Engl. summ. by W. B. Turrill, J. Ecol. 20: 211—213, 1932).
STOCKER, O., – 1963 – Das dreidimensionale Schema der Vegetationsverteilung auf der Erde. Ber. dt. bot. Ges. 76: 168—178.
TANSLEY, A. G., – 1920 – The classification of vegetation and the concept of development. J. Ecol. 8: 118—149.
TANSLEY, A. G. & T. F. CHIPP, eds., – 1926 – Aims and Methods in the Study of Vegetation. Brit. Emp. Veget. Comm & Crown Agents for Colonies, London. 383 pp.
THORNTHWAITE, C. W., – 1931 – The climates of North America according to a new classification. Geogrl Rev., N. Y. 21: 633—655.
THORNTHWAITE, C. W., – 1948 – An approach toward a rational classification of climate. Geogrl Rev., N. Y. 38: 55—94.
TROLL, C., – 1948 – Der asymmetrische Aufbau der Vegetationszonen und Vegetationsstufen auf der Nord- und Südhalbkugel. Ber. geobot. Inst. Rübel, Zürich, 1947: 46—83.
TROLL, C., – 1961 – Klima und Pflanzenkleid der Erde in dreidimensionaler Sicht. Naturwissenschaften 48: 332—348.
TROLL, C. – 1968 – The Cordilleras of the Tropical Americas. Aspects of climatic, phytogeographical and agrarian ecology. (Span. summ.) Colloquium geogr., Bonn 9: 13—56.
WALTER, H., – 1962, 1968 – Die Vegetation der Erde in ökophysiologischer Betrachtung. Fisher, Jena. 2 vols., 538 & 1001 pp.
WARMING, E., – 1909 – Oecology of Plants: An Introduction to the Study of Plant-Communities. Oxford Univ. Press. 422 pp.
WEBB, L. J., – 1959 – A physiognomic classification of Australian rain forests. J. Ecol. 47: 551—570.
WHITTAKER, R. H., – 1953 – A consideration of climax theory: the climax as a population and pattern. Ecol. Monogr. 23: 41—78.
WHITTAKER, R. H., – 1962 – Classification of natural communities. Bot. Rev. 28: 1—239.
WHITTAKER, R. H., – 1970 – Communities and Ecosystems. Macmillan, New York. 162 pp.

WHITTAKER, R. H. & W. A. NIERING, – 1965 – Vegetation of the Santa Catalina Mountains, Arizona. (II.) A gradient analysis of the south slope. *Ecology* 46: 429—452.
WHITTAKER, R. H. & W. A. NIERING, – 1968 – Vegetation of the Santa Catalina Mountains, Arizona. IV. Limestone and acid soils. *J. Ecol.* 56: 523—544.

14 DOMINANCE-TYPES

ROBERT H. WHITTAKER

Contents

14.1	Introduction	67
14.2	The School of CLEMENTS	67
14.3	The Meaning of Dominance-Types	69
14.3.1	The Concept	69
14.3.2	Implications of Species Distributions	69
14.3.3	Limitations	72
14.4	Application	73
14.4.1	Definition of Dominance-Types	73
14.4.2	Grouping of Dominance-Types	73
14.4.3	Subordinate Units	75
14.5	Conclusion	76
14.6	Summary	77

14   DOMINANCE-TYPES

14.1   Introduction

As BEARD observes in the preceding article, formations are often divided into subordinate community-types characterized by their dominant species. Many authors have termed these units 'associations,' and BEARD considers this term suitable for them. Community-types recognized by dominant species differ from the associations of the school of BRAUN-BLANQUET, however, in manner of definition and in scale or inclusiveness of the resulting units; moreover the approach through dominance represents a different perspective on what deserves emphasis in the study of vegetation from the floristic system of BRAUN-BLANQUET (1951, article 20 in this volume). The author (WHITTAKER 1962) has suggested that clarity is served by assigning the formal term 'association' to units of the school of BRAUN-BLANQUET, and designating units defined by dominant species as dominance-types or, simply, types.

Such types are among the most widely used of vegetation units. Dominance-types have been used by some authors in all the regional traditions, but they are particularly characteristic of American and British ecology. They have an important place (under the term 'formation') in Russian classifications (article 17). They have been the preferred units of many practical ecologists, for forestry, range and wildlife ecology, and land management particularly. Dominance-types will be discussed both as part of the formal system of CLEMENTS and as more informally used by plant and animal ecologists not of that school.

14.2   The School of Clements

CLEMENTS (1905, 1916, 1928, 1936) sought to fashion a system for treating the vegetation of a continent then rich in plant communities little disturbed by man. The concept of succession of COWLES (1899, 1901) and others was linked with the formation as a unit and with the analogy of the community as an organism. The formation was seen as a complex organism which arises, grows, matures, and dies; the climax formation is the adult organism for which successional communities are developmental stages (CLEMENTS 1928: 125-6).

The formation was also a regional vegetation unit, covering a definite geographic area for which it was the climax, the climate of which it expressed in the growth-form of its dominant species. CLEMENTS' formations corresponded in general to those of other plant geographers; but the idea of succession was used to elaborate a distinctive system of classification.

Formations were divided into associations usually defined by dominant species; associations differed from one another in response to climatic differences within the region of the formation. Associations were divided into consociations characterized by a single dominant species, and fasciations by combinations of dominant species (WEAVER & CLEMENTS 1938). Societies were subordinate units characterized by subdominants, species of growth-form different from the dominants. These terms applied only to communities regarded as climax in CLEMENTS' (monoclimax) sense, and a set of parallel units (formies, associes, consocies, fascies, socies) were applied to successional communities. CLEMENTS' interpretation distinguished from the climatic climax various stabilized communities (proclimaxes) that were not considered true climaxes. Since there was only one climax community in a given region, the successional units were applied to proclimaxes as well as to clearly unstable communities. The lociation, lamiation, sociation, and other units appeared in the system later (CLEMENTS 1936), to produce perhaps the most complex terminology developed by any school. All units below the formation were recognized by dominant (or subdominant) species.

The Clementsian system was for some time the dominant approach to plant communities in the United States, though it was never accepted by all ecologists. It was also widely influential in British ecology (WHITTAKER 1962). As some of CLEMENTS' ideas spread others, notably the organismic analogy and the basing of classification on often hypothetical successional relations, were increasingly criticized and discarded (WHITTAKER 1957, 1962). The work of BRAUN (1938, 1947, 1950, 1956) on the eastern deciduous forests of the United States represents a development from CLEMENTS' approach, without many of the features for which it has been criticized. BRAUN's approach was less formal and deductive than CLEMENTS'. BRAUN used the 'association' concept of CLEMENTS (though recognizing additional eastern forest associations), but placed much less emphasis on classification as such. Though the associations were treated as regional climaxes, these were seen in a perspective quite different from CLEMENTS'—a perspective accepting the existence of a variety of stable communities in a given area and emphasizing the complexity of species distribution and vegetation pattern

(BRAUN 1956). The eastern forests were furthermore seen in a long-term historic view of evolution from Tertiary time to the present. One highly mixed present community-type, the mixed mesophytic forest of the southern Appalachian Mountains, was regarded as nuclear to the eastern forests and most like the ancestral Arcto-Tertiary forests. Other major community-types might be interpreted as derived from the mixed mesophytic by simplification, as 'association-segregates' (BRAUN 1950, 1956, cf. RICHARDS 1952).

CLEMENTS' system has otherwise been of declining influence in the United States in the last two decades. Neither in American nor in British ecology has it been replaced, for most authors, by any other formal system. A majority of ecologists have used, for classification below the formation, dominance-types as units of convenience, part of no formal hierarchy but serving the purposes of research.

14.3 The Meaning of Dominance-Types

14.3.1 THE CONCEPT

A dominance-type is a class of communities defined by the dominance of one or more species. These species are usually the most important ones of the uppermost stratum of the community, but sometimes of a lower stratum of higher coverage. The concept of 'dominant' itself has no clear boundary. In a forest it may be true that the canopy trees are important above all other species for their effects, through microclimate and soil chemistry, on the environments to which other species must be adapted. In an open community such as a desert scrub such effects of major species on the environments of minor ones are less evident. We generally know little of the real importances of species in the sense of effects on one another's populations. 'Dominant' in most cases means only that the species is conspicuous in the community and is high in one of the importance values (coverage, biomass, density, frequency, etc.) by which we seek to express the contribution of species to community structure. In practice a dominance-type unites communities or samples in which some one species or combination of species has high importance values.

14.3.2 IMPLICATIONS OF SPECIES DISTRIBUTIONS

It is then important for understanding of these units to know how dominant species relate to one another. As indicated above (article

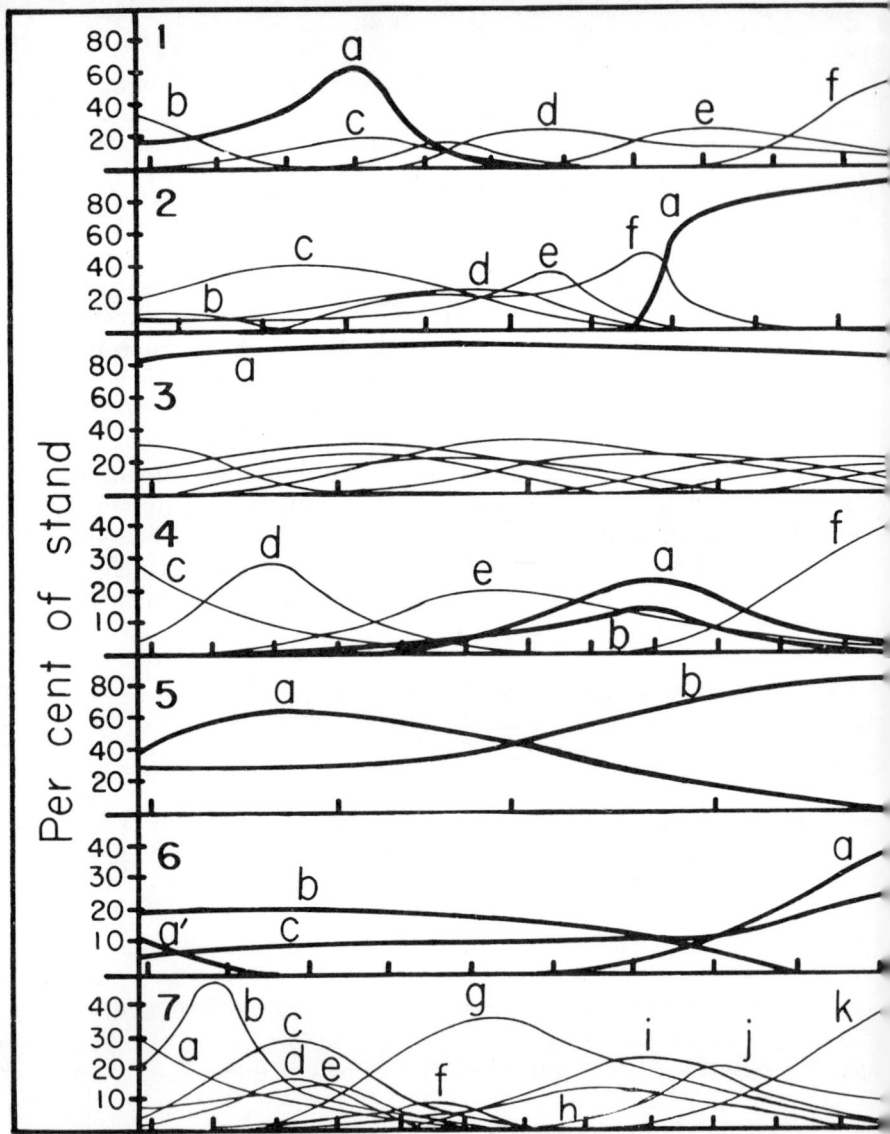

Fig. 1. Distributional relations of species characterizing dominance-types in the Great Sr Mountains (WHITTAKER 1956, 1962). Population distributions of tree species are plotte relative densities (per cents of stems 2 cm and over in stands).

1) Dominance of a single species (*Tsuga canadensis*) with a bell-shaped distribution, the moisture gradient at 1070—1380 m elevation.

2) Dominance of a single species (*Fagus grandifolia*, 'gray' ecotype) with a 'pla distribution, along the elevation gradient in mesic sites.

3) Dominance of a single species (*Abies fraseri*) with a broad distribution along moisture gradient in subalpine forests (1900—2000 m); subordinate species have narr distributions.

4) Dominance of two species (*Quercus prinus*, a and *Castanea dentata*, b) with si distributions in a moisture-gradient transect at 760—1070 m.

5) Dominance of two species (*Abies fraseri*, a, and *Picea rubens*, b) with dissimila broadly overlapping distributions, in a moisture-gradient transect at 1380—1670 m.

6) Dominance in various combinations of three species (*Quercus alba*, aa', *Q. prinus*, b *Castanea dentata*, c) with dissimilar but broadly overlapping distributions, in an elev transect in subxeric sites.

7) Dominance in various combinations of several species, in a moisture-gra transect at 760—1070 m. Major species of the cove forest or mixed mesophytic grou *Batula alleghaniensis* a, *Tsuga canadensis* b, *Halesia monticola* c, *Acer saccharum* d, *Tilia hetero* e, and *Liriodendron tulipifera* f.

3), gradient analysis has provided information on species distributions along environmental complex-gradients. Fig. 1 illustrates for 7 community-types the manners in which their dominant species are distributed along environmental gradients:

1) In a pattern of continuously gradating vegetation, one species (a) has a typical, bell-shaped distributional curve and is clearly dominant in a part of the environmental gradient. Species a defines a dominance-type, but the limits of this type (whether at 50, 40, 30 per cent of canopy stems, etc.) are necessarily arbitrary.

2) A species (a) with a 'plateau' distribution is strongly dominant over a span of the gradient, and its range of dominance is more abruptly separated from that of other species (article 3). The species defines a dominance-type that is relatively discontinuous with other types along one environmental gradient (but may be continuous with other types along other gradients).

3) A dominant species (a) has a very wide amplitude along the gradient, but more narrowly distributed undergrowth species permit the recognition of site-types in the Finnish sense (article 15).

4) Two species (a and b) with similar distributions along the gradient characterize a type in much the same manner as one. The boundaries of the type are again arbitrary, and these two dominants with similar distributions along the moisture gradient are differently distributed in relation to elevation.

5) Two species (a and b) have dissimilar, but broadly overlapping distributions, dominating a wide range of communities separately or together. A classifying ecologist must choose whether to recognize: a single type dominated by either of the species or both; or two types, one dominated by a and one by b; or three types as dominated by species, a, species b, or the combination of a and b.

6) Three species (aa', b, and c) have different distributions but overlap broadly. Either a single type including any combination of the dominants, or some number of types characterized by particular dominants or combinations, may be recognized. (The case illustrated is the Quercus-Castanea association of CLEMENTS, within which CONARD (1935) distinguished two associations on Long Island and WHITTAKER (1956) four major types in the Great Smoky Mountains.)

7) Several species (a to f) have overlapping distributions and often occur together, though in proportions varying from stand to stand. These species are a commodal group in the sense of WHITTAKER (1956); but the classifying ecologist must decide how many species to include in the group, and hence the inclusiveness of the unit the group defines. He may, or may not, choose to recognize as 'segregate' types in the sense of (BRAUN (1950) stands strongly dominated by particular species or combinations.

14.3.3 LIMITATIONS

Dominance-types are thus based on species of most varied distributional relationships. A dominant species may be of very wide range along environmental gradients (case 3); the type it defines then comprises stands that are diverse in habitat characteristics and in composition of species other than the dominant. Some other dominant species are narrowly distributed and define fairly homogeneous, localized types. Community-types defined by dominance consequently can be units of widely different scale and inclusiveness. Dominant species provide no secure basis for objective decisions about identity, scale, and number of units on which ecologists can agree. Decisions must be made, from the preferences of individuals, on what species are important enough to contribute to the definition of dominance-types, on what combinations of species sharing dominance are to be accepted as dominance-types, and on where boundaries, if any, are to be placed in their continuous intergradations.

A dominance-type thus is not, and cannot be, a particular, standardized kind of community-unit (WHITTAKER 1962). Dominance-types are poorly suited to the construction of a formal, hierarchial classification of vegetation. There is convenience in the use of dominance-types because the types can be recognized without intensive study of the full floras of a range of communities. Use of dominance-types often seems appropriate also because they are defined by quantitative relationships and by the species that may be of most interest to ecologists. In comparison with the units of the school of BRAUN-BLANQUET these are not unmixed advantages, for dominance-types are often based on a more limited range of information than are floristic units. There is some justice in the view of members of that school that classification based on dominant species alone is a cruder and more superficial approach to community units than theirs. For systematic classification and for detailed knowledge of community characteristics and relations to environment, the approach through dominance-types has significant disadvantages compared with the system of BRAUN-BLANQUET (article 20).

The choice of dominance-types as the basis of classification should involve acceptance of these limitations. There are reasons many ecologists will continue to accept them and to classify by dominance and physiognomy. For some kinds of study formal classification is not a purpose. Dominance-types may serve well enough to describe some major features of a vegetation pattern, to designate certain communities whose characteristics are being compared, or to summarize some of the results of gradient analysis. Granting the usefulness of dominance-types, some comments on terminology and use of these units follow.

14.4 Application

14.4.1 Definition of Dominance-Types

The preceding implies that dominance-types must be defined by the ecologist's choices of what species and combinations are to characterize types. The lability of the choices provides the means of making the classification work. The options of what combinations or segregations (as in cases 5, 6, and 7) are to be accepted as types permits some adjustment of the number and inclusiveness of the types to suit the classifier's purposes, and to produce units that are reasonably consistent with one another in the ranges of environmental and community variation that each comprises. The process is one of typification, as referred to also by BEARD in the preceding article. In the field of variation that is a vegetation pattern, a sufficient number of types are recognized to represent major, recognized differences in community composition (cf. SCHMID 1952, POORE 1956, WHITTAKER 1962). The characteristics of these types may then be incorporated into the definitions of community-types as classes (article 12). These more formal definitions of classes may in turn be so modified and accommodated to one another as to form a workable classification. This classification should provide a reasonable number of units, reasonably comparable in their inclusiveness, and so defined that most or all samples taken from the vegetation pattern can be assigned to one of the units.

14.4.2 Grouping of Dominance-Types

There are a number of ways these types may be grouped into broader units. Among them:
1) By physiognomy into formations. Formations (or biomes) may be used as higher units, within which dominance-types are lower units (WARMING 1909, see also preceding article). Formations, as community-types defined by dominant growth-forms and broad environmental relations on particular continents, are in turn grouped into formation-types, comprising similar formations of similar environments on more than one continent. Various difficulties affect the definitions of formations themselves. There are cases in which formations and dominance-types do not very naturally accommodate themselves to a hierarchial, 'genus and species' relationship (WHITTAKER 1962). In general, however, the flexibility inherent in the definitions of both will permit the creation of suitable units. It is in some cases appropriate to recognize physiognomically defined subformations as a level of classification between formation

and dominance-type. Thus CLEMENTS (WEAVER & CLEMENTS 1929, CLEMENTS & SHELFORD 1939) includes among the North American temperate grasslands tall-grass prairie, mid-grass, and short-grass plains, and California grassland now dominated by annuals. These are termed 'associations' by CLEMENTS; but their definitions are primarily physiognomic (even though CLEMENTS specified major dominants for them), and each includes a number of dominance-types.

2) By species relationships into collective types. Dominance-types may be variously grouped into larger units that are less inclusive than the formation. Dominance-types that share one or more of their major species, or share major species in various directions although none is common to all, or are floristically similar in their subordinate species, or have taxonomically related dominants, may be grouped together. For such units the term collective-type is suggested (WHITTAKER 1962). Comparable units are called associations by British authors (TANSLEY & CHIPP 1926, TANSLEY 1939, CROCKER & WOOD 1947, BEADLE 1948, BEARD 1955), federations by DU RIETZ (1932), alliances by BEADLE & COSTIN (1952). Some of CLEMENTS' associations are collective types in this sense. For example, the Quercus-Carya association of the eastern deciduous forests and the Pinus-Juniperus association of the woodland formation each includes a wide range of combinations of dominant species, linked together by the genera of the dominants.

3) By successional status and prevalence. The Clementsian system distinguished climax associations and successional associes; but many of the latter were stabilized, proclimax communities. The system of BRAUN-BLANQUET (1951) more realistically distinguishes (i) terminal communities ('Schlussgesellschaften,' climatic, regional, or prevailing climaxes) that are stabilized in relation to the climate of a region or elevation belt, (ii) stable communities ('Dauergesellschaften', edaphic or local climaxes, some of the proclimaxes of CLEMENTS) that are also self-maintaining but are adapted to special biotopes in an area, and (iii) unstable, developmental or successional communities. Many difficulties for the Clementsian system resulted from the effort to force on vegetation a dichotomy between communities that were regional and climax, and communities that were local and therefore presumptively successional. Resolution of the difficulty lies in the distinction of three levels of conditions – regional or prevailing climax, local or edaphic climax, and local successional communities. (The fourth combination of regional prevalence with instability is clearly possible.) Many authors and schools have developed concepts of regional (and climax) community-types (WHITTAKER 1953, 1962). Regional types can be defined as units

on different levels – as formations, subformations, collective-types, and dominance-types. The associations of CLEMENTS ranged across the latter three possibilities from dominance-types (e.g. Fagus grandifolia-Acer saccharum) to collective types (Quercus-Carya) to subformations (some of the grassland associations).

4) By developmental relations into successional series and climax-complexes. A sequence of developmental community-types leading to a stable or climax type in a given kind of habitat is a successional series or sere. A grouping of related successional and climax community-types is a climax-complex (BRAUN-BLANQUET 1951, cf. the formation in the sense of MOSS 1910, TANSLEY 1911, and CLEMENTS 1916, 1936). In a narrower sense, the climax-complex (or succession-al complex) comprises only a particular climax type plus the various communities of primary successions and disturbance developing toward it. In the broader and more frequently used sense, the climax-complex groups the varied climax and successional communities of a region or elevation zone around a prevailing climax community-type; it is then a complex of successional complexes.

5) By environmental relations. Apart from succession, a chain (or continuum) of dominance-types or other units along a complex-gradient of environment is an ecological series (see also articles 5, 15). Communities that are related to one another in response to an environmental pattern (e. g. a landscape pattern) form a community-complex or -mosaic (cf. KRAUSE 1952). (If successional, as well as spatial, relations are in question, the community-complex is also a climax-complex.) In some applications of ecological series, each section or zone of the series forms a pattern of community-types, one of which prevails and characterizes the zone. Thus along such major complex-gradients as elevation in mountains and tide-levels in the littoral, the ecological series is a chain (or continuum) of community-complexes.

14.4.3 SUBORDINATE UNITS

Division of dominance-types into subtypes may be by different combinations of dominants and subdominants: (i) by marked change in importance value relations (or presence and absence) of the two or several dominant species of the type, (ii) by segregations of particular dominant species or combinations from a larger group of species by which the type is defined, (iii) by subdominants as major species of a stratum other than that of the dominant. The latter subtypes may be, but will not necessarily be, sociations. When forest types are divided by undergrowth composition in relation to site properties, the resulting subtypes may be site-types. Dominance-types may also be divided by groups of character-species or differential-species, and the subtypes may then correspond to lower units of the system of BRAUN-BLANQUET. There are thus possibilities for relating a classification employing dominance-types, which may be rather broad units, to the units of other schools.

14.5 Conclusion

The complexities of communities offer many choices of criteria by which ecologists may seek indication of environment or classification of communities. Choice of which community characteristics are to be used as criteria involves balancing together desired properties of these (WHITTAKER 1954, 1962). The criteria should be: (i) accessible – characteristics of the community that are evident in the field or, at least, can be observed or measured without excessive labor, (ii) significant – important as a characteristic of one community distinguishing it from others, strongly correlated with other characteristics of the community and its environment, and (iii) effective – successful in expressing environmental difference or producing units of classification at the level of intensiveness of a given study. (An accessible and significant criterion may be ineffective in this sense; physiognomy, for example, may be a poor criterion to distinguish among local forests and small site differences.) Differences among schools of ecology represent in part different choices of community characteristics to be emphasized, on the basis of different balances among these desired properties, in relation to different kinds of vegetation and perspectives on vegetation research.

The school of BRAUN-BLANQUET's primary criterion is floristic composition, particularly representation of diagnostic species (character-species and differential-species). As a community characteristic for classification and indication, floristic composition is (i) reasonably accessible, though requiring thorough knowledge of and attention to whole-community flora, (ii) significant, in that the utilization of information on a number of species, particularly those that are narrowly distributed or are near the borders of their distributions, produces information strongly correlated with other community and habitat characteristics, and (iii) effective, since choice of species is flexible, different groups of diagnostic species can serve for different levels of hierarchial classification, or different intensiveness of environmental expression. Dominance, as a primary criterion, is (i) more easily accessible, (ii) significant, but often at a lower or uncertain level because a given dominant may have a wide distribution, or may through disturbance be replaced by another dominant with only limited change in the rest of the community, (iii) often less effective for indication and classification than community composition.

Choice of dominance as a criterion is based on a different balance among these properties, emphasizing accessibility and quantitative importance to some sacrifice of significance in the sense just given and effectiveness. The preceding section suggests a range of units on different levels for classification by physiognomy and dominance — formation-type, formation, subformation and collective type when appropriate, dominance-type, and subtype. By no means is this intended as a system in competition with that of BRAUN-BLANQUET. If a formal, hierarchial classification is the objective, dominance-types are not the way. Yet many ecologists, affected by different interests than phytosociologists, are likely to continue to classify their communities by dominance. For those making such a choice, this discussion is intended to offer clarification of the basis, limitations, and possibilities for such classification, and suggestions on units and concepts.

14.6 SUMMARY

In British and American ecology, classification (at levels below the formation) has most often been based on the dominant species of communities. The American author CLEMENTS elaborated a formal classification based on successional relations and dominant growth-forms and species. The major units of this system included the formation, the 'association' as a broadly defined regional community-type within the formation, and successional types (associes, etc.). While CLEMENTS' work was widely influential, his classification has little current use.

Much current work, however, classifies plant communities into units which are defined by dominants (most conspicuous or important species), and which may be termed 'dominance-types.' The distributions of dominant species overlap with one another in most varied ways. Dominant species are consequently combined in varied, and often continuously intergrading, ways into the communities that must be classified. The ecologist must choose what dominant species, at what levels of importance, are to define dominance-types. Dominance offers no simple and objective criterion for classifying communities, and is poorly suited to the construction of a formal, hierarchial classification.

Dominance-types can be used, however, to construct a workable classification for a given area. Dominance-types may be grouped into formations or subformations, and in some cases into collective dominance-types (groups of dominance-types linked together by shared major species or other characteristics). Dominance-types may be divided either into subtypes by different combinations of their major species, or into some of the units of other schools of ecology (sociations, site-types, or associations and other units of BRAUN-BLANQUET).

These units can be used to form a hierarchy – formation-type, formation, subformation and collective type when appropriate, dominance-type, and subtype. Such a hierarchy should not be regarded as a formal classification comparable to that of BRAUN-BLANQUET. As a basis of classification dominance is more convenient than full floristic composition, but may less effectively express community and environmental relations.

REFERENCES

BEADLE, N. C. W., – 1948 – The vegetation and pastures of western New South Wales, with special reference to soil erosion. Tennant, Govt. Printer, Sidney. 281 pp.
BEADLE, N. C. W. & A. B. COSTIN, – 1952 – Ecological classification and nomenclature. With a note on pasture classification by C. W. E. MOORE. *Proc. Linn. Soc. N. S. W.* 77: 61—82.
BEARD, J. S., – 1955 – The classification of tropical American vegetation-types. *Ecology* 36: 89—100.
BRAUN, E. LUCY, – 1938 – Deciduous forest climaxes. *Ecology* 19: 515—522.
BRAUN, E. LUCY, – 1947 – Development of the deciduous forests of eastern North America. *Ecol. Monogr.* 17: 211—219.
BRAUN, E. LUCY, – 1950 – Deciduous Forests of Eastern North America. Blakiston, Philadelphia. 596 pp.
BRAUN, E. LUCY, – 1956 – The development of association and climax concepts: their use in interpretation of the deciduous forest. *Am. J. Bot.* 43: 906—911.
BRAUN-BLANQUET, J., – 1951 – Pflanzensoziologie: Grundzüge der Vegetationskunde. 2nd ed. Springer, Wien. 631 pp. 3rd ed., 1964, 865 pp.
CLEMENTS, F. E., – 1905 – Research Methods in Ecology. Univ. Publ. Co., Lincoln, Nebr. 334 pp.
CLEMENTS, F. E., – 1916 – Plant succession: an analysis of the development of vegetation. *Publs Carnegie Instn* Washington 242: 1—512.
CLEMENTS, F. E., – 1928 – Plant Succession and Indicators: A Definitive Edition of Plant Succession and Plant Indicators. Wilson, New York. 453 pp.
CLEMENTS, F. E., – 1936 – Nature and structure of the climax. *J. Ecol.* 24: 252—284.
CLEMENTS, F. E. & V. E. SHELFORD, – 1939 – Bio-ecology. Wiley, New York. 425 pp.
CONARD, H. S., – 1935 – The plant associations of central Long Island: a study in descriptive plant sociology. *Am. Midl. Nat.* 16: 433—516.
COWLES, H. C., – 1899 – The ecological relations of the vegetation on the sand dunes of Lake Michigan, I. Geographical relations of the dune floras. *Bot. Gaz.*

27: 95—117, 167—202, 281—308, 361—391.
Cowles, H. C., - 1901 - The physiographic ecology of Chicago and vicinity; a study of the origin, development, and classification of plant societies. *Bot. Gaz.* 31: 73—108, 145—182.
Crocker, R. L. & J. G. Wood, - 1947 - Some historical influences on the development of the South Australian vegetation communities and their bearing on concepts and classification in ecology. *Trans. R. Soc. So. Aust.* 71: 91—136.
Du Rietz, G. E., - 1932 (1930) - Vegetationsforschung auf soziationsanalytischer Grundlage. *Handb. biol. Arb Meth.* 11, 5: 293—480.
Krause, W., - 1952 - Das Mosaik der Pflanzengesellschaften und seine Bedeutung für die Vegetationskunde. *Planta* 41: 240—289.
Moss, C. E., - 1910 - The fundamental units of vegetation: historical development of the concepts of the plant association and the plant formation. *New Phytol.* 9: 18—53.
Poore, M. E. D., - 1956 - The use of phytosociological methods in ecological investigations. IV. General discussion of phytosociological problems. *J. Ecol.* 44: 28—50.
Richards, P. W., - 1952 - The Tropical Rain Forest: An Ecological Study. Cambridge Univ. Press. 450 pp.
Schmid, E.,- 1952 - Natürliche Vegetationsgliederung am Beispiel des Spanischen Rif. *Ber. geobot. Inst. Rübel*, Zürich 1951: 55—79.
Tansley, A. G., ed., - 1911 - Types of British Vegetation, by Members of the Central Committee for the Survey and Study of British vegetation. Cambridge Univ. Press. 416 pp.
Tansley, A. G., - 1939 - The British Islands and Their Vegetation. Cambridge Univ. Press. 930 pp.
Tansley, A. G. & T. F. Chipp, eds., - 1926 - Aims and Methods in the Study of Vegetation. Brit. Emp. Veget. Comm. & Crown Agents for Colonies, London. 383 pp.
Warming, E., - 1909 - Oecology of Plants: An Introduction to the Study of Plant-Communities. Oxford Univ. Press. 422 pp.
Weaver, J. E. & F. E. Clements, - 1929 - Plant Ecology. Mc Graw-Hill, New York. 520 pp. 2nd ed., 1938, 601 pp.
Whittaker, R. H., - 1953 - A consideration of climax theory: the climax as a population and pattern. *Ecol. Monogr.* 23: 41—78.
Whittaker, R. H., - 1954 - Plant populations and the basis of plant indication. (Germ. summ.) *Angew. PflSoziol.*, Wien, Festschr. Aichinger 1: 183—206.
Whittaker, R. H., - 1956 - Vegetation of the Great Smoky Mountains. *Ecol. Monogr.* 26: 1—80.
Whittaker, R. H., - 1957 - Recent evolution of ecological concepts in relation to the eastern forests of North America. *Am. J. Bot.* 44: 197—206.
Whittaker, R. H., - 1962 - Classification of natural communities. *Bot. Rev.* 28: 1—239.

15 THE FINNISH SCHOOL AND FOREST SITE-TYPES

TOOMAS E. A. FREY

Contents

15.1	Introduction	83
15.2	CAJANDER and the Site-Type Concept	83
15.2.1	Development of the Concept	83
15.2.2	Community Dynamics and Undergrowth	85
15.3	Applications to Classification	86
15.3.1	Site-Type Classes	86
15.3.2	Site-Types	87
15.3.3	Subordinate Units	92
15.4	Application in Management	93
15.5	Series and Complexes	96
15.6	The Finnish and Other Approaches	99
15.7	Conclusion	102
15.8	Summary	103

15 THE FINNISH SCHOOL AND FOREST SITE-TYPES

15.1 Introduction

The classification of vegetation seems an eternal problem, without final answer. It is so because of the existence of regions differing widely not only in climate, soils, and flora, but also in the dynamics of the vegetation as affected by natural disturbance and increasing civilized violence as well. It is hardly possible that a single classificatory procedure should provide for all the problems and patterns with which students of vegetation in different areas must be concerned. The complexity of the task has been met with an equally complex and varied methodology worked out in various countries and phytosociological schools.

It is to be emphasized that ecological methods tend to reflect the properties of the vegetation being studied, and one has to agree with WHITTAKER (1962) that there exists an ecology of ecologists. Hence, any attempt at a universally acceptable procedure for the establishment of classificatory units of world vegetation must fail. At best, perhaps a common platform could be found for different approaches (see HUSTICH 1960). Two major approaches – the physiognomic and the floristic (of BRAUN-BLANQUET) – have gained widest, but far from universal, acceptance. The approach to forest site-types pioneered by CAJANDER has had in vegetation classification a more limited, but still significant role that is to be discussed here.

15.2 Cajander and the Site-Type Concept

15.2.1 DEVELOPMENT OF THE CONCEPT

The Finnish biocoenotic school was founded by A. K. CAJANDER (1879-1943), from wide field experience in Fennoscandia and Siberia with northern vegetation that is comparatively monotonous and fairly poor in species. Early in the century (1901-1906) these extensive areas, unaffected or only slightly disturbed by human activity, were studied by CAJANDER (1903, 1904, 1909a, 1909b 1913) with results that deserved wide attention.

From these studies CAJANDER acquired an excellent perception not only of geographic and habitat relations, but also of vegetation as a dynamic entity. The changes caused by momentary factors (clear cuttings, forest fires, and burnt–over clearings in woods under cultivation) may be striking, and hence the temporary communities in such stands show successional patterns from early pioneer to more stable stages. As a rule, climatic extremes (severe drought and exceptional cold) repeat themselves more frequently and are of modest influence, so that the changes in communities they produce are mostly of small degree and last only for a few years. The more severe disturbances initiate successions that lead back to communities adapted to the average climatic (and edaphic) conditions of their biotopes or habitats, and these communities can be regarded as almost permanent in their composition and structure.

CAJANDER was not the first to make a clear distinction between temporary and permanent communities. However, he was the first to take the stable communities as the basis for the classification of vegetation. He thus anticipated such concepts as the formation of Moss (1910), TANSLEY (1911), and CLEMENTS (1916), and the climax-complex of BRAUN-BLANQUET (1951). For example, in his early works on the vegetation of the Lena basin (1903, 1904) he recognized the following successional stages for lowland vegetation: (1) sandy barrens, (2) Salicetum viminalis, (3) mixed bushes rich in species, mainly *Salix*, (4) Betuletum odoratae, (5) Picetum obovatae, (6) Piceeto-Lariceetum dahuricae, the last-named being the most stable community on these sites. Within the limits of the given series, the ecological situation does not differ so much as one might suspect from the contrasting composition and structure of the communities; i.e. the real capacity of the biotope or site remains almost unaltered during successional changes in the communities. Therefore, in so far as the forest type is a unit for practical forestry, the forest type should be described in such a manner that one can identify the type regardless of the facies or the seral stage, and this identification will indicate the site's potential productivity. Hills holds a similar viewpoint when he states: 'Since combinations of macroclimate and landform not only constitute the basic potential productivity but exercise a functional control on the integrated development of vegetation, soil and ecoclimate, combined macroclimate landform units provide the basic classification of ecosystems' (HILLS 1960: 39). It is difficult, however, to use the many, continuously varying qualities of climate, landform, and soil to produce biologically appropriate units of classification. The CAJANDER system uses the expression of these qualities in the composition of stable communities as a means of appropriate and useful classification of the communities, and thereby of the sites which they (and related unstable communities) occupy.

15.2.2 COMMUNITY DYNAMICS AND UNDERGROWTH

Where forests are concerned, the difficulty is increased by the longevity and competitive capacity of dominant tree species which may call forth long-lasting differences in communities. Nevertheless, as KALELA (1960: 46) writes: 'We should try to find climax communities and to make first-hand investigations in them. The successional phases in which development has progressed farthest follow in the second place, and only after them the short-lived initial stages of the successions. In limiting stand types – especially among the unstable successional communities – we are forced to comparatively broad and collective types, omitting the most temporary and rarely occurring communities. Otherwise the number of stand types will be so great that it would be quite impossible to deal with them or to gain any comprehensive idea of the system as a whole.'

In definition of site-types the undergrowth must be emphasized. The Scandinavian and Baltic area has relatively few tree species and each of these may occur over a wide range of sites with different undergrowth associates. The Finnish system makes use of the *relative* autonomy of different layers, though it does not divide vegetation into separate unistratal associations or unions as does LIPPMAA (1933, 1935, 1939; see also CAJANDER 1909a, GAMS 1918, BOLLETER 1921, KUJALA 1929, DU RIETZ 1930, CAIN 1936, and article 16 in this volume). Granted that the independence of canopy and undergrowth is only partial, it is still true that the undergrowth provides much more information, more effectively expressing biotope conditions, than the canopy tree species. In some cases one canopy species may be replaced by another in a given kind of site, with little change in undergrowth composition. In other cases the change in undergrowth with canopy species (or age and density) is great, yet the investigator versed in the relations of undergrowth species to site may recognize in this altered undergrowth the characteristics of the stable community that indicates the site-type. It is by knowledge of successional process and of undergrowth species relationships that it is possible, though not always easy, to group the variety of disturbed stands in an area in relation to the stable or normative communities of the site-types.

To put it in other words, the main idea of the Finnish theory of site-types lies in the typification, from the vegetation of a given region, of a limited number of ecosystem-types recognized through the expression of site conditions in stable vegetation, and especially undergrowth composition. Actually the composition and structure of the individual stands representing an ecosystem-type are of a fairly uneven nature and overlap with those of other ecosystems. The site-type as a whole is thus an abstraction based on a series of individual stands which show some divergence caused by climatic, edaphic, cultural and other factors. It should be realized, however, that the variation in disturbed temporary communities is connected with the information about the instability of the ecological situation, whereas the permanent communities carry information on the ecosystem itself.

Variation among stands in a given type of site may be of any degree up to change in the leading dominant species and suppression of undergrowth so that no apparent similarity to a virgin stand remains. But regardless of the degree of disturbance, the given site maintains its basic properties, including its potential productivity and ability to reach the virgin or climax state. Hence, when knowledge of undergrowth and succession have permitted us to assign a disturbed stand to a site-type, we have achieved knowledge of biotope properties, successional direction, productive potential, suitability for different tree species, and appropriate management, that are by no means explicit in the characteristics of the disturbed stand by itself.

We may quote finally another definition given by CAJANDER himself (1926, 1949): 'All those stands are referred to the same forest type the vegetation of which at or near the time of maturity of the stands and provided the stands are normally stocked, is characterized by a more or less identical floristic composition and by an identical ecologico-biological nature, as well as all those stands the vegetation of which differs from that defined above only in those respects which – being expressions of differences due to age, fellings, etc. – have to be regarded as merely accidental and ephemeral or at any rate as only temporary. Permanent differences call forth a new forest type in cases where they are sufficiently well-marked, or a sub-type in cases where they are less essential, but nevertheless, noticeable.'

15.3 Application in Classification

How to ascertain that a difference in the ecological-biological nature of stands is sufficiently well-marked to indicate a difference in site-type? This question naturally arises and refers to an often misinterpreted point in CAJANDER's theory since, at least at the first glance, he did not give clear enough directions to do so.

15.3.1 SITE-TYPE CLASSES

The problem of distinguishing forest site-types of the Finnish school is usually settled in three stages. The first stage or level of classification involves *site-type classes*. CAJANDER himself divided the Finnish forests into five such classes: (1) dry-and -poor (heathlands), (2) fresh-mesic (mossy forests), (3) fresh-and-rich (broad leaved forests), (4) wet-and-rich (inundated forests), and (5) wet-and-poor (bog forests) site-types. This division is clearly based on more or less well-known correlations with environmental features. For practical purposes these site-type classes are very useful and readily recognizable, and of evident importance in making a decision when choosing a forest regeneration and management strategy.

Similar or more detailed treatments of classes of site-types have been applied in, e.g., Sweden (ARNBORG 1960), Canada (HEIMBURGER 1941, KUJALA 1945), the United States (HEIMBURGER 1934, CURTIS 1959), the USSR (POGREBNJAK 1929, SUKACHEV 1930, 1931, 1951, VOROB'EV 1953, etc.), and particularly in Estonia (ILVES 1953, KARU & MUISTE 1958, MASING 1969, LÕHMUS 1971). Here it might be of interest to note that already in the preceding centuries colloquial Estonian included as many as ten special words for site-type classes as we now recognize them. Today they are used in their right as terms (Fig. 1).

15.3.2 SITE-TYPES

The second stage is connected with the definition of *site-types* within site-type classes. Due to the circumstances quoted above, the definition should bear on permanent stands and their ground flora, the latter being the most distinctive attribute of Finnish forest typology. Site-types are best recognized by the whole undergrowth flora, for it is the whole composition of species present (and absent) that best indicates the properties of the biotope or habitat. In this emphasis of 'characteristic species composition' the site-type approach may in principle converge with that of the school of BRAUN–BLANQUET (though abundance and frequency relations of species may be given more emphasis). Undergrowth dominants alone are not, however, a sufficient basis for the definition and identification of site-types. A generally important dominant may on occasion be absent (e.g. *Vaccinium myrtillus* from the Myrtillus-type) from a stand that, as judged from over-all floristic composition, belongs to the site-type it characterizes. Yet in practice, in the species-poor vegetation of the North, undergrowth dominants (including bryophytes and lichens) are emphasized in the characterization and naming of site-types. Also in practice, the vitality (growth rate and height) of trees is much used for recognition of site-types, especially by foresters. In areas such as western Canada with more numerous tree and shrub

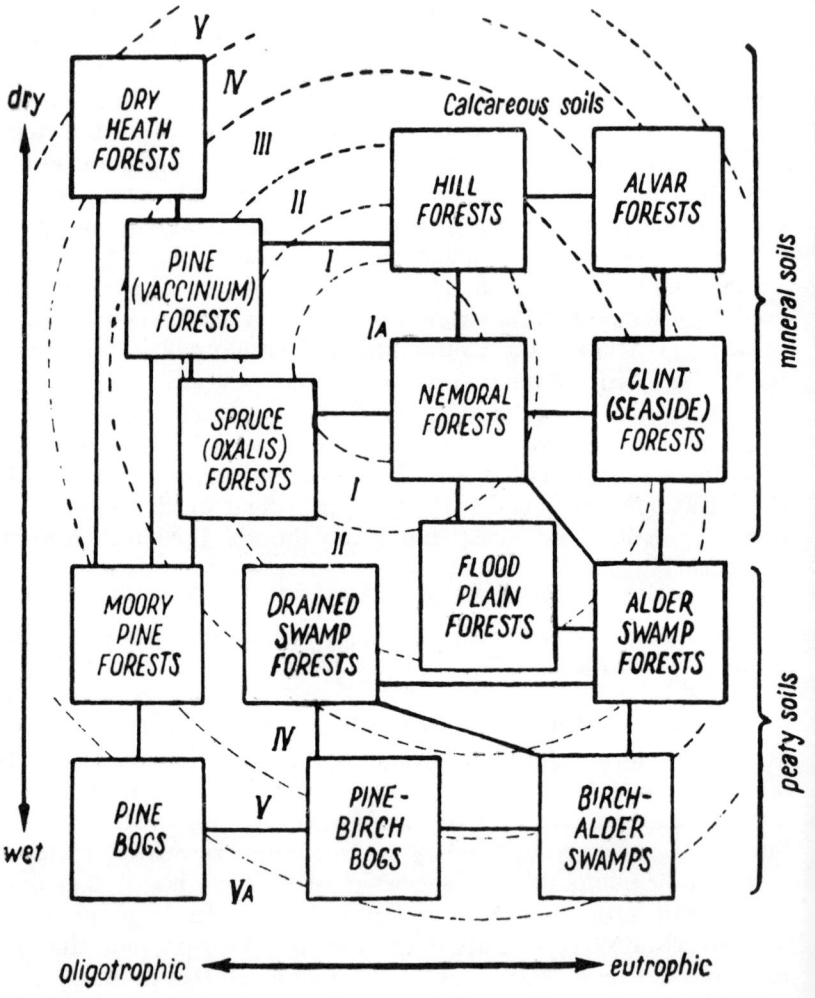

Fig. 1 Estonian forest site-type classes according to Masing (1969). Solid lines indicate the existence in nature of transitional stages between the classes, and interrupted lines the forest quality classes (isobonitets).

species, the indicator value of these has been considered along with that of the lower strata.

CAJANDER in particular was much concerned with dominance and the phenomenon of competition, which he was one of the first to consider. Already in his early papers CAJANDER paid considerable attention to the problem of interspecific competition as expressing the effect of the intermingled ecological factors upon vegetation. He recognized that in the comparatively extreme conditions of boreal forests there are only a few species able to dominate. The other species occurring there must adapt themselves not only to the biotope characteristics but also to the dominant species. From the limited number of dominant species there may result discontinuity in vegetation. Site-types can in any case be characterized, particularly in stable communities whose composition results from competition through a long time, by means of (i) dominant, (ii) constant, (iii) characteristic and (iv) differential species (see CAJANDER 1926: 27, 1930: 27–28). Once established for permanent stands, these species can be readily used in judging the membership of any of the stands representing temporary communities. In secondary forests the tree layer is usually much degraded, and centuries may be needed before the balance is reached again. The ground vegetation, however, generally establishes itself within decades; and in spite of the strong effects of dominant trees upon the other layers, the latter always consist of enough species that this assemblage, especially on better and mesic sites, is a more sensitive indicator than the tree layer itself.

Using dominant undergrowth species or groups of prevailing species (e.g the extraordinary role of lichens in the North), site-types are easily recognizable in poor site-type classes. In his early work CAJANDER (1909a) recognized five site-types for the first class:

(1) Cladina-type (ClT),
(2) Myrtillus-Cladina-type (MClT),
(3) Calluna-type (CT),
(4) Empetrum-Myrtillus-type (EMT),
(5) Vaccinium-type (VT).

Usually there is no problem in identifying types (1), (3) and (5), but the other two are of somewhat mixed nature. The distinctions between these types do not produce real difficulties, however, because in extreme conditions the boundaries are usually well-marked. CAJANDER attributes this phenomenon to the strong competitive relations between the dominant species. The types for dry-poor sites described in areas other than Finland fit in fairly well with CAJANDER's units, regardless of the degree of cultural disturbance in a particular country.

Reasonable agreement between students of vegetation in different areas is easily obtained at the other extreme as well, considering the site-types of mires and bogs. But, the greater the number of species the smaller the role of dominant species; and as soon as one considers mesic conditions, the task turns out to be rather more complicated. The third site-type class was divided by CAJANDER into the following categories:

(1) Geranium-type (GT),
(2) Dryopteris-type (DT),
(3) Oxalis-Maianthemum-type (OMaT),
(4) Sanicula-type (ST),
(5) Aconitum-type (AT),
(6) Vaccinium-Rubus-type (VRT),
(7) Lychnis diurna-type (LT).

All these site-types are rich, of high species diversity and variability within and between stands; and as a result it is difficult to distinguish them. Therefore, the site-types themselves are to some extent speculative in nature. Their boundaries are often gradual, fine-grained and topography-dependent with wide transition zones, so that it is hard to reduce the role of subjective decisions. As a rule, not only dominant and constant but also characteristic and differential species or/and groups of species must be used to define and recognize these site types. It is not very surprising that the fresh-and-rich site types proposed by CAJANDER for the boreal forests of Finland do not coincide with the types suggested for use in the hemiboreal forests of Estonia (Fig. 2). Differences result from both the vegetation itself and the criteria applied. The former point may be illustrated by the fact that LINKOLA (1929), who first applied the Finnish theory to Estonian forests, was forced to introduce a sixth class of site-types into CAJANDER's typology, viz. – (6) dry-and-rich (alvar forests), lacking in Finland but frequent in Estonia as well as in southern Sweden.

Additional difficulties concern the temporary communities of the fresh-and-rich site-type class. It appears that in all types there is a general tendency to form communities poor in species at the early and middle successional stages. The high productivity results in closed and overlapping canopies under which in almost all cases only *Oxalis acetosella* can survive. Other species, excluding mosses, occur in a sparsely scattered manner; and sometimes it is impossible to deduce the basic site-type from the dominance of *Oxalis* and the absence of other herbs. Therefore in Estonia a special site-type, the Oxalis-type, is used. However, it should be borne in mind that an Oxalis-type site does not usually denote more than a site which with age will change into a site of the Oxalis-Maianthemum-type, or passing through the Vaccinium-Rubus-type may acquire the properties of the Dryopteris or even the Sanicula-type (in CAJANDER's terminology).

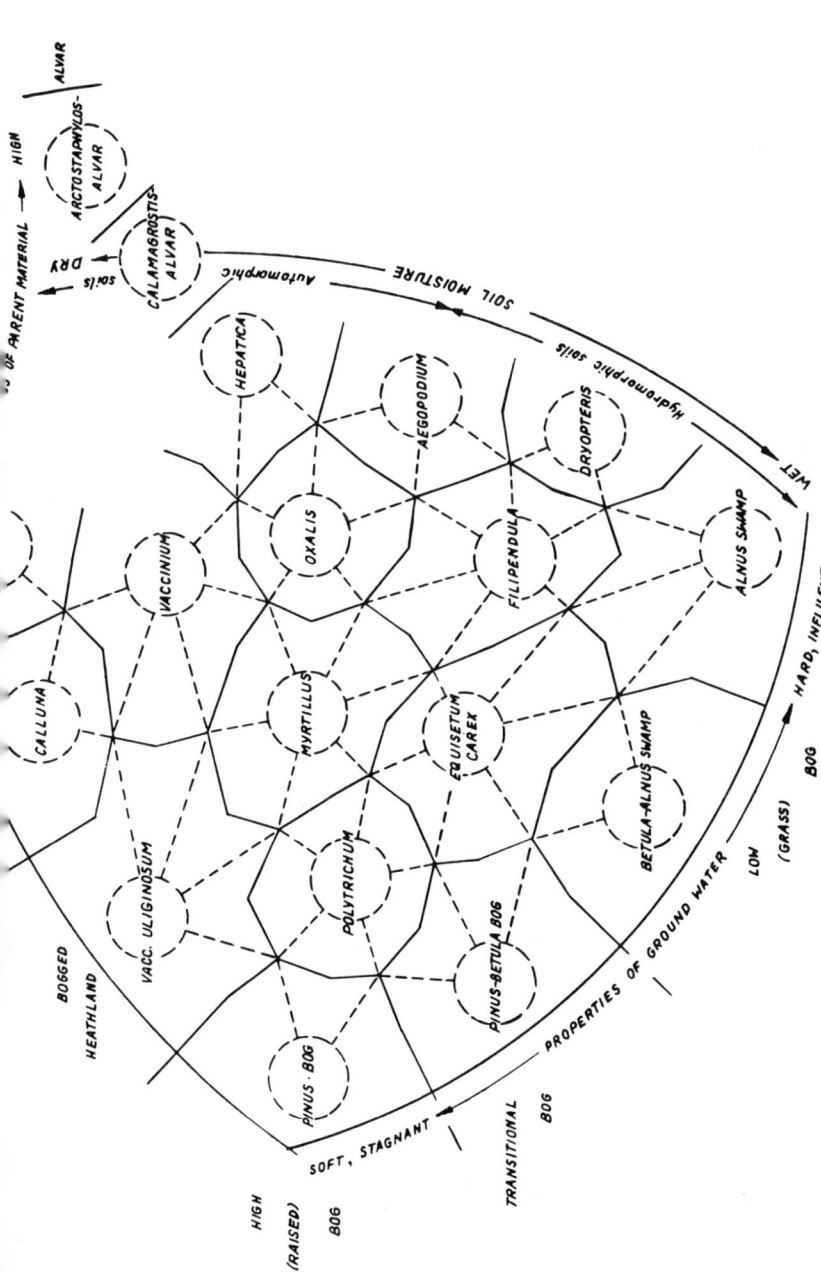

Fig. 2. A chart of forest site-types in relation to one another and leading soil factors in Estonia (Lõhmus 1971). The circles denote the site-type gravity centers defined both by ground vegetation (using the presence of prevalent species weighted by their average cover) and by soil properties. Solid lines indicate the distribution along environmental gradients of the site types, and dotted lines of subtypes, respectively.

15.3.3 Subordinate Units

The third stage of classification involves change in the dominant tree species within the limits of a site-type. In extreme site-types there may be no replacement of the normally dominant tree species by other species. Mesic and rich sites, however, show a variety of dominants or, to put it otherwise, one can find there a great number of cover types or *forest types*. Forest types are thus (unlike site-types) to be distinguished by tree species. CAJANDER (and the Russians) termed forest types distinguished by both canopy species and undergrowth 'associations'; but the international term for vegetation units defined by the dominant species of the different strata is now *sociation* (articles 17 and 18). For example, in sites having undergrowth of the Myrtillus site-type, four forest types or sociations are distinguished by dominant tree species, the more stable Pinetum myrtillosum and Piceetum myrtillosum, and the temporary Betuletum myrtillosum and Populetum myrtillosum. The Mercurialis-Aegopodium site-type is characterized by as many as six dominant tree species, the corresponding sociations being mainly temporary. For example, *Betula verrucosa* may be replaced by *Populus tremula* before *Picea abies* replaces *Populus* to take its rightful dominance approximately one hundred years after felling.

On the other hand, within the dry-and-rich site-type class (alvar forests) there are four site-types (KARU & MUISTE 1958):

(1) Arctostaphylus-Cladonia-type with only *Pinus silvestris*,
(2) Vaccinium-type with *Pinus* or/and *Picea abies*,
(3) Oxalis-Corylus-type with *Pinus*, *Picea*, or *Quercus robur*,
(4) Sesleria-type with the previous three plus *Betula verrucosa* as dominant tree species.

Thus alvar forests include altogether ten forest types or sociations in four site-types.

To sum up the leading principles of the Finnish method of forest site-types, it should be emphasized that the classification of forests into site-types is based on all the similar and distinguishing characters of the stands under consideration (e.g. TUOMIKOSKI 1942, KALELA 1939, 1954, 1960). Usually the question of how widely or narrowly the site-types are to be delimited has been regarded as having only a secondary importance. However, where they are broadly defined CAJANDER as well as his followers have taken advantage of differentiating them into subtypes (cf. the definition of the site-type above) and variants and/or facies. These units too cannot escape some subjectivity in definition and decision about limits. Yet it should be emphasized that such divisions are based on objective data – the tabular summaries of the stand samples first proposed by CAJANDER (1903) independently of the Central-European investigators.

15.4 Application in Management

It is a fair test of the system that its site-types, using undergrowth as an indicator of the biotope, should predict the forest growth the biotope will support. The relation of site-type to forest growth was investigated by ILVESSALO (1920, CAJANDER & ILVESSALO 1921) and many others. ILVESSALO observed especially the effects of relative favorableness of site on stand density and stand growth. For example, in South Finland major site-types (excluding the hydric) may be arranged in a series of decreasing favorableness – Oxalis-type (OT), Oxalis-Myrtillus-type (OMT), Myrtillus-type (MT), Vaccinium-type (VT), Calluna-type(CT), and Cladonia-type (CIT). When densities of trees by diameter size classes are plotted, for stands of the same age in each of these site-types, a striking relationship appears (Fig. 3). The more favorable the site, the smaller

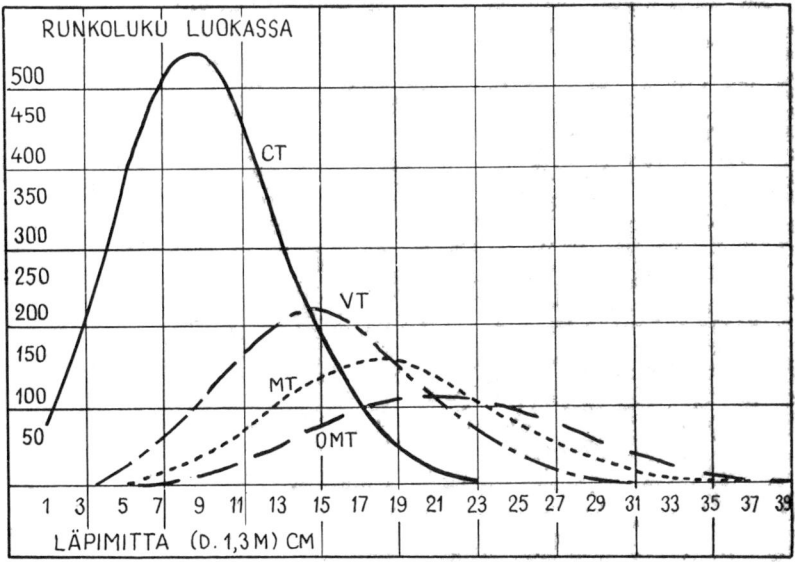

Fig. 3. Mean frequency distributions of tree stems by diameter classes, for 70-year-old *Pinus silvestris* in four forest site-types (CAJANDER & ILVESSALO 1921, Fig. 6). The vertical scale is numbers of stems in 4-cm classes per hectare, the horizontal scale is stem diameter in cm at breast height.

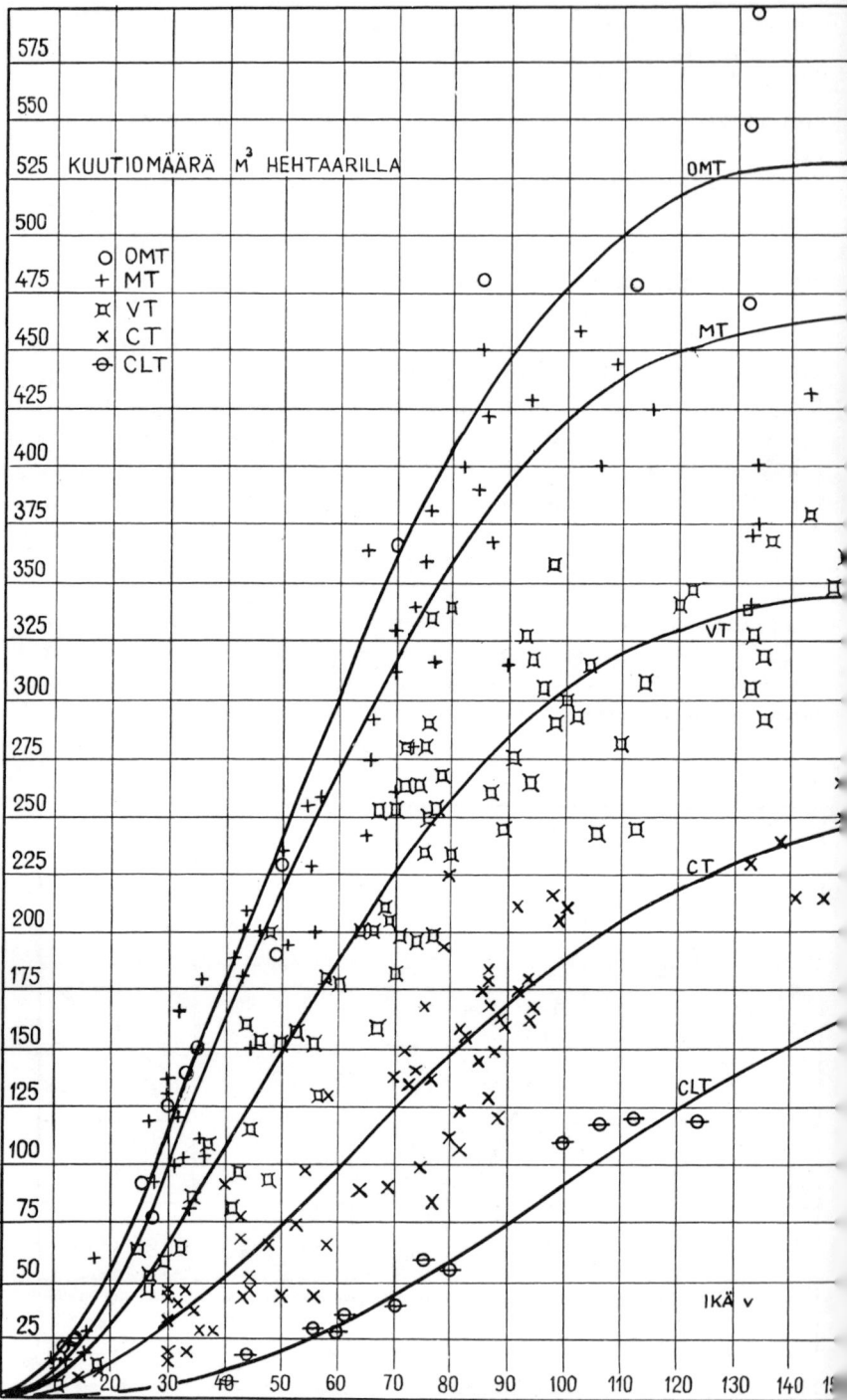

Fig. 4. The relationship of tree growth in *Pinus silvestris* to age in five forest site-typ (CAJANDER & ILVESSALO 1921, Fig. 9, see also CAJANDER 1949, Figs. 2-4). The vertic scale is accumulated stem wood volume in cubic meters per hectare, the horizont scale is stand age in years.

the number of trees, the larger the mean diameter, and the broader the dispersion of diameters from the mean. The inverse relation of favorableness to density reflects a widely significant principle of plant competition (YODA et al. 1963, HARPER 1967). The difference in tree sizes must result from difference in growth rate, as represented in Fig. 4. There is dispersion in the growth of different stands of a given site-type at a given age; such dispersions are almost inescapable in ecological phenomena. Yet the essential requirement is met: The site-type designation gives a prediction (within probability limits) of productive potential of the habitat.

CAJANDER thought much less effective a site indication based only on rate of height growth (site quality or 'Bonität' classes). The quality class can only be determined for a given species growing in the biotope at an age and condition suitable for measurement. The site-type, in contrast, can be determined independently of the present tree stand in the biotope. A quality class for one species in a site may offer no prediction of the growth rate of another species there, for there is no fixed proportion between the growth rates of different species in different sites. A site-type designation, however, offers a prediction of growth potential in that site for any species that might be planted there, if the growth rate of that species in the site-type has been measured. A quality class measurement for one species in a site may give little indication of what other species would be suitable to the site. The site-type classification provides an indication of which tree species may reasonably be planted in the site and which of these may best realize the site's potential for forest production at what stocking density. The site-types have much to offer for the skills of forest management.

The objectives of the system for management are those stated outside the Finnish school, by KRAJINA (1960: 55): 'In order that an ecological classification may be used by the forester, silvicultural characteristics might be included, having respect to: productivity and quality of trees; capacity for conservation of soil, water, and other habitat factors; methods of cutting and logging; methods of slash disposal; methods of reforestation; stocking, and methods and time of thinning and pruning; relation to pests; possibility for ranch management; wildlife management. Only the accurate synecological knowledge of forests, concretely registered and mapped for every forest region, will form the sound foundation for any forest management, silvicultural or genetical'. This summary fairly characterizes the nature and practical application of the Finnish theory of site-types in the Scandinavian and Baltic countries, especially in Finland and Estonia.

15.5 Series and Complexes

ILVESSALO'S (CAJANDER & ILVESSALO 1921) work implies that in many areas major site-types can be arranged in a sequence of decreasing quality from the fresh-and-rich to the dry-and-poor. Such a sequence is an *ecological series* in the sense of CAJANDER (1903) and Russian authors (Fig. 5). A further result of interest in ILVESSALO'S (1922) work is the relation of species diversity to the series from the Oxalis-type to the Cladonia-type in South Finland (Table I). In

TABLE I. Species diversities in forest samples with stands of intermediate ages (ILVESSALO 1922). Samples are mostly 0.2 or 0.25 hectare, but some down to 0.1 ha. Species diversities decrease and proportions of non-vascular plants increase along the moisture gradient from the Oxalis-Type to the Cladonia-Type.

Types:	Oxalis	Oxalis-Myrtillus	Myrtillus	Vaccinium	Calluna	Cladonia
Number of quadrats	21	58	53	38	37	5
Total number of higher plant species	107	105	86	58	28	9
Mean number of higher plant species	37.5	32.0	28.7	18.4	7.6	4.2
Standard deviation, species per quadrat	9.0	8.7	7.3	5.8	3.1	1.7
Per cents of all species,						
higher plants	86.3	84.7	81.9	75.3	63.6	45.0
mosses	10.5	9.7	11.4	14.3	15.9	25.0
lichens	3.2	5.6	6.7	10.4	20.5	30.0

this area species diversity declines with decreasing site quality and productivity (see also CAJANDER 1909a, PALMGREN 1915-17 LINKOLA 1916, 1917, LAKARI 1920), though in some other areas maximum diversity is in the middle of a moisture gradient, or productivity and diversity are inversely related (WHITTAKER 1969). ILVESSALO (1922) gives data also on effects of age of stand and species of canopy dominant on species diversity. PALMGREN (1928) studied the relation of bird communities to a site-type series.

The use of ecological series of site-types as a basis of vegetation interpretation was developed further by KUJALA (1938, 1945). KUJALA considered that, with two major site-types in each of CAJANDER's three classes excluding the hydric, six site-types would in many areas form a series representing the main axis of community response to habitat. For south Finland such a series (from dry to mesic) would be: (1) Calluna-Cladonia-type, (2) Vaccinium-type, (3) Myrtillus-type, (4) Oxalis-Myrtillus-type, (5) Oxalis-Majan-

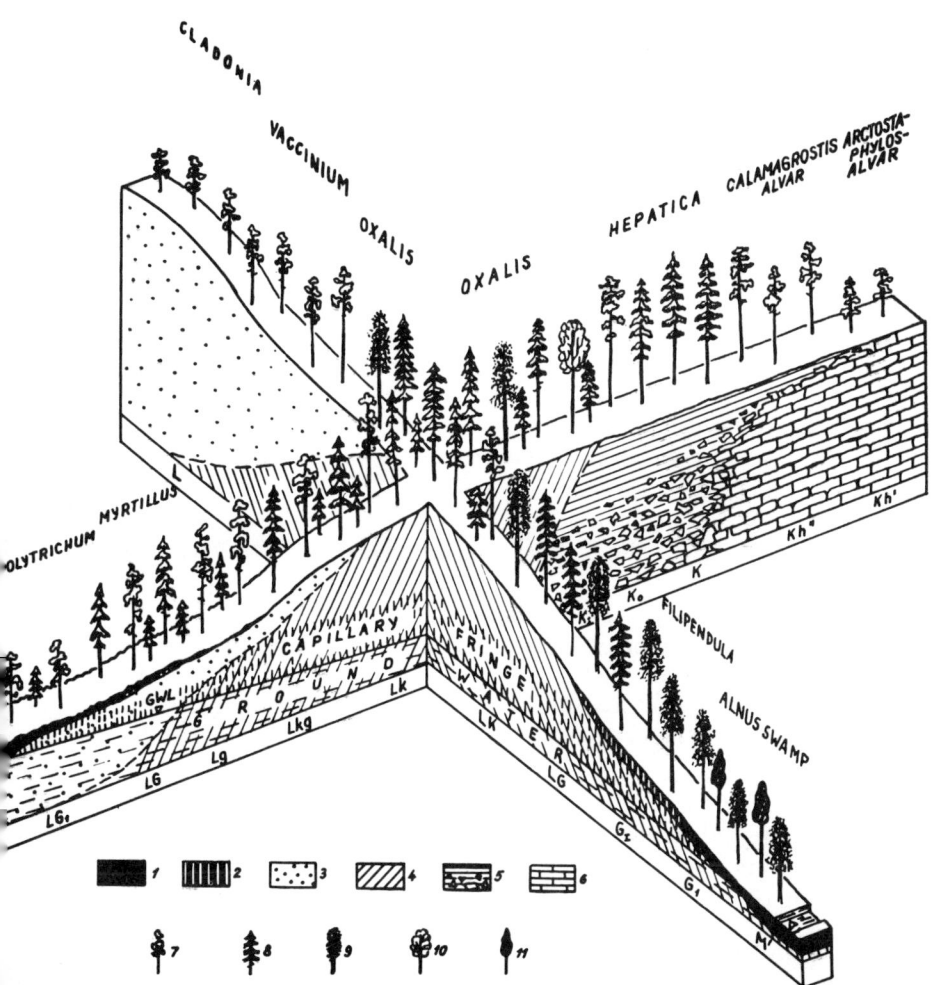

ecoclines showing the interrelations of Estonian forest site types and their ecological position ork of complex habitat factors (LÕHMUS 1971):1) peat, 2) organic-mineral (peaty) horizon and, 4) noncalcareous moraine, 5) calcareous moraine, 6) limestone; 7) *Pinus silvestris*, 8) *Picea la verrucosa, B. pubescens*, 10) *Populus tremula*, 11) *Alnus glutinosa*: L – podzol, Lg – gleyed podzol, podzolic, Lkg – gleyed soddy podzolic, LG – glay podzol, LG_1 – peaty podzol, R' – shallow $_1$–podzolic glay, G_1 – peaty glay, M'–shallow low moor, K_1– brown pseudopodzolic, K_0– brown – brown lessive, Kh'' – shallow limestone rendzina, K' – very shallow limestone rendzina soil.

themum-type, and (6) Oxalis-Filices-type. When one proceeds to a different geographic area, each of these site-types may be replaced by another, related site-type; thus from south to north Finland the Oxalis-Myrtillus-type is replaced by the Dryopteris-Myrtillus-type as the fourth member of the series. These two ecologically equivalent communities are *vicariant* site-types. In north

Finland the types given for south Finland are replaced by vicariants to form the series: (1) Calluna-Myrtillus-Cladina-type, (2) Empetrum-Vaccinium-type, (3) Hylocomium-Myrtillus-type, (4) Dryopteris-Myrtillus-type, (5) Geranium-Dryopteris-type and (6) Filices-type. These two series, comprising corresponding or vicariant types, are *parallel type-series*. KALELA (1961) distinguished six or seven type series each characterizing a forest region in Finland.

A type-series permits one to characterize some main features of the vegetation pattern of an area of northern forest. (Such series may not, however, include hydric or bog communities, and in some areas the vegetation pattern is further complicated by fertility and other soil differences.) Parallel type-series permit comparison of the vegetation patterns of different areas. Thus in his treatment of the Canadian boreal forests KUJALA (1945) compares parallel type-series across the continent. An eastern (Gaspé Peninsula)series resembles the Finnish – (1) Kalmia-Cladonia-type, (3) Vaccinium canadense-type, (4) Cornus-Aralia-type, (5) Oxalis-Cornus-type, (6) Oxalis-Dryopteris-type, (7) Impatiens-Circaea-Athyrium-type. A western (Vancouver Island, maritime-climate) series comprises only – (3) Gaultheria-type, (4) Achlys-Gaultheria-type, (5) Achlys-Tiarella-Aspidium munitum-type. The latter is geographically and floristically distant from the Finnish series and shows an apparently smaller degree of type differentiation along the moisture gradient. KUJALA (1945) further observes that the middle members of the type-series are normally best developed and most extensive in an area and express its climate. These middle members, generally (3) and (4), are consequently *leading-types* (*Leittypen*) for the characterization of regional vegetation; the concept of leading-type converges with that of climatic climax (CLEMENTS 1916) and, perhaps more closely, prevailing or regional climax (WHITTAKER 1953, BRAUN 1956). The characterization of vegetation patterns through type-series may be compared also with that of mountain vegetation through topographic coenoclines (WHITTAKER 1956, 1960, WHITTAKER & NIERING 1965, article 3).

KALELA (1939, 1954, 1958, 1960, 1961, 1962) also developed CAJANDER's system further for the treatment of vegetation patterns, emphasizing both the dynamical and geographical aspects of typology in a regional or zonal approach that is now of major importance in the Finnish school. Each forest site-type, regional or local, comprises a climax community-type and various successional communities developing towards it. The stands belonging to a larger topographic unit form a *stand complex*, i.e. 'the parts of plant cover which are more or less clearly distinguished from the surroundings and composed of numerous stands representing diverse stand types of various sizes and alternating in different ways–being mosaic, belt-like or reticular in shape' (KALELA 1960: 42). Just as stands may be grouped into stand types, he suggests the grouping of *stand complexes* into their *types*. Within each *vegetation region* (KALELA 1939, 1954) the macroclimate is homogenous in broad features and the region is characterized by the prevalence of the same complex-types in ecologically similar localities (see also AHTI et al. 1968). As a rule, the stand types of a vegetation region can be divided into regional (climatic, zonal) and local (edaphic, azonal) plant communities. Phytosociologists have not always paid sufficient attention to this division. In a given region prevailing climate influences the development of certain kinds of soils. The communities that are most successful on the normal and non-extreme soils in a region form together its *regional plant communities*. KALELA thus develops, in the Finnish school, ideas comparable to the community-complex or pattern, landscape-type, and regional or prevailing climax community of other schools.

15.6 The Finnish and Other Approaches

The principles of forest typology proposed by CAJANDER (1909a, 1926, 1930, 1949, CAJANDER & ILVESSALO 1921) were applied, in some cases with modifications, by AALTONEN (1919), ILVESSALO (1920, 1922, 1929, 1967), LAKARI (1920), KUJALA (1921, 1929a, 1929b, 1936a, b, 1945, 1960), LINKOLA (1924, 1929, 1930), KALELA (1941, 1952, 1954, 1958, 1960, 1961), KALLIOLA (1943), KELTIKANGAS (1945, 1959), PERTTULA (1950), JALAS (1950), KALLIO (1957), HÄMET-AHTI (1963), HÖGNÄS (1966), MÄKIRINTA (1968) and others in Finland and elsewhere. Site-types for hydric or bog communities were considered by CAJANDER (1913), LUKKALA (1929), AARIO (1932), PAASIO (1937, 1940, 1941), RUUHIJÄRVI (1960), and EUROLA (1962). The system has been influential outside Finland also (see KALELA 1960, KRAJINA 1960, WHITTAKER 1962, TRASS 1964, ALEKSANDROVA 1969). Closely related approaches in the Scandinavian and Baltic area include the work of MALMSTRÖM (1926, 1936) and NIHLGÅRD (1970) in Sweden, BORNEBUSCH (1923-5, 1929) in Denmark, KIRSTEIN (1929) and MALLNER (1944) in Latvia, and LINKOLA (1929, 1930) in Estonia. Forest site-types were studied in Central Europe by CAJANDER (1909a, 1949, LINKOLA (1924), and KUJALA (1936). North American applications have included ILVESSALO (1929), HEIMBURGER (1934, 1941), KUJALA (1945), SPILSBURY & SMITH (1947), HUSTICH (1949, 1950), and CRANDALL (1958),

AHTI (1959), KALELA (1962), and HÄMET-AHTI (1965). Patagonian forest site-types were studied by KALELA (1941).

Russian forest phytocoenology developed in partial independence of the Finnish school, but was influenced by it. Russian work, from the early writings of MOROSOV (1928) to the recent of SUKACHEV (1944, 1954, 1960) emphasizes the forest stand as a biogeocoenose, a complex whole of organisms and environment and their interrelations. The forest types of SUKACHEV (1928, 1932) and others give equal weight to the canopy dominant and the undergrowth; these are not site-types in the sense of CAJANDER but sociations (termed by the Russians 'associations'). They are, however, treated in terms of ecological series as in Finnish work, though the emphasis has been on series in relation to a number of gradients (see article 17, Fig. 1) rather than on a single series of principal site-types (KUJALA 1945). Russian authors have used ideas of CAJANDER on biologically equivalent sites (SUKACHEV 1934) and twin associations (sociations differing in one of their stratal dominants or unions) (KATZ 1929, 1930). ALEKSANDROVA (1969) has pointed out how CAJANDER's ideas have influenced, or coincided with, the ecological series of SUKACHEV (1925, 1931, 1934, 1938), SOKOLOWA (1935, 1936), and SOKOLOV (1937, 1962), the groups of SMAGIN (1950), the circles of KATS (1936), the 'ingregations' of BYKOV (1962), the sections of MIRKIN (1965), and other Russian interpretations.

A Swedish site-type system was developed by ENEROTH (1936), ARNBORG (1940, 1953) and LINDQUIST (1954). The forest types of this system are sociations, and these are grouped in relation to forest succession in a manner resembling the Finnish. A forest site-type comprises a whole series of plant communities of which only one stage – present under certain conditions, particularly in normal density of the tree layer – is normative for the type. A number of different undergrowth communities are thus grouped together in relation to a kind of site and a normative community which occurs in such sites under certain conditions (ARNBORG 1940). The concept is close to CAJANDER's but as observed by WHITTAKER (1962) 'The site-type approach is thus carried to its logical conclusion; it is a classification not simply of communities, but of sites as characterized by groups of related communities (cf. KALELA 1954).' The Swedish system converges also with the Russian in treatment of stratal communities (unions) as these are variously combined into forest types (sociations), with the latter arranged in ecological series (Fig. 6).

The influence of CAJANDER spread also into Middle Europe (WHITTAKER 1962), but the system has little current use there. The Austrian system of AICHINGER (1951, 1952, 1967, WENDELBERGER 1951) employs a different unit, the *vegetation-development-type*. Devel-

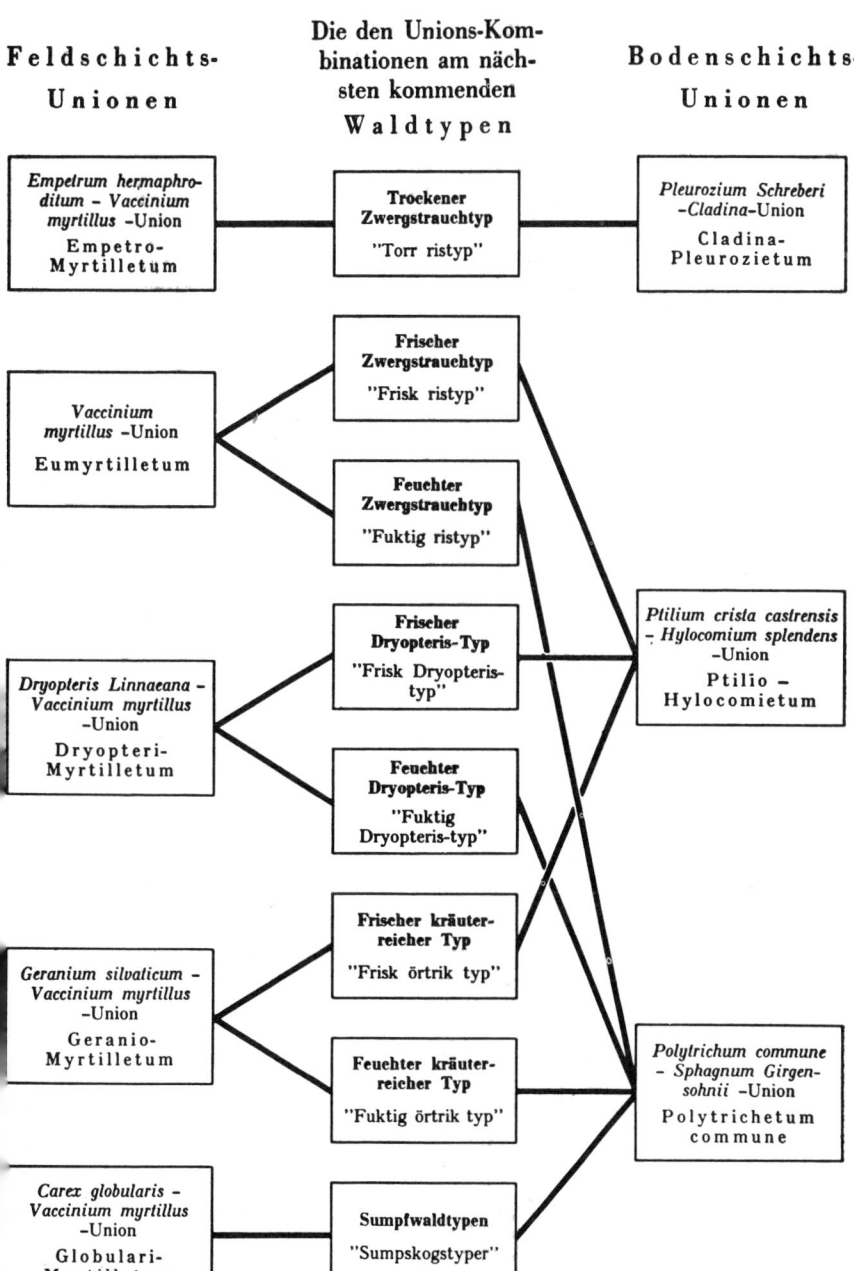

Fig. 6. A system of forest types for a north-Swedish forest (ARNBORG 1940). The middle column gives forest types as formed by different combinations of undergrowth unions those of the field (herb and shrub) layer on the left and the ground (moss and lichen) layer on the right. Forest types and unions are arranged in relation to the moisture gradient from the dry dwarf-shrub type (*Trockener Zwergstrauchtyp*) through fresh (*Frischer*) and moist (*Feuchter*) types to the wet swamp forest types (*Sumpfwaldtypen*).

opment-types are described by stratal dominance in the manner of sociations, but are defined also by their place in successions. The system, with its strong emphasis of successional relations, is adapted to vegetation heavily and variously influenced by man. Extensive work in applied forestry in middle and southern Europe has applied the BRAUN–BLANQUET system; such work is to be discussed in a succeeding part of the *Handbook*. In general the CAJANDER system has had continued influence only in application to boreal and subalpine forests, the taiga formation-type.

15.7 Conclusion

In historic perspective, CAJANDER himself was influenced by several Scandinavian research workers – e.g. by VAINIO's (1878) deep concern with habitat factors, NORRLIN's (1870) idea of the interdependence of the vegetation and its habitat, NILSSON's (1901) principle of vegetation series based primarily on the character of undergrowth, BLOMQVIST's (1872, 1879) forest growth classes, etc. In developing these and other ideas into a system for the treatment of northern vegetation, CAJANDER influenced and is representative of the whole Northern Tradition (WHITTAKER 1962, TRASS 1964). Many of the ideas of the Finnish school correspond to those of other schools; but the school is distinctive, characterized by its own nuclear concept: the site-type as a grouping of successionally related communities around a mature or normal community-type which indicates site quality and which is to be recognized and related to the successional communities by undergrowth composition. From this central concept have radiated the means of classifying communities, practices of forest management, treatment in terms of ecological series, applications to animal communities, and interpretation of vegetation dynamics and landscape patterns.

It seems broadly true that the emphasis in forest structure shifts downward with increasing latitude, from tropical forests with their floras strongly concentrated in the canopy trees and their epiphytes, through warm-temperate forests with both canopy and undergrowth well developed and mixed in composition, to boreal forests with one or few tree species and with the flora (and in some cases maximum coverage) concentrated in the lower strata. The Finnish and related approaches are suited to the latter part of this spectrum, and the emphasis of stratal relations in the northern schools contrasts with that of floristic composition (or canopy dominance) in warm-temperate and of physiognomy in tropical ecology. We may wonder (see KALELA 1941) whether the Finnish system is not appropriate also for cold-temperate climates of the Southern Hemisphere – for, say, the *Nothofagus* forests of New Zealand. For the Scandinavian and Baltic area at least, CAJANDER's ideas stimulated the development of an approach to vegetation adapted to the ecology of northern forests, and were productive as a means for research on and management of them.

15.8   SUMMARY

The Finnish School developed under the leadership of CAJANDER, with the forest *site-type* as a central concept. A site-type groups together mature forest stands of similar undergrowth composition expressing similar site qualities, and unstable stands of the same kinds of sites. In northern forests the canopy tree species less effectively expresses site conditions than does the undergrowth. Site-types are consequently defined largely by the undergrowth of mature or stable stands, and knowledge of undergrowth communities is used to relate unstable stands to stable ones.

With the site-type as the basic unit, vegetation of an area may be classified into (a) three to six site-type classes expressing broad relationships to soil moisture and fertility, (b) within each class one to several site-types and subtypes, each expressing narrower site differences in distinctive undergrowth communities, and (c) within each site-type or subtype a number of forest types or sociations, defined by dominant species of both canopy and undergrowth and including both stable and unstable communities.

The site-type provides a biological expression of biotope or site characteristics. Recognition of the site-type for a forest stand (which may be disturbed) thus provides indication of the quality of the site, the mature community and the direction of succession toward it, the potential productivity in that site of different tree species, and consequently the appropriate forest management.

Site-types may be arranged in ecological series along environmental gradients. Several site-types (in many cases about six in one region, excluding hydric sites) thus arranged along the moisture gradient may express major features of vegetational variation in an area. Site-types may also be arranged into patterns including ecological series in relation to more than one environmental gradient. The site-types most widespread in a landscape (those in the middle of the moisture-gradient series) may be considered the leading-types or regional plant communities.

The Finnish approach has been extensively used in the Scandinavian and Baltic area and has influenced Russian forest phytocoenology, but has not been much used in warm-temperate areas. The approach seems best suited to boreal and subalpine forests; for these it is an effective means of ecological investigation and practical management.

ACKNOWLEDGMENT

The author thanks A. KALELA for his comments on the article and R. H. WHITTAKER for his contributions to it during revision.

REFERENCES

AALTONEN, V. T., – 1919 – Kangasmetsien luonnollisesta uudistumisesta Suomen Lapissa. I. (Germ summ.: Über die natürliche Verjüngerung der Heidewälder im Finnischen Lappland. I). *Comm Inst. forest. fenn.* (*Metätiet. Tutkimuslait. Julk.*) 1: 1–319+1–56.
AARIO, L., – 1932 – Pflanzentopographische und paläogeographische Moorunter- suchungen in N-Satakunta. *Fennia* 55(1): 1–179.
AHTI, T., – 1959 – Studies on the caribou lichen stands of Newfoundland. *Ann. Bot., Soc. Zool. -Bot. Fenn. 'Vanamo'* (*Suomal. eläin-ja kasvit. Seur. van. kasvit. Julk.*) 30(4): 1–44.
AHTI, T., L. HÄMET-AHTI, & J. JALAS, – 1968 – Vegetation zones and their sections in northwestern Europe. *Ann. bot. fenn.* 5: 169–221.
AICHINGER, E., – 1951 – Vegetationsentwicklungstypen als Grundlage unserer land- und forstwirtschaftlichen Arbeit. *Angew. Pfl Soziol.*, Wien 1: 17–20.
AICHINGER, E., – 1952 – Fichtenwälder und Fichtenforste als Waldentwicklungs- typen. *Angew. Pfl Soziol.*, Wien 7: 1–179.
AICHINGER, E., – 1967 – Die Waldentwicklungstypen im Raume von Kirch- leerau. (Engl. & French summs.) *Veröff. geobot. Inst. ETH, Stiftg Rübel*, Zürich 39: 187-270, 283, 293.
ALEKSANDROVA, VERA D., – 1969 – Classification of Vegetation. Principles of Classification and Classification Systems of Various Phytocoenological Schools (in Russian) Publ. House 'Nauka,' Leningrad. 274 pp.
ARNBORG, T., – 1940 – Der Vallsjö-Wald, ein nordschwedischer Urwald. *Acta phytogeogr. suec.* 13: 128-154.
ARNBORG, T., – 1953 – Det nordsvenska skogstypsschemat. 3rd ed. Svenska Skogsvardsför. Förlag, Stockholm. 20 pp.
ARNBORG, T., – 1960 – Can we find a common platform for the different schools of forest type classification? *Silva fenn.* 105: 13-15.
BLOMQVIST, A. G., – 1872 – Tabeller framställande utvecklingen af jemnårig och slutna skogsbestånd af tall, gran och björk. Helsingfors.
BLOMQVIST, A. G., – 1879 – Eine neue Methode, den Holzwuchs und die Star t- ortsvegetation bildlich darzustellen. *Bidr. Känn. Finl. Nat. Folk*, 31: 145-15?.
BOLLETER, R., – 1921 – Vegetationsstudien aus dem Weisstannental *Jb. St. Gall. naturw. Ges.* 57(2): 1-140.
BORNEBUSCH, C. H., – 1923-5 – Skovbundsstudier I-III, IV-IX. (Engl. summ.) *Forstl. Fors Vaes. Danm.* 8: 1-148, 181-288.
BORNEBUSCH, C. H., – 1929 – Danmarks skovtyper. *Acta for. fenn.* 34(11): 1-18.

BRAUN, E. LUCY., – 1956 – The development of association and climax concepts: their use in interpretation of the deciduous forest. *Am. J. Bot.* 43: 906-911.
BRAUN-BLANQUET, J., – 1951 – Pflanzensoziologie: Grundzüge der Vegetationskunde. 2nd ed. Springer, Wien. 631 pp. 3rd ed. 1964, 864 pp.
BYKOV, B. A., – 1962 – On the composition of some formations of the ingregations (in Russian). *Trans. Bot. Inst. Kazakh S. S. R. (Trudy alma-atin. bot. Sada)* 13: 3-27.
CAIN, S. A., – 1936 – Synusiae as a basis for plant sociological field work. *Am. Midl. Nat.* 17: 665-672.
CAJANDER, A. K., – 1903 – Beiträge zur Kenntnis der Vegetation der Alluvionen des nördlichen Eurasiens. I. Die Alluvionen der unteren Lena-Thales. *Acta Soc. Sci. fenn.* 32(1): 1-182.
CAJANDER, A. K., – 1904 – Studien über die Vegetation des Urwaldes am Lena-Fluss. *Acta Soc. Sci. fenn.* 32(3): 1-40.
CAJANDER, A. K., – 1909a – Ueber Waldtypen. *Acta for. fenn.* 1(1): 1-175.
CAJANDER, A. K., – 1909b – Beiträge zur Kenntnis der Alluvionen des nördlichen Eurasiens III. Die Alluvionen der Tornio- und Kemi-Thäler. *Acta Soc. Sci. fenn.* 37(5): 1-224.
CAJANDER, A. K., – 1913 – Studien über die Moore Finnlands. *Acta for. fenn.* 2(3): 1-208.
CAJANDER, A. K., – 1926 – The theory of forest types. *Acta for. fenn.* 29(3): 1-108.
CAJANDER, A. K., – 1930 – Wesen und Bedeutung der Waldtypen.*Silva fenn.* 15(1): 1-66.
CAJANDER, A. K., – 1949 – Metsätyypit ja niiden merkitys. Forest types and their significance. (Finn. & Engl.) *Acta for. fenn.* 56(4): 1-69, 1-71.
CAJANDER, A. K., & Y. ILVESSALO, – 1921 – Über Waldtypen II. *Acta for. fenn.* 20(1): 1-77.
CLEMENTS, F. E., – 1916 – Plant succession: an analysis of the development of vegetation. *Publs Carnegie Instn,* Washington 242: 1-512.
CRANDALL, DOROTHY L., – 1958 – Ground vegetation patterns of the spruce-fir area of the Great Smoky Mountains National Park. *Ecol. Monogr.* 28: 337-360.
CURTIS, J. T., – 1959 – The Vegetation of Wisconsin: An Ordination of Plant Communities. Univ. Wisconsin, Madison. 657 pp.
DU RIETZ, G. E., – 1930 – Vegetationsforschung auf soziationsanalytischer Grundlage. *Handb. biol. ArbMeth.* 11, 5: 293-480.
ENEROTH, O., – 1936 – Om skogstyperna och deras praktiska betydelse. *Handl. LandbrVeck,* Stockholm 1936: 821-837.
EUROLA, S.,–1962 –Über die regionale Einteilung der südfinnischen Moore. *Ann. Bot., Soc. Zool.-Bot. Fenn. 'Vanamo' (Suomal. eläin-ja kasvit. Seur. van. kasvit. Julk.)* 33: 1-243.
GAMS, H., – 1918 – Prinzipienfragen der Vegetationsforschung. Ein Beitrag zur Begriffsklärung und Methodik der Biocoenologie. *Vjschr. naturf. Ges. Zürich* 63: 293-493.
HÄMET-AHTI, L., – 1963 – Zonation of the mountain birch forests in northernmost Fennoscandia. *Ann. Bot. Soc. Zool., Bot. Fenn. 'Vanamo' (Suomal. eläin-ja kasvit. Seur. van kasvit. Julk.)* 34(4): 1-127.
HÄMET-AHTI, L., – 1965 – Notes on the vegetation zones of western Canada, with special reference to the forests of Wells Gray Park, British Columbia. *Ann. bot. fenn.* 2: 274-300.
HARPER, J. L., – 1967 – A Darwinian approach to plant ecology. *J. Ecol.* 55: 247-270.
HEIMBURGER, C. C., – 1934 – Forest-type studies in the Adirondack region. *Mem. Cornell Univ. (N.Y.) agric. Exp. Stn.* 165: 1-122.
HEIMBURGER, C. C., – 1941 – Forest-type classification and soil investigation on Lake Edward Forest Experimental Area. *Canada Dept. Mines Resources, Domin-*

ion Forest Serv., *Silvic. Res. Note* 66: 1-60.

HILLS, G. A., – 1960 – Comparison of forest ecosystems (vegetation and soil) in different climatic zones. *Silva fenn.* 105: 33-39.

HÖGNÄS, B., – 1966 – Undersökning av skogstyper och beståndsutvickling på Åland. (Engl. summ.) *Acta for. fenn.* 78(2): 1-127.

HUSTICH, I., – 1949 – On the forest geography of the Labrador peninsula. A preliminary synthesis. *Acta Geogr., Helsingf.* 10(2): 1-63.

HUSTICH, I., – 1950 – Notes on the forests on the East coast of Hudson Bay and James Bay. *Acta Geogr., Helsingf.* 11(1): 1-83.

HUSTICH, I., ed., – 1960 – Symposium on forest types and forest ecosystems. *Silva fenn.* 105: 1-142.

ILVES, A., – 1953 – Eesti NSV arumetsatüübid (Russ. summ.: Estonian upland forest types). In, 'Loodusuurijate Seltsi juubelikoguteos,' (Jubilee Symposium of the Estonian Naturalists Society), pp. 11-49. Tallinn.

ILVESSALO, Y., – 1920 – Tutkimuksia metsätyyppien taksatoorisesta merkityksesta nojautuen etupäässä kotimaiseen kasvutaulujen laatimistyöhön. (Germ. summ.: Untersuchungen über die taxatorische Bedeutung der Waldtypen). *Acta for. fenn.* 15(3): 1-157, 1-51, 1-32, 1-26.

ILVESSALO, Y., – 1922 – Vegetationsstatistische Untersuchungen über die Waldtypen. *Acta for. fenn.* 20(3): 1-73.

ILVESSALO, Y., – 1929 – Notes on some forest (site) types in North America. *Acta for. fenn.* 34(39): 1-111.

ILVESSALO, Y., – 1967 – Luonnonnormaalien metsiköiden kehityksestä Kainuussa ja sen lähiympäristössä. (Engl. summ.: On the development of natural normal forest stands in south-eastern North-Finland). *Acta for. fenn.* 81(5). 1-85.

JALAS, J., – 1950 – Zur Kausalanalyse der Verbreitung einiger nordischen Os- und Sandpflanzen. *Ann. Bot., Soc. Zool.-Bot. Fenn. 'Vanamo'* (*Suomal. eläin-ja kasvit. Seur. van. kasvit. Julk.*) 24(1): 1-362.

KALELA, A., – 1939 – Über Wiesen und wiesenartige Pflanzengesellschaften auf der Fischerhalbinsel in Petsamo Lappland. *Acta for. fenn.* 48(2): 1-523.

KALELA, A., – 1941 – Über die Entwicklung der herrschenden Bäume in den Beständen verschiedener Waldtypen Ostpatagoniens. *Ann. Acad. scient. fenn.* (*Suomal. Tiedeakat. Toim.*)Ser. A, 4 Biol., 3: 1-65.

KALELA, A., – 1952 – Kainuun alueen metsätyypeistä. (Germ. summ.: Über die Waldtypen des Kainuugebietes zwischen Mittel- und Nordfinnland.) *Comm. Inst. forest. fenn.* (*Metsätiet. Tutkimuslait. Julk.*) 40(26): 1-17.

KALELA, A., – 1954 – Zur Stellung der Waldtypen im System der Pflanzengesellschaften. *Vegetatio* 5/6: 50-62.

KALELA, A., – 1958 – Über die Waldvegetationszonen Finnlands. *Bot. Notiser* 111: 353-368.

KALELA, A., – 1960 – Classification of the vegetation, especially of the forests, with particular reference to regional problems. *Silva fenn.* 105: 40-49.

KALELA, A., – 1961 – Waldvegetationszonen Finnlands und ihre klimatischen Paralleltypen. *Arch. Soc. zool.-bot. fenn. 'Vanamo'* (*Suomal. eläin-ja kasvit. Seur. van. Tiedon.*) 16 (Suppl): 65-83.

KALELA, A., – 1962 – Notes on the forest and peatland vegetation in the Canadian Clay Belt Region and adjacent areas. *Comm. Inst. forest. fenn.* (*Metsätiet. Tutkimuslait. Julk.*) 53(33): 1-14.

KALLIO, K., – 1957 – Käenkaali-mustikkatyypin kuusikoiden kehityksestä Suomen lounaisosassa. Taksatoris-liiketaloudellinen tutkimus. (Engl. summ.: On the development of spruce forests of the Oxalis-Myrtillus site type in the south-west of Finland. Forest mensuration and management research). *Acta for. fenn.* 66(3): 1-155.

KALLIOLA, R., – 1943 – Porajärven seudun metsatyypeistä. (Germ. summ.: Über die Waldtypen der Gegend von Porajärvi in Ostkarelien). *Comm. Inst. forest fenn. (Metsätiet. Tutkimuslait. Julk.)* 31(2): 1-15.

KARU, A. & L. MUISTE, – 1958 – Eesti metsakasvukohatüübid. (Estonian forest site-types). Tallinn. 44pp.

KATZ, N. J., – 1929 – Die Zwillingsassoziationen und die homologen Reihen in der Phytosoziologie. *Ber. dt. bot. Ges.* 47: 154-164.

KATZ, N. J., 1930 – Die grundlegenden Gesetzmässigkeiten der Vegetation und der Begriff der Assoziation. *Bettr. Biol. Pfl.*, 18: 305-333.

KATS, N. YA., – 1936 – On plant communities of the Caucasian State Reserve in light of the species affinities (in Russian). *Zemlevêdénie* 38(3): 247-280.

KELTIKANGAS, V., – 1945 – Ojitettujen soitten viljavuus eli puuntuottokyky metsätyyppiteorian valossa. (Swed. & Engl. summs.: The fertility of drained bogs as shown by their tree producing capacity, considered in relation to the theory of forest types). *Acta for. fenn.* 53(1): 1-237.

KELTIKANGAS, V., – 1959 – Suomalaisista seinäsammaltyypeistä ja niiden asemasta Cajanderin luokitusjärjestelmässä. (Engl. summ.: Finnish feather-moss types and their position in Cajander's forest site classification). *Acta for. fenn.* 69(2): 1-266.

KIRSTEIN, K., – 1929 – Lettlands Waldtypen. *Acta for. fenn.* 34(33): 1-20.

KRAJINA, V. J., – 1960 – Can we find a common platform for the different schools of forest type classification. *Silva fenn.* 105: 50-55.

KUJALA, V., – 1921 – Havaintoja Kuusamon ja sen eteläpuoliston kuusimetsäalueiden metsä- ja suotyypeistä. (Germ. summ.: Beobachtungen über die Wald- und Moortypen von Kuusamo und der südlich von dort gelegenen Fichtenwaldgebiete). *Acta for. fenn.* 18(5): 1-65.

KUJALA, V., – 1929a – Die Bestände und die ökologischen Horizontalschichten der Vegetation. *Acta for. fenn.* 34(17): 1-26.

KUJALA, V., – 1929b – Untersuchungen über Waldtypen in Petsamo und in angrenzenden Teilen von Inari Lappland. *Comm. Inst. forest. fenn. (Metsätiet. Tutkimuslait. Julk.)* 13(9): 1-125.

KUJALA, V., –1936a – Tutkimuksia Keski- ja Pohjois-Suomen välisestä kasvillisuus rajasta (Germ. summ.: Über die Vegetationsgrenze von Mittel- und Nord-Finnland). *Comm. Inst. forest. fenn. (Metsätiet. Tutkimuslait. Juik.)* 22(4): 1-95.

KUJALA, V., – 1936b – Waldvegetationsstudien im östlichen Mitteleuropa (Finn. summ.) *Comm. Inst. forest. fenn. (Metsätiet. Tutkimuslait. Julk.)* 22(6): 1-115.

KUJALA, V., – 1938 – Metsätyyppien parallelisuudesta. (Germ. summ.: Über Parallelität der Waldtypen). *Comm. Inst. forest. fenn (Metsätiet. Tutkimuslait. Julk.)* 27(1): 1-17.

KUJALA, V., – 1945 – Waldvegetationsuntersuchungen in Kanada mit besonderer Berücksichtigung der Anbaumöglichkeiten kanadischer Holzarten auf natürlichen Waldböden in Finnland. *Ann. Acad. scient. fenn. (Suomal. Tiedeakat. Toim.)* Ser. A, 4 Biol., 7: 1-434.

KUJALA, V., – 1960 – Can we find a common platform for the different schools of forest type classification? *Silva fenn.* 105: 56-59.

LAKARI, O. J., – 1920 – Tutkimuksia Pohjois-Suomen metsätyypeistä (Germ. summ.: Untersuchungen über die Waldtypen in Nordfinnland). *Acta for. fenn.* 14(4): 1-85.

LINDQUIST, B., – 1954 – Ein Waldtypenschema für die skandinavischen Buchenwälder. *Angew. PflSoziol.*. [Wien], Festschr. Aichinger 2: 965-970.

107

LINKOLA, K., – 1916 – Studien über den Einfluss der Kultur auf die Flora in den Gegenden nördlich vom Ladogasee. I. Allgemeiner Teil. *Acta Soc. Fauna Flora fenn.* 45(1): 1-432.

LINKOLA, K., – 1917 – Itä-Karjalan metsätyyppejä koskevia havaintoja. *Acta for. fenn.* 7: 224-245..

LINKOLA, K., – 1924 – Waldtypenstudien in den Schweizer Alpen. *Veröff.geobot. Inst. Rübel*, Zürich 1: 139-224.

LINKOLA, K., – 1929 – Zur Kenntnis der Waldtypen Eestis. *Acta for. fenn.* 34(40): 1-73.

LINKOLA, K., – 1930 – Über die Halbhainwälder in Eesti. *Acta for. fenn.* 36(3): 1-30.

LIPPMAA, T., – 1933 – Taimeühingute uurimise metoodika ja Eesti taimeühingute klassfikatsiooni põhijooni. (Germ. summ.: Grundzüge der pflanzensoziologischen Methodik nebst einer Klassifikation der Pflanzenassoziationen Estlands). *Acta Inst. Hort. bot. tartu.* 3 (4):1-169, and *Loodusuur. Seltsi Aruanded* 40(1-2): 1-169.

LIPPMAA, T., – 1935 – La méthode des associations unistrates et le système écologique des associations. *Acta Inst. Hort. bot. tartu.* 4 (no. 1-2, art. 3): 1-7, and *Loodusuur. Seltsi Aruanded* (1934), 41(3-4): 205-211.

LIPPMAA, T., – 1939 – The unistratal concept of plant communities (the unions). *Am. Midl. Nat.* 21: 111-145.

LÕHMUS, E., – 1971 – Uusi andmeid Eesti metsatüüpidest (Some new data on Estonian forest types). In: '*Eesti Metsad.*' (in press).

LUKKALA, O. J., – 1929 – Tutkimuksia soiden metsätaloudellisesta ojituskelpoisuudesta. (Germ. summ.: Untersuchungen über die waldwirtschaftliche Entwässerungsfähigkeit der Moore). *Comm. Inst. (quaest.) forest. fenn.* (*Metsatiet. Tuttkimuslait, Julk.*) 15(1): 1-301.

MÄKIRINTA, U., – 1968 – Haintypenuntersuchungen im mittleren Süd-Häme, Südfinnland. *Ann. bot. fenn.* 5(1): 34-64.

MALLNER, F., – 1944 – Vorgeschichte und Werdegang des lettländischen Waldtypensystems. *Z. Forst. u. Jagdw.* 76: 105-136.

MALMSTRÖM, C., – 1926 – The experimental forests of Kulbäcksliden and Svartberget in North Sweden. 2.Vegetation. *Statens Skogsförsöksanst.* [Sweden] *Exkurs.-Ledare* 11: 27-87.

MALMSTRÖM, C., – 1936 – Norrlands viktigaste skogstyper. *Sver. Nat.* 27: 21-41.

MASING, V., – 1969 – Structural analysis of plant cover and classification problems. In 'Plant Taxonomy, Geography and Ecology in the Estonian S.S.R,' pp. 49-59. Tallinn.

MIRKIN, B. M., – 1965 – On the ecological classifications of meadow vegetation in floodplains (Russ. with Engl. summ.). *Bot. Zhur. SSSR.* 50: 324-334.

MOROSOW, G. F., – 1928 – Die Lehre vom Walde. Neuman, Neudamm. 375 pp.

MOSS, C. E., – 1910 – The fundamental units of vegetation. Historical development of the concepts of the plant association and the plant formation. *New Phytol.* 9: 18-53.

NIHLGÅRD, B., – 1970 – Vegetation types of planted spruce forests in Scania, Southern Sweden. *Bot. Notiser* 123: 310-337.

NILSSON, A., – 1901 – Sydsvenska ljunghedar. Stockholm.

NORRLIN, J. P., – 1870 – Bidrag till sydöstra Tavastlands flora. *Notiser ur Förhandl., Sällsk. Fauna Flora fenn.* 11 (N.S.8, 1871): 73-196. (Transl. as: Beiträge zur Flora des südöstlichen Tavastlands. *Acta for. fenn.* 23(4): 15-52, 1923.)

PAASIO, I., – 1937 – Suomen nevasoiden tyyppijärjestelmää koskevia tutkimuksia. (Germ. summ.: Untersuchungen über das Typensystem der Weissmoore Finnlands). *Acta for. fenn.* 44(3): 1-129.

PAASIO, I., – 1940 – Zur Kenntnis der Waldhochmoore Mittelfinnlands. (Finn.

summ.) *Acta for. fenn.* 49(2): 1-41.
PAASIO, I., – 1941 – Zur pflanzensoziologischen Grundlage der Weissmoortypen. (Finn. summ.) *Acta for. fenn.* 49(3): 1-84.
PALMGREN, A., – 1915-17 – Studier över lövängsområdena på Åland. I-III. *Acta Soc. Fauna Flora fenn.* 42(1): 1-634.
PALMGREN, P., – 1928 – Zur Synthese pflanzen- und tierökologischer Untersuchungen. *Acta zool. fenn.* 6: 1-51.
PERTTULA, U., – 1950 – Kasvillisuudesta ylisellä Syvärillä sekä siihen etelässä rajoittuvalla Juksovon seudulla. (Germ. summ.: Über die Vegetation am oberen Lauf des Flusses Swir nebst der im Süden anschliessenden Gegend von Juksowo). *Ann. Bot. ,Soc. Zool.-Bot. Fenn. 'Vanamo'* (*Suomal. eläin-ja kasvit. Seur. van. kasvit. Julk.*) 23(6): 1-204.
POGREBNJAK, P. S., – 1929 – Über die Methodik der Standortsuntersuchungen in Verbindung mit den Waldtypen. *Verh. Int. Kongr. forstl. Versuchsanstalten* 2: 455-471.
RUUHIJÄRVI, R., – 1960 – Über die regionale Einteilung der nordfinnischen Moore. *Ann. Bot., Soc. Zool.-Bot. Fenn. 'Vanamo'* (*Suomal. eläin-ja kasvit. Seur. van. kasvit. Julk.*) 31(1): 1-360.
SMAGIN, V. N., – 1950 – An attempt towards a classification of forest types of the southern taiga subzone (in Russian). *Byull. mosk. Obshch. Ispȳt. Prir.*, Ser. Biol. 55(3): 86-89.
SOKOLOV, S. Ya., – 1937 – Advances of forest geobotany (in Russian). *Sov. Bot.* 6: 3-23.
SOKOLOV, S. Ya., – 1962 – Taxonomy of forest associations (in Russian). *Problemȳ Bot.* 6: 110-123.
SOKOLOWA, L. A., – 1935 – Die geobotanischen Bezirke der Onega-Dwina Wasserscheide und der Onega Halbinsel. (Russ. with Germ. summ.) *Trudȳ bot. Inst. Akad. Nauk SSSR*, Ser. 3 (Geobotanica) 2: 9-80.
SOKOLOWA, L. A., – 1936 – Die Vegetation des Rayons der Louchi-Kestenga Chaussee (Karelien). (Russ. with Germ. summ.) *Trudȳ bot. Inst. Akad. Nauk SSSR*, Ser. 3 (Geobotanica) 3: 241-306.
SPILSBURY, R. H. & D. S. SMITH., – 1947 – Forest site types of the Pacific Northwest: a preliminary report. *Tech. Publ. Forest Serv. Br. Columb.* (Dept. Lands & Forests), T. 30: 1-46.
SUKATSCHEW, W. N., – 1925 – Pflanzenassoziationen und Forstbestandestypus (Russian with Germ. summ.). *Trudȳ leningr. lesotekh. Akad.* 32: 39-56.
SUKACHEV, V. N., – 1927 – A Manual for Study of Forest Types (in Russian). 2nd ed., 1930, Moscow-Leningrad. 318 pp.
SUKACHEV, V. N., – 1928 – Principles of classification of the spruce communities of European Russia. *J. Ecol.* 16: 1-18.
SUKACHEV, V. N., – 1931 – The fundamental and leading ideas in studying forest types (in Russian). *Mitt. Leningrad Inst. wissenschl. Forsch. Gebiet Holzindust.* 18(3): 51-71.
SUKATSCHEW, W. N., – 1932 – Die Untersuchung der Waldtypen des osteuropäischen Flachlandes. *Handb. biol. ArbMeth.* 11, 6: 191-250.
SUKACHEV, V. N., – 1934 – The phytocoenosis – what is it? (in Russian). *Sov. Bot.* 5: 4-18, 47-50.
SUKACHEV, V. N., – 1938 – The main concepts of the science of vegetation cover (in Russian). In 'The Vegetation of the USSR,' Moscow-Leningrad 1: 15-37.
SUKACHEV, V. N., – 1944 – On principles of genetic classification in biocenology. (Russ. with Engl. summ.) *Zh. obschch. Biol.* 5(4): 213-227.
SUKACHEV, V. N., – 1951 – The basic principles of forest typology. Proceedings of

the Symposium on Forest Typology, Moscow, pp. 7-19.
SUKATSCHEW, W. N., – 1954 – Die Grundlagen der Waldtypen. *Angew. PflSoziol.* [Wien], Festschr. Aichinger 2: 956-964.
SUKACHEV, V. N., – 1960 – The correlation between the concept 'forest ecosystem' and 'forest biogeocoenose' and their importance for the classification of forests. *Silva fenn.* 103: 94-97.
TANSLEY, A. G. ed., – 1911 – Types of British vegetation, by members of the central committee for the survey and study of British vegetation. Cambridge Univ. Press. 416 pp.
TRASS, H., – 1964 – Geobotaanika. Ajalugu ja kaasaegsed arengusuunad. (Geobotany: History and Contemporary Trends of Development). MS, Tartu State Univ. 400 pp.
TUOMIKOSK:, R., – 1942 – Untersuchungen über die Vegetation der Bruchmoore in Ostfinnland. 1. Zur Methodik der pflanzensoziologischen Systematik. (Finn. summ.) *Ann. bot., Soc. Zool.-Bot. Fenn. 'Vanamo' (Suomal. eläin-ja kasvit. Seur. van. kasvit. Julk.)* 17(1): 1-203.
VAINIO, E. A., – 1878 – Kasvistonsuhteista Pohjais-Suomen ja Venäjän-Karjalan rajaseuduilla. (On the flora of the borderland districts between northern Finland and Russian Karelia). *Meddn. Soc. Fauna Flora fenn.* 4(1): 1-161.
VOROB'EV, D. V., – 1953 – Forest Types of the European Part of the U.S.S.R. (in Russian) Kiev. 450 pp.
WENDELBERGER, G., – 1951 – Das vegetationskundliche System Erwin Aichingers und seine Stellung im pflanzensoziologischen Lehrgebäude Braun-Blanquets. *Angew. PflSoziol.* [Wien] 1: 69-92.
WHITTAKER, R. H., – 1953 – A consideration of climax theory: the climax as a population and pattern. *Ecol. Monogr.* 23:41-78.
WHITTAKER, R. H., – 1956 – Vegetation of the Great Smoky Mountains. *Ecol. Monogr.* 26: 1-80.
WHITTAKER, R. H., – 1960 – Vegetation of the Siskiyou Mountains, Oregon and California. *Ecol. Monogr.* 30: 279-338.
WHITTAKER, R. H., – 1962 – Classification of natural communities. *Bot. Rev.* 28(1): 1-239.
WHITTAKER, R. H., – 1969 – Evolution of diversity in plant communities. *Brookhaven Symp. Biol.* 22: 178-196.
WHITTAKER, R. H. & W. A. NIERING, – 1965 – Vegetation of the Santa Catalina Mountains, Arizona. (II) A gradient analysis of the south slope. *Ecology* 46: 429-452.
YODA, K., T. KIRA, H. OGAWA, & K. HOZUMI, – 1963 – Self-thinning in overcrowded pure stands under cultivated and natural conditions. *J. Biol. Osaka Cy Univ.* 14: 107-129.

16 SYNUSIAL APPROACHES TO CLASSIFICATION

Jan J. Barkman

Contents

16.1	Introduction	113
16.2	Historical Development	114
16.2.1	Early Sources	114
16.2.2	The Influence of Gams	115
16.2.3	More Recent Conceptions	117
16.2.4	The Attitude of the Braun-Blanquet School	119
16.2.5	Ranks of Synusial Units, Du Rietz and Lippmaa	122
16.3	Discussion	124
16.3.1	Criteria of Synusiae	125
16.3.2	Proposals on Concepts and Terms	128
16.4	Applications	130
16.4.1	Vascular Plants on Land	130
16.4.2	Epiphytic Societies	132
16.4.3	Aquatic Communities	136
16.4.3.1	Characteristics	136
16.4.3.2	Errant Hydrophytes	137
16.4.3.3	Adnate Hydrophytes	139
16.4.3.4	Larger Hydrophytes	141
16.4.4	Soil-Inhabiting Communities	142
16.4.5	An Example: the Dicrano-Juniperetum	146
16.4.5.1	Sampling Methods	146
16.4.5.2	Synusial Description	147
16.4.5.3	Relations of Synusiae to Associations	148
16.5	Evaluation	150
16.5.1	Arguments against the Synusial Approach	151
16.5.2	Arguments for the Synusial Approach	152
16.5.3	Conclusion	155
16.6	Summary	157

16 SYNUSIAL APPROACHES TO CLASSIFICATION

16.1 Introduction

Most of the classifications discussed in this volume apply in principle to communities or biocoenoses — i.e. the whole assemblages of organisms that live together in a particular site or biotope. Since the plants make up by far the largest part of the biomass, as well as the essential structure of land communities, it is reasonable to base our classifications upon the plants. Yet the plant community itself is often a complexly organized system, comprising groups of plants on different levels, differently adapted to light and substrate, differing in seasonal timing, and manner of resource use and function. It is a general observation from the study of plant communities that these groups of plants, adapted to different environmental factors within the community, also may differ in their distributions in relation to habitat factors and climatic gradients. These groups of plants show 'individuality' in their distribution and response to environment, both within and between communities, that justifies study of their composition, ecology, and dynamics, and may even justify a separate classification.

A particular plant community, in its particular local or topographic environment or *biotope*, is a *phytocoenose*. A grouping or class of phytocoenoses that are similar in species composition, or other significant characteristics, is a *community-type* or *coenon*; when the standard units of the school of BRAUN-BLANQUET are in question, community-types are *syntaxa* (associations, etc.). A subdivision of a particular phytocoenose that is based primarily upon life-form, secondarily on position or manner of function of its species in the community, is sometimes termed a *society*.

A phytocoenose may be subdivided in two ways (BARKMAN 1970). The first, abstract approach groups the species of the phytocoenose according to (i) their structure into growth-form types, (ii) their reaction upon limiting factors of the environment into life-form groups, (iii) their ecological range into ecological groups, (iv) their sociological affinity (association index) into sociological groups. It is also possible to distinguish species groups according to (v) dissemination types, (vi) taxonomic groups, (vii) areal types, etc. Each grouping may be expressed in a spectrum, indicating the percentage share of all the specific groups present in the particular phytocoenose. This spectrum may then be generalized for the syntaxon in question.

A second, concrete approach divides phytocoenoses according to (i) the ecological aspect (microhabitats) into microcommunities, (ii) the spatial aspect into layer communities, (iii) the temporal phases into aspection (seasonal) communities, (iv) on the base of i, ii (and iii) into *societies*. The former approach is synthetic (the species groups must first be established on the basis of comparison with other species and other phytocoenoses), the latter is direct and analytical. It can be carried out within a single phytocoenose. The resulting units are not necessarily spatially distinct in the former case, as the species of one group are often scattered throughout the phytocoenose; but they are usually spatially distinct in the latter case, each society occupying a restricted space within the phytocoenose. Examples of such societies are: the group of tree species populations that make up a forest canopy, a group of herbs in a forest that grow from bulbs or rhizomes below the soil surface to bloom in the spring, a group of lichens that occur together on the bark of a tree trunk between the moss-clad base and the crown. The abstract units into which similar societies (in a number of phytocoenoses) are classified are *synusiae*, and these are the principal concern of this paper.

The study of societies and synusiae is an important alternative and complement to the study of phytocoenoses and syntaxa. I shall consider here the history of the synusial approach, the concepts and terms of greatest usefulness in it, its application to different kinds of communities, and the evaluation of this as an approach to research on communities.

16.2 **Historical Development**

16.2.1 EARLY SOURCES

The recognition and analysis of whole plant communities or phytocoenoses started before that of partial communities. LORENZ (1858) and KERNER (1863) were the first to see the importance of epiphytic and layer communities as fundamental structural elements of plant communities. LORENZ called them 'Komplexe,' and KERNER wrote, 'eine Pflanzenformation ist eine Verkettung von Beständen.' Later he called these 'Bestände' 'Genossenschaften.'

HULT (1881) was the first author to stress the relative independence of layer societies. In a forest with a uniform tree layer for instance two different moss societies may occur side by side, forming a mosaic. The uniform layer he called uniting society ('föreningsbestånd'), the societies of the heterogeneous layer alternating societies ('alternatbestånd'); the whole community is then a twin formation ('tvillings-formation'). He emphasized that any layer may function as uniting society. These ideas were followed by Russian geobotanists. Thus KORJINSKY (1888) spoke of 'horizons,' and KATZ (1929) classified twin formations into parallel and homologous series. Among other early plant geographers WARMING (1909) used the term 'subordinate communities' and DRUDE (1919) 'Elementarassoziationen.' There is difference of opinion on the meaning of the latter; BRAUN-BLANQUET (1928) considered them to correspond with his subassociations or facies, DU RIETZ (1932) with synusiae.

These terms all referred to spatially distinct units within the plant community. CLEMENTS (1905), on the other hand, divided plant communities into seasonal aspects and called these 'societies.' In 1907 he narrowed this conception down to spatially distinct units by introducing the condition that societies should have 'a principal subdominant species.' This necessarily restricted his notion to layer communities, but only to subordinate layers, for a dominant species according to his definition occurs only in the uppermost layer. 'In a forest societies are found only beneath the primary layer of trees.' According to his climax theory CLEMENTS (1916) divided these units into societies proper (i.e. parts of a climax association) and socies (i.e. parts of a seral unit or associes). In 1928 these notions were further elaborated and a distinction was now made between (i) aspect or seasonal, (ii) layer and (iii) cryptogam societies. As to (iii) CLEMENTS remarked that 'distinctions into ground societies, parasitic societies (i.e. those mostly on leaves and herbaceous stems...) and bark societies (...) are convenient, but of minor importance. A distinction based upon life-form, i.e. moss, liverwort, lichen, and fungus is probably of greater value.' (One notes that this is a subdivision based more upon taxonomy (taxocoenoses) than upon life-form.) Apparently spatial isolation, so important a criterion for his layer communities, was not considered by him of much significance for the cryptogam synusiae. Few, if any, other authors have accepted this viewpoint.

16.2.2 THE INFLUENCE OF GAMS

The term synusia (German: 'Verein') was first proposed in 1917 in a lecture by RÜBEL (cf. RÜBEL 1925) and defined as an *edaphic* unit within a plant association (= 'ökologischer Verein' in the sense of

WARMING), 'for instance a layer community or an epiphytic community.' Modern development of the synusial approach really stems, however, from a most thorough and original treatment of the vegetation science of that time by GAMS (1918). Gams very rightly remarked that we should distinguish between concrete and abstract communities (community-types) and he used the term synusia only in the latter sense. Stands of synusiae were called 'Bestände.' Unfortunately later authors used 'synusia' in both concrete and abstract senses[1]).

GAMS distinguished three types of synusiae, which RÜBEL (1925) considered 'eher hemmend als fördernd für die allgemeine Einführung.' Those of the lowest level ('synusiae of the first grade') were defined as communities of plants or animals, the independent components of which belong to the same life-form and within one (biogeographical) district to the same species. Those of the second grade (level) were defined as communities, the independent components of which belong to different species of the same life-form class and essentially the same temporal aspect. Later authors restricted the term 'synusia' to those of the second grade. GAMS' synusiae of the third grade, finally, are not partial communities at all, but are biocoenoses.

The word 'independent' in GAMS' definitions implies that a synusia of any grade may include dependent organisms of quite different life-forms and vegetation layers, as GAMS pointed out himself. In a pine wood for instance the pine trees form a synusia of the first grade together with their parasitic higher plants (*Viscum album*) and their parasitic and mycorrhizal fungi, as well as phytophagous and gall-forming insects. The synusia would exclude, however, some epiphytic mosses, not because they are wholly autotrophic organisms but because they also occur on soil and rocks (!), thus being independent of the tree bark as a substratum. One notes the difficulty here: the moss *species* is independent of the pine synusia but the epiphytic moss *individual* is not; furthermore, species show varying degrees of dependence on others for support, and/or nutrition.

The inclusion of so many different dependent organisms makes GAMS' synusiae rather impractical. GAMS' life-form classes also were rather too broadly conceived. The Chamaephyta for instance in-

1) Favourable exceptions were DE VRIES and SCHEYGROND (SCHEYGROND 1931) who used the terms 'enkelvoudig verband' (concrete) and 'associatie' (abstract). These terms also, however, are very misleading. The Dutch word 'Verband' also exists in German, where it means 'alliance' (a higher abstract vegetation unit of entire phytocoenoses, Dutch: 'Verbond'); whereas 'associatie' was, at that time, being used in many different meanings (CLEMENTS, BRAUN-BLANQUET, DU RIETZ), but always for entire phytocoenoses (concrete or abstract).

clude both perennial bryophytes and dwarf shrubs. The clearly distinct moss layer and dwarf shrub layer in a heath or coniferous wood, therefore, belong to the same synusia (second level). This procedure has not been followed by later authors.

It is most curious that few of the later authors (for instance Du Rietz 1932 and Poore 1955) have read Gams' definitions well. When citing Gams, they omit the word 'independent' and therefore exclude the dependent organisms of other life-forms and strata from their synusiae. This misinterpretation may have been caused by Gams' shorter, less clear definition in his table on p. 421 and in his summary on p. 475 (and by the fact that Gams himself quite wrongly suggested, p. 429, that Clements' aspect and layer societies are nothing else than his own synusiae of the first and second grade). But the oversight may also have been influenced by a choice of practicality. The later authors chose the more useful definition of the synusia as a group of species of the same or closely related life-forms within a community. A synusia can, in principle, be a group of 'independent' plants (e.g. the canopy of a forest), or a group of epiphytic species (e.g. mosses or lichens on the bark of the trees), or a group of species that are partly or wholly dependent in nutrition (e.g. the mycorrhizal fungi of the forest floor).

16.2.3 More Recent Conceptions

Braun-Blanquet (1928) defined synusiae as 'communities of species of one life-form group and with uniform habitat requirements'. In this the explicit reference to the habitat of the community is new. Temporal differences are omitted from the definition; yet seasonal aspects do belong to different synusiae for Braun-Blanquet, partly on the basis of difference in life-form. Within a Mediterranean association, for instance, he distinguished a hiemal synusia (hepatics, winter therophytes), a vernal synusia (geophytes and spring therophytes) and an aestival synusia (chamaephytes and hemicryptophytes). In the second (1951) and third (1964) edition of his '*Pflanzensoziologie*' Braun-Blanquet gave almost the same definitions as in 1928. Even in the 1964 edition the term 'life-form group' is not explained, and the chapter on life-forms uses the term in two different meanings, (i) as a subdivision of a life-form class (p. 145) and (ii) as identical with a life-form class (p. 146). His examples on p. 167 show, however, that, in so far as synusiae are concerned, the term 'life-form group' is applied in the latter sense (i.e. identical with the large life-form classes of Raunkiaer). However, the author is not consistent, for he considers moss layer and dwarf shrub layer to belong to different synusiae.

Contrary to most authors BRAUN-BLANQUET's synusiae are based upon life-form and seasonal aspect only, irrespective of floristic composition. DU RIETZ (1930), in contrast, incorporated floristic composition and 'sociological affinity' between the component species into his definition, but neglected life-form and seasonal aspect. His synusiae are identical with layer communities (stratocoenoses in the sense of BALOGH 1938), but the epiphytes are considered to form separate synusiae, the epilithic species not. In 1936 DU RIETZ defined synusiae simply as 'one-layer communities', but in fact his concept had become narrower, for soil-inhabiting bacteria and fungi were considered to constitute two separate synusiae, although living in the same layer (level). In 1957 his definition was more, although not fully, adjusted to his own use, 'a plant community .... either forming one layer or growing epiphytically on other members of the biocoenose.' DU RIETZ (1965) also distinguished various zoosynusiae, some of which inhabit the same vegetation layer and differ only in life-form and in their way of exploiting the habitat. Thus the concept of 'niche' introduced by zoologists was tacitly incorporated in the synusial concept. BARKMAN (1968) used the term 'synusia' in a sense very close to DU RIETZ (1965). However, not only the epiphytes, but also epilithic communities as well as all terrestrial communities on special substrata were considered to be distinct synusiae. In 1970 he also distinguished seasonal synusiae.

ELLENBERG (1956), CAIN & CASTRO (1959) and DAUBENMIRE (1968) used definitions closely similar to BRAUN-BLANQUET's, but ELLENBERG took a narrower view (mosses and lichens separated), whereas CAIN & CASTRO held a slightly broader view, defining a synusia as 'a social aggregate consisting of one *or a few closely related* life-forms occurring together and having a similar ecology' (italics by the present writer). SCAMONI (1963) cherished a quite different opinion, excluding, as he did, the layer communities and restricting synusiae to parts of a plant community inhabiting special microhabitats (including all layers?). His notion seems to be identical with KORCHAGIN's (1964) microcoenosis. Within the phytocoenose KORCHAGIN distinguished not one, but three different levels of structural elements: coenopopulations, synusiae, and microcoenoses. According to him phytocoenoses may be divided vertically into (i) layer synusiae and these may be subdivided into (ii) intralayer synusiae. Among the former he distinguished (ia) layers characterized by the component biomorphs (dominant life-form?), (ib) sublayers differing by the average height of plants within one layer (tall and low herb layer for instance?), (ic) 'canopies', represented by groups of the same species, but of various heights (due to different age classes). Intralayer synusiae may be (iia) societies within a layer, differing in floristic composition, density, etc., (iib) seasonal aspects or aspect societies, (iic) epiphytes. Inhomogeneous phytocoenoses (mosaic complexes etc.) may be divided horizontally into microcoenoses or coeno-elements, differing in composition, etc. Within each microcoenose the layer synusiae are represented by special societies, differing from those of adjacent microcoenoses. Clearly these microcoenoses can be identical with the narrowly defined sociations of DU RIETZ et al. if the component societies have one dominant species each, but this need not be the case. BARKMAN (1968) used the terms 'Mikrogesellschaft' and (1969, 1970) 'microcoenosis' in exactly the same sense as KORCHAGIN without being familiar with his publication. BARKMAN (1970) defined synusiae as parts of a microcoenose (or phytocoenose, if no microcoenoses can be distinguished), formed by plants that belong to the same stratum and do not differ fundamentally in either periodicity or way of habitat exploitation. As a general term covering both microcoenoses and synusiae the word 'merocoenosis' was introduced by MÖRZER BRUINS (1947). Although being applied to (both plant and animal) synusiae only, his definition is wider and includes both horizontal and vertical partitions of biocoenoses, for he speaks only of 'spatially distinct parts of biocoenoses.' Stratocoenoses (the 'horizontaal complex-verband' of DE VRIES in SCHEYGROND 1931) are the vertically separated counterparts of the microcoenoses (the 'verticaal complex-verband' of DE VRIES, l.c.). The former are often identical with synusiae but may also consist of more than one synusia. A survey of the elements of biocoenoses, their interrelations, and current synonyms is given in table I.

16.2.4 THE ATTITUDE OF THE BRAUN-BLANQUET SCHOOL

In the school of BRAUN-BLANQUET a growing tendency can be observed towards giving both stratal and seasonal synusiae more independent syntaxonomical status, although BRAUN-BLANQUET (1964:168) himself has still ignored the problem of taxonomic independence: 'Es besteht natürlich ohne weiteres die Möglichkeit, die Synusien den floristisch gefassten Vegetationseinheiten einzugliedern.'
Originally all synusiae occurring in one particular site were considered parts of one association, the only exception being the epilithes and epiphytes. These were considered to form dependent communities ('abhängige Gesellschaften'), but they were not included in the relevés nor in the association tables of forests, and

Table I. Elements of the Biogeocoenose and their Relations

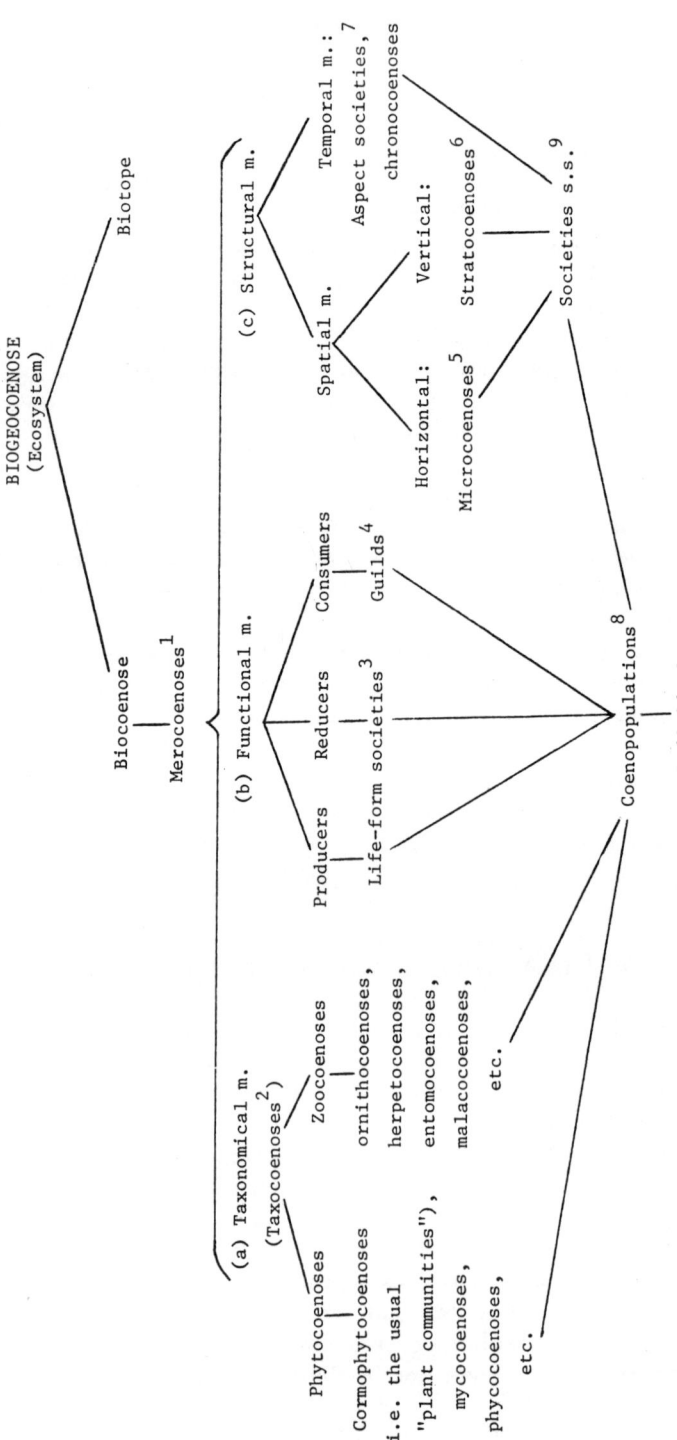

they were described as separate associations. ALLORGE (1922) was the first to publish epiphytic associations, both of woods and wayside trees, followed by OCHSNER (1928) and many others. Later epilithic communities were treated in the same way, starting with MOTYKA (1925) and KLEMENT (1931, 1955). SCHUMACHER (1944), followed by PHILIPPI (1956, 1963), described bryophyte communities on mineral soil in woods as separate associations, whereas the moss communities on humus were still considered part of the forest association. KOPPE (1955) rightly remarked that this is inconsistent; in his opinion none of these bryophyte communities can be considered associations, but all are socions or societates in the sense of DU RIETZ. PIRK & TÜXEN (1949) described a fungus association occurring on cow dung in meadows.

In the BRAUN-BLANQUET school the description of synusial associations was not confined to cryptogam communities, however. TÜXEN (1950) described a number of 'Schleiergesellschaften', which overgrow other plants like a veil, called them associations, and even united then into a separate order (Convolvuletalia sepium). However, these communities root in the same soil as the plants which they overgrow or use as a support, and they cannot be separated from the latter otherwise than as synusiae (BARKMAN 1968,

---

Table 1. (opposite)
Elements of the biogeocoenose are arranged with increasing narrowness of content downward on the vertical, manners of division on the horizontal. Three primary kinds of relationships—taxonomic, functional, and structural (including temporal)—may be used to divide the biocoenose into merocoenoses or community-fractions, which may in turn be subdivided into narrower merocoenoses until one reaches the coenopopulations (of individual species in the biocoenose), each occupying its niche and, finally, the individual organisms of which coenopopulations are composed. To every merocoenose an abstract type or class corresponds; these are in general designated by substituting the suffix -coenon for -coenose (e.g. merocoenon, see Table II). Abstract classes of *societies* in the strict sense, however, are *synusiae* in the sense of this article. References indicated: 1) MÖRZER BRUINS (1947); 2) VAN DER MAAREL (1965); 3) = Synusiae in the sense of BRAUN-BLANQUET, ELLENBERG, DAUBENMIRE, CAIN & CASTRO, not in the sense of GAMS; 4) ROOT (1967); 5) KORCHAGIN (1964), BARKMAN (1969, 1970) = Synusiae sensu SCAMONI (1963); 6) BALOGH (1938) = Synusiae sensu DU RIETZ (1930) = Layer synusiae KORCHAGIN (1964); 7) = Society sensu CLEMENTS (1905); 8) KORCHAGIN (1964); 9) Synusiae sensu DU RIETZ (1965), WILMANNS (1962), BARKMAN (1970) = Intra-layer synusiae KORCHAGIN (1964).

WESTHOFF 1968, WESTHOFF & DEN HELD 1969). Yet, the two last-named authors maintain the order Convolvuletalia, 'only because there is no better solution'. KOCH & TÜXEN (1954, in OBERDORFER 1957) distinguished the floating duckweed communities from the rooting water plants with which they are associated, as a separate class, Lemnetea. Aquatic and epiphytic societies offer special problems which we shall discuss separately below.

A new development in the BRAUN-BLANQUET school was the introduction of seasonal associations. Although PIRK & TÜXEN (1949) observed seasonal aspects in a fungus association on cow dung which were totally different as to species composition, they rejected the idea of splitting this association. OBERDORFER (1954), however, distinguished several phanerogamic weed associations in the same arable field, succeeding each other during the year. HÜBSCHMANN (1960) distinguished, within certain types of arable land, a summer association of vascular weed plants, and two bryophytic associations, developing in autumn and winter respectively. If this trend is to be continued, we might eventually find ourselves in a situation where one cannot see the wood for the trees, as is clearly suggested by recent developments in taxonomy of aquatic communities (DEN HARTOG & SEGAL 1964). It is therefore not surprising that this led to reactions both from opponents of the BRAUN-BLANQUET school (SCHMID 1941) and from within this school (HÖFLER 1959 and ZOLYOMI, WILMANNS, BARKMAN).

16.2.5 RANKS OF SYNUSIAL UNITS, DU RIETZ AND LIPPMAA

GAMS was of the opinion that every society (synusia) has its own ecology and its own type of succession. He therefore considered his synusiae (1st and 2nd grade) as the fundamental vegetation units and wrote a monograph based on these units (GAMS 1927). REGEL (1923) called them 'associations,' a term around which a chaos of confusion existed at that time, and termed types of phytocoenoses 'association complexes.' Until 1930 DU RIETZ, too, worked upon the basis of synusial (one-layer) associations, but he then (1930, 1932) proposed to describe both layer communities and phytocoenoses and to unite them into two independent hierarchical systems, although he still considered the former to be the fundamental units. The ranks of (abstract) synusiae proposed were: (1) socion, (2) consocion, (3) associon, (4) federion, (5) subformion, (6) formion, and (7) panformion. A consocion is a layer community with one constant dominant species or two codominant species, and may be combined with any dominant species in the other layers; a socion differs by being restricted to one particular sociation, i.e. a fixed set of dominants in all layers. Therefore a socion is a layer community associated with one particular socion in each other layer, whereas a consocion may be associated with different socia in the other layers. In single-layer phytocoenoses socion, consociation, and sociation are identical. An associon has various dominant species or no dominants at all, but a specific floristic composition. The same applies to the federion. Subformions and formions have a wide geographical range and hence a great floristic variation. They consist of a number of federions connected by gradual transitions and by having a number of species in common. Panformions are characterized by genera or families only. The seven levels of synusial units therefore correspond more or less with (1) sociations, (2) consociations, (3) associations (in the sense of BRAUN-BLANQUET) (4) alliances, (5) orders, (6) classes and (7) class groups („Klassengruppen'). It is interesting that DU RIETZ confined all these units to synusiae with a stable character, a viewpoint not shared by other authors.

The development of LIPPMAA's ideas shows exactly the opposite trend. Until 1933 he used the term 'association' both for unistratal and pluristratal community-types, but in 1935 he strongly advocated the use of unistratal associations only and applied this principle to the forest vegetation of Estonia. When a community has more than one layer, not the community as a whole, but each layer should be described and named separately. His arguments were (i) the relative autonomy of the layers, being independent topologically and synsystematically (not ecologically, as he pointed out himself), (ii) the great differences in minimum area of the various layers. LIPPMAA also offered a full hierarchy of synusial units. Within a union (associon) LIPPMAA (1938, 1939) recognized geographic 'facies.' For classification above the union LIPPMAA (1933b, 1939) proposed grouping the unions of a given area that are dominated by the same life-form (or two similar life-forms) and that are ecologically related (i.e. in habitats), into an 'Assoziationsgattung' or genus of unions. Higher units were suggested (association-family, -series, and -class) up to the 'division' comprising all the unions of the earth in which the same life-form or two closely related life-forms are dominant.

At the International Botanical Congress of Amsterdam, 1935, LIPPMAA's ideas were 'rejected' and DU RIETZ' ideas (two independent systems) adopted. Later even LIPPMAA (1938, 1939) was converted. At this congress DU RIETZ (1936) also proposed some changes in terminology, calling both his socions and consocions 'societies' (societas), the associon 'union' and the federion 'federation.' These new terms found wide acceptance among students of moss and lichen communities in Sweden, for instance WALDHEIM (1944, 1947), VON KRUSENSTJERNA (1945) and ALMBORN (1948). On two later occasions (1954, 1965) DU RIETZ again stressed the need for two independent hierarchical systems, one for the phytocoenoses, one for the synusiae. In the discussion following DU RIETZ 's 1965 lecture SCHMITHÜSEN and BARKMAN advocated the analysis of all synusiae, including the epiphytic, when describing a phytocoenose. The present writer has now applied this method to his study of juniper scrub (see below). In 1969 (in WESTHOFF & DEN HELD) he proposed to use, within the synusial system, the units union and federation of DU RIETZ as equivalents of association and alliance. The terms 'subformion, formion and panformion' (which were introduced in 1930 by DU RIETZ only as provisional terms and not maintained in his 1936 publication) were rejected by the present author, both because they suggest a relationship with formations, i.e. physiognomic-structural units, which do not fit into a vegetation system based on floristics, and because they have never been used by other authors. However, the terms order and class proposed by WILMANNS (1962), are already in use for phytocoenose units (syntaxa), so the terms 'ordulus' and 'classicula' were proposed instead (cf. BARKMAN 1970).

16.3 Discussion

The treatment of terms and interpretations of synusial units in the preceding may or may not have exhausted the reader, but it by no means exhausts the literature (see also 16.4.1 and 16.4.2). The proliferation of terms is a source of difficulty for any review attempting a synthetic perspective on synusial approaches. It is as if the students of synusiae had sought to emphasize the importance of their approach by creating a literature of terms exceeding that for phytocoenoses. We shall attempt some selection from this literature as regards (i) the criteria by which synusiae are to be recognized as such and (ii) terms and concepts that might be accepted as having the widest usefulness.

16.3.1 CRITERIA OF SYNUSIAE

Among the many and confusing interpretations of synusiae we may recognize the following criteria: (i) life-form, (ii) spatial separation, (iii) temporal separation, (iv) microenvironment, (v) habitat exploitation and function, (vi) dependence, (vii) composition and dominance, and (viii) taxonomy. The appropriateness of these may be discussed in this sequence.

(i) Life-form was introduced as a primary criterion by GAMS (1918) and has been widely used since. Life-forms can be clearly distinguished, and group together species that are related in their major adaptations and functions in the community. Life-forms are more appropriate for separating synusiae than the growth-form concept of physiognomic ecology (article 13). Synusiae defined by life-form have the disadvantage of changing with change in life-form concepts (WILMANNS 1962). Also, we have to choose between different life-form systems (RAUNKIAER 1934, IVERSEN 1936, etc.). So far nearly all synusiae have been based on RAUNKIAER's classification. This system groups terrestrial plants primarily by vertical position of the vegetative buds in the unfavourable season (if there is any) in relation to the ground (substratum) surface. The system has disadvantages: (a) It is partly a growth-form system, (b) it is not very useful in ever-humid and warm (equatorial) climates, (c) it is not well adapted to aquatic plants, bryophytes, and lichens, (d) it does not cover the whole plant kingdom, (e) some of its life-form classes are too broadly conceived for our purpose (e.g. the Chamaephyta which comprise a.o. both bryophytes and dwarf shrubs), whereas the subdivisions of these classes are often too narrowly conceived (Chamaephyta reptantia and velantia for instance). These points (a–e) can be amended and partly have been amended by several authors. (f) If the life-form of the plant species is unknown, no synusiae can be distinguished in the growing season, whereas the unfavourable season (winter in the cold temperate region, summer in the Mediterranean etc.) is hardly suitable for vegetation analysis.

The use of IVERSEN's biological plant types has the disadvantages (a) that experimental work must be carried out before the species can be classified (b) that it is based on a single factor, important though it is, viz. the adaptation of plants to water (moisture, hydrature).

Both life-form systems have the disadvantage that in a forest all trees from seedlings to full-grown specimens would belong to the same synusia, despite their occurrence in layers above the ground with widely different microclimates and also in different root horizons often with very dissimilar physical and chemical soil properties. Besides seedlings and adult plants may have different demands upon the environment. In juniper scrub various shrub species always die as seedlings, whereas *Juniperus communis* itself is often present as a tall shrub only, being unable to germinate under its own canopy. It would seem hardly appropriate to unite these species into one 'shrub synusia.'

(ii) Spatial separation, particularly in a vertical sense (layering), has often been applied as a criterion. It sometimes meets, however, with difficulties, as layers cannot always be clearly distinguished (tropical rain forests (RICHARDS 1952), wet alder woods in the temperate zone, etc.). Distinct stratification is found especially in planted and cultivated woods and one might ask whether it is logical to base a system of layer synusiae upon such communities rather than using undisturbed phytocoenoses as a standard. Clear layering can, however, also be observed in natural subalpine and boreal (subarctic) forests (the latter being often less disturbed than the so-called 'primeval' tropical rain forests!) and in many semi-natural phytocoenoses. Another disadvantage is that two layers that are distinct in one phytocoenose, may form one layer elsewhere (see 16.5.1). Usually, however, this phenomenon is restricted to sublayers, e.g. tall and low herb layer.

(iii) Temporal separation is a more controversial basis of distinction since, if the whole life-cycle is considered to represent a species, different seasonal aspects have the same floristic composition. Yet such groups as early-flowering bulb plants and summer grasses, or summer and winter therophytes, are so distinct as to periodicity that separation on at least a synusial level seems appropriate. All the more this holds true of such autonomous partial communities as moss societies developing on ploughed arable fields in winter and societies of therophytes developing only once in a number of years in open spaces of perennial plant communities (e.g. in Australian semideserts, DOING, verbal comm.).

(iv) Microenvironment is an obvious criterion, but needs some comment. The organisms of special substrata like tree trunks, rotten logs, leaves and herbaceous stems, decaying fungi, excrements, game tracks, rabbit holes, molehills, and outcropping rocks are distinctive and deserve recognition in any system. Yet each of these microbiotopes or microhabitats may support an assemblage of organisms of different life-forms and strata. They cannot be considered synusiae by any definitions except SCAMONI's (1963), but must be regarded as higher structural units as pointed out by KORCHAGIN (1964) and BARKMAN (1968), i.e. as microcommunities or microcoenoses each of which may include a number of synusiae. Microbiotopes thus do not define, but may contribute to the delimitation of, synusiae.

(v) Habitat exploitation and function is a more questionable criterion. It is included in the criterion of life-form (i) and partly in those of spatial separation (ii) and periodicity (iii). If life-forms are not used to distinguish synusiae, yet such groups as autotrophic plants, symbiotic (mycorrhiza), saprophytic, and parasitic fungi should be considered to belong to separate synusiae.

(vi) Independence. Syntaxonomic independence (one synusia occurring in various syntaxa) was one of the reasons for Du Rietz and Lippmaa to distinguish synusiae (even as associations). It is a practical reason indeed, but it should not imply that dependent synusiae are not to be accepted, if otherwise they form a clear-cut element in a phytocoenose. The criterion is therefore superfluous.

Ecological dependence was a reason for Gams for not giving the organisms in question the status of synusia, whereas for Clements ecologically 'independent' societies (uppermost layer) were not recognized as such. Neither of these opposing viewpoints seems reasonable, since (a) none of the societies in a phytocoenose is completely independent of the others, (b) there are all possible intergradations from wholly dependent (parasites) to almost independent (tree society), (c) one and the same synusia may occur in both a 'dependent' and an 'independent' position, for example moss societies on both living tree bark and on dead stumps, rocks and soil within a forest and even outside the forest.

(vii) Composition and dominance are alternative (but for some simple communities and for many societies nearly synonymous) criteria of units in any community classification (articles 14 and 20). Once the synusia concept has been defined on whatever basis, composition and dominance come in as important criteria for their delimitation and classification. However, we are here concerned only with the definition of the synusia as a general concept, not of particular unions. Obviously the above criteria (i), (ii), (iii) and (v) happen to imply a difference in floristic composition as well, although in some cases different strata may consist of the same species. Floristic differences are often also involved in criterion (iv). If, however, in one stratum of a phytocoenose, in an apparently homogeneous microbiotope societies can yet be observed that are consistently and significantly different in composition and/or dominance, these will be distinguished as separate societies whether or not they also differ in life-form, periodicity, or function. Composition therefore is a primary criterion, regardless of the others.

(viii) A few authors distinguished synusiae on the basis of taxonomic groups. Klement (1955) and Ellenberg (1956) for instance separated bryophyte and lichen synusiae. This procedure is not generally accepted and it is not recommended. Higher taxonomic groups are seldom, if ever, homogeneous as to life-form, periodicity, or habitat requirements, nor do they as a rule belong to a single stratum or a single ecological or sociological group. Even moss and herb layer do not coincide with taxocoenoses, for vascular plants like *Tillaea muscosa, Sagina procumbens,* and *Cassiope hypnoides* distinctly belong to the moss layer, wherever they occur. Synusiae based on taxonomic groups only, should rather be called taxocoenoses, a term introduced by VAN DER MAAREL (1965) for animal groupings. For purely practical reasons it may be necessary to study taxocoenoses first (we must not forget that all 'plant communities' are in fact taxocoenoses), but our final aim must be to study biocoenoses and synusiae.

16.3.2 PROPOSALS ON CONCEPTS AND TERMS

It will be evident from the discussion in 16.3.1 that some criteria have to be rejected and others meet with serious objections or difficulties in application, with the exception of microhabitat and floristic composition. None of them is in itself a satisfactory basis for the definition of synusiae. Even microhabitat and floristic composition, which need to be incorporated in any definition, only define microcommunities if applied as the sole criteria. Synusiae should be based on these criteria either in combination with life-form or in combination with spatial and temporal separation and habitat exploitation. To the writer's mind the latter solution is the better. A synusia is thus defined as a structural part of a phytocoenose inhabiting (a) a special microhabitat, with (b) a specific floristic composition and consisting of species that (c) belong to the same stratum and that do not differ fundamentally in either (d) periodicity or (e) way of exploitation of their environment. The preceding has also indicated the need for distinction between microcommunities and synusiae. If we omit c, d and e from the above definition, it covers the notion of microcommunity. A microcommunity or microcoenose is thus defined by a special microhabitat and a specific floristic composition (a-b). Horizontal combinations of synusiae of the same stratum (b-c) are called 'stratocoenoses.' If a consistent nomenclature is desired, aspect societies (b-d) could be called 'chronocoenoses.' For taxonomically limited fractions of phytocoenoses the term 'taxocoenose' is already in use. A general term for any of the above partial communities is 'merocoenose,' introduced by MÖRZER BRUINS (1947). All these terms (except synusia) refer to individual stands. Any of these kinds of partial communities can be classified into abstract types. Much confusion has been aroused by using the same word in a concrete and an abstract sense. In analogy with the term 'coenon' (for an abstract whole-community unit or community-type) VAN DER MAAREL (1965) proposed the terms 'merocoenon,' 'stratocoenon,' and 'taxocoenon,' BARKMAN (1970) the term 'microcoenon.' If need be, we could also speak of 'chronocoena.' Unfortunately the term 'synusia' is still being used in both the concrete and abstract sense. The word 'Bestand' used by GAMS for concrete stands of synusiae cannot be recommended because it is not international, and because it is frequently used for entire phytocoenoses (also as its English equivalent, 'stand'). The author and the editor of this chapter suggest that 'synusia' should be applied to abstract units, as GAMS intended, and that 'society' should be revived as the English term for a concrete stand of a synusia (in spite of the different meaning given to 'society' by DU RIETZ (1936), cf. 16.2.5). The German equivalent, 'Verein,' has been more widely used in both abstract and concrete senses, but there would be real advantage in limiting it to the concrete, while 'synusia' and other more formal terms apply to the abstract units. For the principal synusial category in English the term 'union' has been established, especially by DU RIETZ (1936), LIPPMAA (1939) and CAIN & SHARP (1938); it may well be accepted in other languages as an international term for the basic synusial unit, hence paralleling 'association' as the basic unit for phytocoenoses. 'Union' may best apply to broader synusial groupings defined by total species composition, in parallel with the manner in which associations are defined. A subdivision of a union, characterized by differential species, can be called 'subunion' (cf. subassociation). It is proposed here to continue the use of the terms variant, subvariant, vicariant, and subvicariant also for synusial groupings and not to create new terms. A subdivision of a union defined by dominance of (or limited to) a single species, can be called 'socion' (paralleling 'sociation'). If consisting of the dominant species only, the 'socion' is identical with KORCHAGIN's 'coenopopulation.' For higher categories grouping unions I suggest 'federation,' 'ordulus,' and 'classicula' (these being in parallel with alliance or 'Verband,' order, and class for syntaxa) (BARKMAN 1970).

In the effort to bring some order out of confusion, with a number of terms that is sufficient but not excessive, Table II has been prepared after discussion of the problems between author and editor. Table II is void of hierarchial terms for taxo-, micro-, strato-, and chronocoena. The temptation to fill these spaces with new sets of terms has been firmly resisted. It would, for one thing, complicate our system of vegetation units too much. And it is quite possible and sufficient to use simple designations or to describe such merocoena in terms of combinations of synusiae. An additional advantage of the latter method is that it provides us with a flexible system to cope with all problems (cf. 16.4.5).

TABLE II.

Some major concepts and terms for communities and fractions thereof.

	Level of treatment	Concrete objects of study	Abstract units, on any level	Abstract units, hierarchial categor
A.	Whole communities	Community, biocoenose	Community-type, coenon	Various units (see a
B.	Plant communities	Plant community, phytocoenose, stand	Syntaxon (for school of BRAUN-BLANQUET)	*association* higher units–allian order, class, class g lower units–subasse variant, facies; soc
C.	Any subdivision of the whole community or of the plant community	Subcommunity, merocoenose	Merocoenon	
C. 1.	Taxonomic groupings of species	Taxocoenose	Taxocoenon	
C. 2.	Microbiotope groupings	Microcommunity, microcoenose	Microcoenon	
C. 3.	Layer divisions	Layer community, stratocoenose	Stratocoenon	
C. 4.	Seasonal groupings	Aspect society, chronocoenose	Chronocoenon	
C. 5.	Combinations of C2-4	Society	Synusia	*union* higher units–feder ordulus, classicula lower units–subun variant, vicariant;

16.4 Applications

16.4.1 VASCULAR PLANTS ON LAND

The synusial approach to whole plant communities on land leads from the proposals of GAMS (1918) and application in the Alps by BOLLETER (1921) and GAMS (1927), to Russian work (see also KORCHAGIN 1964, ALEKSANDROVA 1969, and article 17), and the distinctive Estonian school of LIPPMAA (1931, 1933a, 1933b, 1935a, 1935b, 1935c, 1939), PASTAK 1935, SIRGO 1936, TOMSON 1937 and VAGA 1940.

GAMS, as discussed above, recognized three levels or degrees of synusiae, and distinguished these as 'ecological' units from 'topographic' units, such as the association, formed usually from combinations of synusiae. The approach was applied by BOLLETER (1921, cf. GAMS 1927, 1936), who treated communities both in terms of ecological units or synusiae ('Vereine') and in terms of spatial or topographic units ('Gemeinde') often consisting of several of these. The 'Gemeinde' were some of the more important, among the many possible, combinations of unions and were characterized by the dominant layer and component unions. In the Soviet Union the synusial approach was applied to desert and steppe communities by KELLER (1932) and GROSSHEIM (1930). KELLER recognized synusial units as 'Genossenschaften'. GROSSHEIM (1930, GROSSHEIM & PRILIPKO 1929) further elaborated GAMS' grades of synusiae with assumptions on the evolution of communities (from loosely organized and simple 'aggregations' (synusiae of the first degree) to more stable, simple 'agglomerations' (of second degree), to pluristratal 'semiassociations' in which the strata retain independence, and pluristratal 'associations' in which the strata have lost independence (the last two are synusiae of the third degree)).

LIPPMAA (1933b, 1939) considered that the elemental units of vegetation were 'associations unistrates,' later called 'unions,' confined to a single stratum, with one or two closely related life-forms dominant. A union might consist of many species related by life-form and ecology, or of only one or two species. These single-layer associations were thought to show evident independence of one another, and to combine themselves in a quite variable fashion into forest stands, in which environmental relations were better revealed by the lower strata than by the canopy (LIPPMAA 1935a, 1935b, 1935c, cf. article 15). Such complex vegetation as that of a forest was regarded by LIPPMAA (1935b) as a complicated assemblage of superimposed 'associations,' with each stratum composed of one or more elemental associations or unions. Since species are differently distributed, floristic composition alone was considered inadequate for the definition of unions. LIPPMAA felt that classification of communities must be ecological, and that the union must be characterized by habitat and life-forms as well as floristic composition (LIPPMAA 1931, 1933a, 1939). Unions could be grouped and divided to form a full hierarchy of synusial units (see 16.2.5).

LIPPMAA (1938, 1939) discussed as an example the Galeobdolon-Asperula-Asarum union of European forest herbs in its distribution, relations to forest communities, division into geographic facies and probable history. LIPPMAA's unions were considered by him different and narrower units than the synusiae of GAMS.

The union differed also from the stratal components of the sociation (Du Rietz 1930, 1932), the 'socion' and 'consocion'; for the unions were broader units that might be of varied floristic composition (cf. the 'associon' of Du Rietz 1932). The approaches of the Uppsala and Estonian schools were otherwise closely related, since both treated communities in terms of stratal or synusial units and also of 'topographic' units built up of these; the schools differed in which of the units received primary emphasis (see also article 18). Approaches based on stratal units, and complexes of these, were applied also by Regel (1923, 1927) in Lithuania, and de Vries (1926, 1932) in The Netherlands.

The synusial approach has not, however, been widely applied to the vascular plants of land communities. Gams, Bolleter, and Lippmaa et al. appear to have demonstrated an interesting possibility which, however, the great majority of phytosociologists have been content to leave as such while concentrating their primary emphasis on the composition (or stratal structure) of whole phytocoenoses. Thus the synusial approach has faded from active use for the vascular plants of land communities, while at the same time having growing influence on the study of epiphytes, water plants, and soil organisms.

16.4.2 Epiphytic Societies

Until recently epiphytic societies have either been ignored in vegetation analysis or described as autonomous syntaxa (associations etc.) even by those who otherwise support the idea of pluristratal associations. The first has been done by most students of terrestrial vegetation of all schools, the second procedure was followed by moss and lichen specialists with vegetational interests. Until recently the latter usually made only brief mention of the forest types in question, without even indicating their syntaxonomic status. Even when both epiphytic and terrestrial associations were fully described, there was sometimes no mention of their connections (e.g. Allorge 1922, Jovet 1949). Distributions of lichen species were studied in various forest associations by Sulma (1935), and along a continuum of forest types by Hale (1955); but neither of them described either the lichen communities or the forest types. Koskinen (1955) published full vegetation records of lichens on trees, grouped according to forest types, host species and microhabitats on the trees; but he did not name the lichen communities nor did he describe the forest types, which were based on structure and site-type.

A distinct improvement was the work by CAIN & SHARP (1938) who, within each forest association, studied one stand and fully described both qualitatively and quantitatively its tree layer and the terrestrial (soil-), epilithic (rock-), epixylic (wood-), and corticolous (bark-surface) bryophyte societies. However, they did not indicate whether there was any ecological or topographical connection between a particular epilithic society and a particular epixylic or corticolous society. The herb enumeration was incomplete and without quantitative data. A similar study was done by SZCZAWINSKY (1953), and COOKE (1955) who described lichen, moss, and fungus societies for forest associations. BARKMAN (1958b) was primarily concerned with epiphytic community-types. He only briefly indicated the forest alliances and associations in which they were found. WILMANNS (1962) gave detailed enumerations of the terrestrial associations, including grasslands, weed communities, bogs etc., in which the epiphytic communities occurred. Neither of them published full joint descriptions of epiphyte and forest communities nor did they fully analyse the epiphytes of one forest stand or association. The latter has been done by RYDZAK (1961) who compared even different subassociations of forests, but he refrained from describing the epiphytic community-types. His vegetation records are often heterogeneous, comprising as they do the whole vegetation on the tree trunks from 0 to 3 m, including all sides and both fissures and bark plateaux. WILMANNS & BIBINGER (1966), too, analysed all epiphytes of each forest association, but they also described the epiphytic unions as such and determined their distribution on the tree boles (height, aspect) and their frequency percentages on the trees per forest stand. IWATSUKI & HATTORI (1957) in Japan were the first to publish full synoptical tables of records including both terrestrial plants and epiphytes; the same was done by SCOTT & ARMSTRONG (1966) in New Zealand. In the latter study all trees of the sample plot were taken into account, even if they were devoid of epiphytes, and the whole epiphytic vegetation up to 2 m was recorded. If a forest stand is to be analysed as a whole and characterised by its epiphytic vegetation, this is probably the most correct method; but if we want to describe the epiphytic societies as such, smaller and more homogeneous sample plots should be chosen, and the method of WILMANNS & BIBINGER (1966) is to be preferred.

All authors cited above, as well as a number of others agree that certain epiphyte synusiae (as community-types) are restricted to a particular tree species, irrespective of the forest association, whereas others are restricted to a special forest association or alliance, irrespective of the tree species. Only very few are more narrowly restricted — to one tree species in one association. Many, however, occur on both a number of tree species and in a variety of forest associations. Whenever a special linkage is observed, either to trees or forest associations (and above all if the latter are concerned), this linkage is usually of a local nature. Complete linkage of an epiphytic community-type throughout its area is extremely rare. It is only found in endemic unions with very small areas (cf. BARKMAN 1958b). WILMANNS (1962) studied the linkage between epiphytic unions and forest syntaxa, restricting her analysis to southwest Germany. She found that some unions are restricted to (are 'Charakter-Union' of) one association or even subassociation, others to one alliance or order, whereas some are not specific at all. Following GAMS (1918) she called these unions stenocoenotic and eurycoenotic respectively. These terms may coincide with stenoecious or stenotopic and euryoecious or eurytopic respectively, but not necessarily so: a particular forest association may be euryoecious itself and so is in that case the stenocoenotic epiphytic union strongly linked with that association. As a rule forest alliances and suballiances are well characterised by their epiphytic unions, forest associations are usually not. In addition to the 'character-unions,' some epiphytic unions and species can serve well as 'differential-unions' and differential-species for forest alliances, thus contributing to better delimitation of these vegetation types (WILMANNS 1962).

As stated above, it has been a tradition in the BRAUN-BLANQUET school from the very beginning to analyse, name and classify epiphytic societies as separate from forest and other terrestrial phytocoenoses. They were generally considered 'associations,' 'alliances,' etc. CAIN & SHARP (1938), however, called them 'unions' divided into 'facies' and opposed the use of the suffix -etum for these units, since they are not equivalent to the pluristratal associations of which they are part. It is therefore illogical that they had no objection to uniting these unions into unistratal alliances with the ending -ion. WALDHEIM (1944, 1947) and ALMBORN (1948) were more consistent, using the suffixes -etum and -ion although they considered these epiphytic synusiae unions and federations respectively. RICHARDS (1939) on the other hand used the term 'associule' and did not use the suffix -etum. HOSOKAWA (1954) rejected this term except for special studies on epiphyte succession, since it involves a dynamic element in the sense of CLEMENTS (1936). ŞTEFUREAC (1941), PHILLIPS (1951) and BARKMAN (1958b) considered epiphyte societies to be classified into synusiae, yet called the latter 'associations.' FELFÖLDY (1941) who based his units on characteristic species combinations held similar views. Yet he speaks of 'sociations' or 'microsociations' which are in fact pluristratal phytocoenoses, based on dominant species. WILMANNS (1958, 1959) still considered epiphytic communities as associations etc.; but in 1962 she changed her ideas and from then on called them synusiae, although still using the suffixes -etum, -ion, -etalia. In 1966 she extended this interpretation to epilithic and terrestrial cryptogamic synusiae.

Quite different opinions were offered by some Japanese authors, but scarcely accepted by other phytosociologists. HOSOKAWA (1951) found that in tropical forests even synusiae of vascular epiphytes are not correlated with those of vascular terrestrial plants. For this reason he kept them apart and created a series of new terms by adding the prefix aero- to the synusial units of DU RIETZ (1930, 1932), e.g. aerosocion, aeroassocion, etc. OMURA (1952, 1953) distinguished forest associations and within these 'epiphyte-associations' divided into 'epiphyte-sociations' without using the suffix -etum. Both were considered to be one-layer communities. HOSOKAWA (1954) argued that these terms are misleading since they suggest reference to complete phytocoenoses. Curiously enough, he also rejected his own terms ('aeroassocion' etc.) as well as DU RIETZ's terms from which they were derived, because (i) one aeroassocion may occur in different forest associations, (ii) one forest association or even sociation often has several epiphyte associons (unions), even on a single tree, i.e. units of a higher hierarchical level than the corresponding forest syntaxon itself. The concrete epiphyte societies are in his opinion part of a concrete forest phytocoenosis (ecological dependence), whereas the abstract epiphytic units are not part of the abstract forest syntaxa (taxonomic independence). This, of course, was what had been pointed out by DU RIETZ, LIPPMAA and others and the reason why DU RIETZ proposed two independent hierarchies and coined his synusial terms. Most authors, including the writer, therefore see no necessity for a set of new terms such as 'epies,' 'epilia,' and 'epido' for societies, unions, and federations of epiphyte societies, as was proposed by HOSOKAWA (1954, see, however NAKANISHI 1970). The terms introduced by HOSOKAWA (1951) and OMURA (1952) are equally superfluous because the epiphytic synusiae are not fundamentally different from other synusiae, except that they comprise adnate societies on (the mostly dead bark of) other living plants. But the same unions may be found on dead trees and even on rocks. Epilithic moss and lichen synusiae are adnate too, and the epiphytic ones described by these Japanese authors also contain flowering plants.

There are two reasons other than those of HOSOKAWA for questioning whether epiphytic communities are really synusiae. (i) They often consist of different life-forms as conceived by BARKMAN (1958b; cf. his table XVIII, p. 194). But if life-forms are conceived in a broader sense, for instance RAUNKIAER's, this objection is invalidated. Besides, we have seen that for good reasons some authors have based synusiae upon spatial location, especially layering, and not upon life-forms. (ii) Epiphytic synusiae may themselves be several-layered, having for instance a fruticose and a foliaceous or crustaceous lichen layer. These layers may in turn show distributional (and hence, potentially, syntaxonomic) independence (BARKMAN 1958b, p. 331). This implies that, according to our definition, such epiphyttic community types are in fact microcoena rather than synusiae (BARKMAN 1969). I have, however, refrained from treating them as such since most epiphytic communities are one-layered, and in others the separation of the layers is obscure, or the layers if distinct are seldom superposed, but form a 'horizontal' mosaic. It is even questionable whether such mosaics are not rather to be treated as successional stages or mixtures of different units.

Summarizing we may state that to-day's trend in phytosociology is to classify epiphytic societies (even if they may be conceived as microcoenoses) into synusiae, and to give them names – also for practical reasons – according to the procedures followed in the classification of phytocoenoses. There is near-unanimous opinion on the last point because: (i) the use of other suffixes than -etum, -ion, etc. would necessitate hundreds of alterations in names of synusial units firmly established in the literature. (ii) it would be necessary to do the same for terrestrial synusiae some of which are unistratal (e.g. Salicornietum strictae) and therefore identical with syntaxa; it would be a quite arbitrary decision which suffix should then be used. (iii) the name of the name-giving species will usually make it clear whether one has to do with a synusia or a syntaxon.

16.4.3 AQUATIC COMMUNITIES

16.4.3.1 *Characteristics*

Aquatic communities offer special problems because they usually have a lower level of organization, and their synusiae are less tightly connected and more often developed as local facies or as pure stands of single species. This is due in part to the fact that differences in specific environments are less stable than in terrestrial communities. The latter show marked vertical gradients both above the soil (microclimate) and in the soil (horizons), whereas water has no vertical gradient except for light and temperature. Water currents tend to obliterate both horizontal and vertical differences in water chemistry, whereas soils may have distinct and permanent horizontal patterns of physical and chemical factors[1]). Finally, floating communities may be driven by wind and currents from one rooting plant community to another. For these reasons the synusial approach has been more widely used here than in terrestrial communities, although as DEN HARTOG & SEGAL (1964) pointed out, all plant and animal communities of a water body belong to the same nutritive cycle. ALMQUIST (1929), who otherwise dealt with biocoenoses only, almost confined himself to the description of synusiae when dealing with fresh-water vegetation. DU RIETZ (1965) followed the same procedure.

Before going into detail it is useful to distinguish with LUTHER (1949) between haptophytes (adnate or non-rooting hydrophytes), rhizophytes (radicant or rooted hydrophytes), planktophytes, and pleustophytes (i.e. microscopic and macroscopic errant or mobile hydrophytes respectively). The distinction of plants into adnate, radicant, and errant life-forms by GAMS is most fundamental and useful in the study of both terrestrial and aquatic vegetation.

16.4.3.2 *Errant Hydrophytes*

The planktophytes and smaller pleustophytes (i.e. the Lemnids and Ricciellids in the sense of DEN HARTOG & SEGAL 1964) differ from all other water plants by forming mobile synusiae, able to shift from one phytocoenose to another. They are also easily and frequently infested with diaspores of other species supplied by wind and birds. For this reason these synusiae are generally considered separate associations, even in the BRAUN-BLANQUET school. Curiously enough, this has been done for planktophytes well before the pleustophyte associations were recognised as such. TÜXEN (1937) for instance still incorporated societies of Lemnaceae in the order Potametalia and even considered *Lemna gibba* a faithful species of the Myriophyllo-Nupharetum.

On the other hand, ALLORGE (1922) distinguished already associations of planktontic algae within aquatic vascular plant communities. He was followed by many others. The reasons for this difference in treatment might be, (i) the fact that planktontic organisms often show vertical movements in addition to horizontal

[1]) Yet VILLERET (1965) claims that the aquatic environment has a mosaic structure and is far from homogeneous. This is likely to hold in particular for shallow stagnant waters, for VILLERET (1954) found an extreme short-distance variation of habitat factors in water at the edge of *Sphagnum* carpets in oligotrophic pools of Brittanny. I have had the same experience in shallow Dutch bog pools.

ones (but *Stratiotes* does the same), (ii) the frequent occurrence of seasonal aspects of these short-lived organisms which may be completely different floristically, (iii) the fact that they often form communities of their own (in large, deep-water basins), (iv) the difference in life-form, (v) the difference in reaction to ecological factors, resulting in syntaxonomic independence, (vi) the completely different methods of investigation, and (vii) the fact that they are studied by other specialists. The last point, although seemingly irrelevant, plays an important role. Many hydrobiologists are not familiar with vascular plant communities, and hardly any phytosociologist is familiar with (microscopic) algae. It is evident from points (vi) and (vii) that in the first stage of investigation planktonic algal communities have to be analysed and synthesized separately, i.e. as taxocoenoses. The same applies to adnate microphyte communities. In a later stage it may be useful to see in how far these communities are linked with and can be incorporated in other water plant communities (SCHMITZ 1965).

Although none of the above points is in itself decisive in requiring distinction of separate plankton phytocoenoses and syntaxa, all contribute to the fact that it may be practical to do so. BRAUN-BLANQUET (1964) is most apodictic in this respect: 'es handelt sich um Biozönosen.' MARGALEF (1947, 1949, 1955), however, considers the whole living community of a water basin as one biocoenosis, but he divides these biocoenoses into 'associations,' some of which are planktontic and comprise microscopic forms, both plants and animals. For him, points (iv) and (v) are obviously not relevant. SCHMITZ (1965), dealing with (mostly adnate) microphytic water communities (microscopic algae), considers them to be synusiae or rather 'microphytocoenoses,' and comments that 'it is hardly recommendable to distinguish several associations in the same locality, yet often impossible to attribute a microphytic community to a particular macrophytic association.' The names of his synusiae are formed in the same way as those of associations (-etum). VILLERET (1965), who speaks of algal associations, doubts their validity as they have been delimited: It is extremely rare, he says, to make a record that fits into one of these abstract units, with the exception of associations of extreme conditions (thermal waters, snow, rapids, salt water sources, etc.). One of the great difficulties in the study of microphytic vegetation is of course that one cannot see the plants. Consequently it is impossible to judge whether the microvegetation is homogeneous or not, nor can we determine the minimum area in the customary way. These difficulties are comparable with those of the study of large algae communities of deep water and the societies of fungi, soil inhabiting or small animals, and animals with nocturnal activity.

As to the communities of smaller pleustophytes, their autonomy has been recognized since KOCH & TÜXEN (in OBERDORFER 1957) created the class of Lemnetea, with one alliance, Lemnion minoris. DEN HARTOG & SEGAL (1964), however, divided this alliance into a Lemnion minoris s.s. of floating pleustophytes (lemnids) and a Lemnion trisulcae of swimming, submerged pleustophytes (ricciellids). Naturally an enormous difference exists between the water surface and the purely aquatic (submerged) environment as habitats. But the same degree of contrast applies to the microclimate of the canopy of a dense forest and its terrestrial moss layer, and likewise to the often acid humus of the latter and the purely mineral, often alkaline subsoil in which the trees root. Such layers should be distinguished, but not as associations (phytocoenotic units).

16.4.3.3 *Adnate Hydrophytes*

Another special group of water plants is formed by the haptophytes. In the terrestrial environment the haptophytic or adnate communities are widely different according to their substratum (living herbs and tree leaves, dead bark, decaying wood, stones and rocks). In the aquatic environment the floristic difference in response to substratum quality may be much less important, probably because the adnate plants absorb water and nutrients over their entire surface from the surrounding water. For practical reasons a distinction is usually made between the haptophytic communities of microscopic algae (see above) and those of larger algae, bryophytes and lichens. Both have, for a long time, escaped the attention of phytosociologists, in particular in the school of BRAUN-BLANQUET. They do not even figure in the survey of water plant communities by LOHMEYER c.s. (1962), nor in that by DEN HARTOG & SEGAL (1964), Yet RÜBEL (1933) already distinguished a separate order of haptophytes, the Encyonematetalia, which, however, only comprised (three) algal communities (as nomina nuda). Macroscopic adnate algal communities have mainly been studied in the littoral zone of sea coasts and brackish waters, often as biocoenoses including sessile animals (barnacles etc.) and semi-sessile animals (molluscs) (for literature see for instance DEN HARTOG (1959) and BRAUN-BLANQUET (1964)).

The first detailed phytosociological description of such communities was published by KORNAŚ & MEDWECKA-KORNAŚ (1950). They distinguished a Fuceto-Furcellarietum which was later split up into a mud-inhabiting rhizophytic Zostero-Furcellarietum and a stone-inhabiting haptophytic Fuco-Furcellarietum (KORNAŚ 1959). Whenever mud and stones or shells are mixed, even in fine-grained patterns, the distinction of their communities on the phytocoenose level seems justified. VAN DEN HOEK (1960) followed the same procedure. DEN HARTOG (1959) is of the opinion, however, that on pebbly bottoms such a distinction is impossible. His associations of rocky sea-coasts, although primarily based upon the dominant alga species of the upper layer, do comprise all layers from large Fucaceae down to crustose red algae and lichens. He stressed the ecological and syntaxonomical independence of some layers and illustrated this by a very instructive scheme (l.c., fig. 14). As DEN HARTOG also pointed out, it is quite possible to classify the separate layers or socions instead of (or in addition to) the often three- or four-layered algal phytocoenoses.

KLEMENT (1955) was of the opinion that lichen communities should be kept strictly apart. When making vegetation records, he even omitted the bryophytes and algae. Thus his marine Verrucarietum maurae contains only lichens, whereas the association table published by DEN HARTOG (1963) also contains four algae. KLEMENT stands rather isolated in this opinion. Fresh-water lichen associations were first described by MOTYKA (1925, cf. KLEMENT 1955). They are all of an amphibious nature. As to the adnate bryophyte communities of fresh water, OCHSNER (1952) still treated them as part of the rhizophytic Potametea, but KRAJINA (1933) and KOCH (1925) considered them autonomous associations and alliances and VON HÜBSCHMANN (1957) even created three classes of fresh-water bryophyte communities (including lichens and mixed communities) and one class of marine associations. Algae were not excluded from the records.

In swift mountain streams bryophytes may actually form pure societies, but in low-land streams of slow currents they are often mixed with phanerogams. *Fontinalis antipyretica*, which VON HÜBSCHMANN considered faithful to the low-land alliance Fontinalion, is also frequent in stagnant water and may then be freely floating or attached to the submerged part of reed culms; the same applies to *Leptodictyum riparium*. In the latter case one could better speak of synusiae within phanerogamic associations. Besides, there is a gradual transition from permanently submerged moss associations to those submerged in winter only, which were also included in the order Fontinaletalia. Now the latter associations, especially VON HÜBSCHMANN's (1957) association of *Leskea paludosa* and *Leptodictyum riparium*, is closely allied to the Chiloscypho-Mnietum which even contains *Fontinalis antipyretica* (BARKMAN 1958b). This, however, is an epiphytic synusia occurring on tree bases that are inundated daily in willow marshes of fresh-water tidal areas. It seems illogical to consider one unistratal community an association and a closely allied unistratal community a synusia, which would mean that they form part of different hierarchical systems. In the writer's opinion this shows that all fresh-water bryophyte 'associations' (phytocoenoses) are best treated as 'unions' (synusiae). These synusiae may or may not be combined with other synusiae into phytocoenoses.

16.4.3.4 *Larger Hydrophytes*

Finally we come to perhaps the most difficult problem, viz. the communities formed by rhizophytes and large pleustophytes. DEN HARTOG & SEGAL (1964) made a strict separation between associations formed by these life-forms. They distinguished 9 classes, 3 of which were new. Apart from the Lemnetea, which form a rather isolated group, three classes of large pleustophytic plants and five classes of rhizophytes were distinguished. This is, however, subject to criticism (cf. BARKMAN 1968, and the reaction by SEGAL in the same issue). Rhizophytes and large pleustophytes have in common that they are not or are hardly mobile. They share the same water mass, the only difference being that the former root in a sand, mud or peat bottom (often a very loose substratum and hardly different from the water itself!), whereas the latter do not. This difference is smaller than it seems, for both groups of plants absorb most of their nutrients from the water, not from the bottom, as was emphasized by DEN HARTOG & SEGAL (1964: 378). Moreover, there are transitions and species that can behave both ways, for instance *Elodea canadensis* and *Myriophyllum spicatum*. In a system of plant communities based primarily on floristic composition, a sharp syntaxonomic separation of these groups as syntaxa would only be justified if it coincided with a clear-cut floristic demarcation. Mixtures do, however, occur: 'it can be a uniform mixture that can hardly be disentangled,' 'vegetations in which rhizophytes and pleustophytes coexist, *have to be* regarded also as mixtures of two vegetation units, even when the vegetation at first glance seems to be homogeneous.' The sharp floristic difference ('the communities of rhizophytes... and pleustophytes ... have not a single species in common') is then a product of the method they use, separating the groups *a priori*. They suggest also as a new criterion success of sexual reproduction, with unwelcome consequences for classification. WESTHOFF & DEN HELD (1969) rejected the classes Ceratophylletea, Utricularietea and Stratiotea. They maintained the Charetea, already proposed by FUKAREK (1961), because in deep lakes the bottom synusiae of Characeae may be the only synusiae present, thus forming phytocoenoses, and because they form a distinct layer, when participating in other phytocoenoses. This too can be criticized by the arguments given above.

16.4.4 SOIL-INHABITING COMMUNITIES

Although plant sociologists have generally neglected the study of soil-inhabiting organisms, there does exist a respectable number of publications on this topic, mostly, it is true, by non-sociologists. Yet this field of study is by far less advanced than that of the autotrophic macrovegetation, and still less is known about the connections between particular soil communities and the corresponding higher vegetation. There are several reasons for this state of affairs. (i) The study requires special knowledge of these plant groups, which only a few specialists have. None of them is familiar with all groups (bacteria, soil-inhabiting algae, slime moulds, micromycetes, larger fungi). (ii) The number of known species is tremendous. APINIS (1970) mentioned the occurrence of more than 400 species of soil-inhabiting microfungi in a total of five littoral communities, poor in higher plants. MEISEL-JAHN & PIRK (1955) recorded 33 species of flowering plants, 21 bryophytes and 121 macrofungi (mushrooms) in spruce forests of the Sauerland (Germany). (iii) Many species have yet to be described. (iv) Determinations of those described are often difficult. Identification often needs the preparation of special cultures in the laboratory (bacteria, algae) or is impossible when the plants are sterile (fungus mycelia), and efforts to induce the latter to form fruit-bodies generally fail. Many of these plants (bacteria, algae) cannot well be preserved in herbaria, others (higher fungi) have often been badly preserved (not dried, no macroscopical description added) or not preserved at all. The literature is scattered, often hard to obtain and expensive, and often also confusing. (v) The plants are invisible in the field (or appear only as fruiting bodies that may be ephemeral), so special sampling methods are needed.

Yet it is important to include them in community analyses. Our knowledge of communities should be based on full species composition, and not only do the fungi alone greatly outnumber the vascular plants and bryophytes in many communities, but they also form an essential part of the biocoenose as humus decomposers. Their role in the biocoenose is highly important in a quantitative sense; in a deciduous wood in England biomass and energy turnover of the fungi are estimated to exceed those of all other lower plants and all animals, taken together, by factors 10 and 4 respectively (SATCHELL, unpublished report).

With regard to higher fungi (macromycetes) it may be remarked that the irregular and often sporadic appearance of fruiting bodies necessitates the sampling of one quadrat at least 6–8 times a year during at least three successive years (LANGE 1948), or every ten days during two years (KALAMEES 1968). However, 5 to 10 years is likely to be a better approximation, according to the present author's experience. The data obtained so far by mycosociologists show that the minimum area for higher fungi is likely to be much greater than for higher plants, i.e. in the order of 1000–5000 sq. m. in a wood, for instance. This makes it difficult to compare the respective synusiae, as it is often hard to find a homogeneous area of woodland of that size. On the other hand, even a large forest area that is homogeneous in its phanerogamic and bryophytic vegetation may be quite heterogeneous for fungi. Their patchy occurrence, and the existence of 'mushroom deserts' and 'mushroom oases' in an otherwise homogeneous forest is well known to mycologists. The effort required to study fungus communities effectively is clearly of a different order of magnitude from that for the usual phytosociological relevés.

Moreover, the limits between fungus and autotrophic communities often do not coincide (cf. for instance LANGE 1948 and TOMILIN 1964), probably because the former are independent of light and strongly dependent on the composition of the humus or the presence of special host species, whereas the latter are more dependent on light and minerals. The relative independence of different synusiae that is one of the themes of this account applies also to fungi vs. green plants (and, probably, to different groups of fungi). GRAHAM (1927) found the same fungus community in two different though neighbouring associations. Both were advanced stages in the succession and grew on soil rich in humus. It is likely that in pioneer communities on more mineral soil fungal synusiae and phanerogamic associations are more closely correlated (ANDERSSON 1950). It is therefore not surprising that HUECK (1953) advocated studying fungus communities separately, at least for the present, for both practical and theoretical reasons (different specialists and methods, different community limits). But he also wrote, 'A synthesis of fungal and autotrophic communities in a later phase of the investigation appears to be desirable, if not necessary.'

In the light of what has been said above and in view of the frequent treatment of moss and lichen synusiae as 'associations,' it is rather surprising that this point of view has not been adopted for fungus communities, except for those on special substrata. Many authors have investigated soil-inhabiting microfungi (for literature see TÜXEN 1964, 1966), but none of them described these as separate associations. As to macrofungi, HÖFLER (1937) regarded them as members of independent communities, because of the lack of topographical correlation with community-types of higher plants, but UBRISZY (1940, 1956) did not recognize the existence of independent fungus associations. FAVRE (1948) used the term 'associations fongiques,' but he explained that it is preferable to consider them synusiae within the larger phytocoenoses, since fungi are heterotrophic and therefore ecologically dependent on higher plants. At least, he says, this should be done until the fungus associations are better known! WESTERDIJK (1949), dealing with lower fungi, however, proposed the term 'association' to designate characteristic combinations of species occurring on certain definable substrata. HUECK (1953) dealing with fungus communities generally, shared this view, stating that 'as the fundamental definition of the term' (i.e. association) 'is unaltered, there appears to be no need to create a new term.' One might remark that synusia and union were not exactly new terms in 1953, and if unions are based on the same (floristic) criteria as associations, this does not imply that the two are identical! SCHWEERS (1949) and ANDREAS (1950) described the mycoflora of certain types of moist to wet, acid, poor grasslands that are rich in bryophytes. In their opinion the special, very characteristic floristic composition of this fungus community justifies the status of association ('*Hygrophorus*-meadow'). This point of view is highly exceptional, however. KALAMEES (1965) is of the opinion that fungal groupings in a forest do not form separate communities. They are part of the biocoenose, but they differ from synusiae by having an entirely different rhythm of development. According to him a new term has to be created, but for the time being they are best regarded as 'consortia' (a term introduced by RAMENSKY).

In the Braun-Blanquet school (cf. TÜXEN et al. 1957) fungus associations are only described as such if they occur on special substrata without other plants, for instance the Fometum igniarii Pirk 1952 on willow trunks, the Trametetum gibbosae Pirk et Tüxen 1957 on rotten beech stumps and the Coprinetum ephemeroidis Pirk et Tüxen 1949 on fresh cow dung. According to this principle PIRK (1950), who published a table of vegetation records of fungi on burnt soil, considered these merely part of the Funarion hygrometricae. Some of the fungi were even considered faithful species of that moss alliance. The practice of naming these fungal societies without higher plants or mosses as 'associations' seems questionable. After all, the beech stumps form part of the forest biocoenose, and the cows form an essential part of the meadow biocoenose. If each fungus society on its special substratum in the forest is to be named an 'association,' the number of these can become overwhelming. It would be more logical and practical to treat fungi through synusiae and to distinguish the latter within a single phytocoenose, according to substratum, position, and function.

Mycorrhizal fungi and parasitic fungi often form synusiae quite independent from other fungi. They are much more linked with certain host species, whereas the saprophytic fungus synusiae seem to be more linked with community-types. However, some mycorrhizal fungi are not at all host-specific (*Lactarius rufus*, *Amanita rubescens*, *A. muscaria* etc.), others are also dependent on other factors, for instance acid soil or limestone, or association variants that occur on wet soil or are poor in grasses (TOMILIN 1964, KALAMEES 1968). Some mycorrhizal-fungi of *Quercus dahurica* are even absent where this tree only develops a shrubby form (TOMILIN 1964). The same applies to parasites. In Eastern Siberia *Uromyces heimerlianus* occurs on various species of *Vicia*. On *Vicia multicaulis* it is only found where it grows in meadows with *Carex schmidtii* and *C. pallida* (TOMILIN 1964). Parasites that are strictly dependent on one host species, may yet occur in various plant associations when the host has a wide habitat range. According to TOMILIN (1964) most saprophytic micromycetes have a wide ecological and sociological range. But saprophytic higher fungi may be very specific, for instance those occurring only on fallen pine cones or beach leaves.

The difference between parasitic and saprophytic synusiae is not always sharp, as parasites may continue their life as saprophytes after the death of their host plant. The soil fungi may form an intergrading spectrum from the purely parasitic, through various kinds of mutualistic and commensal relations, to the saprophytic wholly independent of any living plant; and in the field one cannot see to which group a terrestrial mushroom belongs. It is thus not easy to draw boundaries between the different synusiae for fungi. As a first approach to these, however, it may be reasonable to recognize six broad synusial groupings of fungi in a given forest or scrub phytocoenose: (i) permanently subterranean synusiae of micromycetes, (ii) terrestrial synusiae of saprophytic and mycorrhizal macrofungi with overground fruit-bodies, occurring on humus, (iii) saprophytic fungi on animal excrements and corpses, (iv) parasitic and saprophytic macrofungi on tree bases, decaying stumps and buried wood, (v) epiphytic synusiae of mostly resupinate fungi on the bark of living and dead trunks and branches, (vi) parasitic synusiae on living herbaceous stems, twigs and leaves. Fungi occurring on alien substrata like paper, shoes and other rubbish should be listed separately. In phytocoenoses without woody plants only the groupings (i), (ii) ,(iii) and a fraction of (vi) may be present. In one phytocoenose each category may contain several unions, for instance unions representing different succession stages on decaying stumps. Clearly, further division of these groupings (which seem excessively broad compared with those used for epiphytes) is possible with further study and further knowledge as will be shown below.

16.4.5 An Example: the Dicrano-Juniperetum

16.4.5.1 *Sampling methods*

Vegetation synthesis is dependent on vegetation analysis. Elements that have not been distinguished during the analysis, naturally cannot be described or classified separately. In the Braun-Blanquet school it is customary to use relatively large sample plots (although there is a growing tendency towards smaller plots), and to list the species of each layer separately. On the basis of such relevés it is therefore possible to ordinate or classify the separate layers as well as the whole phytocoenoses, but not the microcoenoses or those societies that form a mosaic within one layer. Hence it is somewhat surprising that in this school layers as a whole have hardly ever been described as separate community-types, whereas microcoena and synusiae within one layer often have. (The reasons why this is so, will be discussed later.)

In order to distinguish such partial communities we must use a different sample procedure with smaller plots. If we want to investigate all elements of a horizontal vegetation mosaic, we can either make vegetation charts, or use line transects, or divide the sample plot into smaller subquadrats of appropriate size (that can be determined by a refined minimal area method, Barkman 1968). The technique has been applied by the present author and his collaborators to investigate herbaceous and bryophyte synusiae within the juniper scrub association on poor sandy soils in the Netherlands and northwest Germany, the Dicrano-Juniperetum (Barkman 1970). On the basis of 1100 small quadrats (30×30 cm) 12 microcoena, comprising 18 unions of herbs, bryophytes, lichens, and algae could be distinguished. These were also studied pedologically and microclimatologically. In addition, the synusiae (unions) of fungi, myxomycetes and epiphytes were investigated. The area of each bryophyte or herb society was much too small to have a representative mycoflora. The synusiae of the latter therefore had to be investigated differently, i.e. by noting all fungus species present in the biocoenoses and indicating, for each species. the associated bryophytes and herbs as well as the substratum of the individual fruiting bodies. This led to the distinction of one union of myxomycetes and 15 terrestrial fungus unions.

In total more than 600 plant species were found in this association alone. Some of the main results will be dealt with here, because: (i) They have been published in Dutch; (ii) this is the only study so far of all synusiae (except bacteria and microfungi) within a plant association following the BRAUN-BLANQUET method; (iii) scrub vegetation is highly complex, often more so than even forest vegetation, and therefore most suitable to demonstrate the complex relations between a phytocoenose and its microcommunities and societies.

16.4.5.2 *Synusial Description*

Juniper scrub generally consists of a mosaic of dense shrubs and small open spaces. There are five microbiotopes – Nm, Sm, Sn, dm, and dn. The open spaces differ according to aspect (north and south side of the shrubs). The north side has a moss vegetation (Nm), the south side a narrow strip without vegetation (only fallen needles, Sn) close to the shrub and a broader strip with bryophytes (Sm) at a greater distance from the shrub margin. In the dense scrub moss patches (dm) alternate with patches of bare needles (dn). The latter may have flowering plants and fungi, however. The moss vegetation is different in Sm, dm and Nm. Field measurements show that the floristic differences between these five biotopes are probably due to differences in (i) light intensity, (ii) temperature maxima, (iii) precipitation (throughfall), (iv) potential evaporation, (v) rate of litter fall.

The 18 unions of terrestrial bryophytes, lichens, and herbs are partly controlled by the above microclimatic factors, partly (different synusiae within one of the five biotopes) by pedological factors (pure sand, humus, decaying wood buried in the soil, admixture of oak litter, differences in mineral content of the soil). Each union contains a number of species and is characterised by one species or by a combination of species. Each of the six herb unions is more or less linked with one or more of the 12 moss unions. There are 12 combinations forming microcoena. These microcoena are all linked together by a common shrub union and a common union of lianas. The unions of fungi are considered distinct from the moss synusiae, (i) because of their heterotrophy and quite different periodicity,

(ii) because they either inhabit special microhabitats or form a distinct stratum (The mycelia live either in the rhizosphere of the shrubs and trees, or in the uppermost 2–5 cm of the humus, whereas the herbs root deeper and the bryophytes and lichens live on the very soil surface, as adnate plants.)

There are two mycorrhiza unions, both alien to the juniper scrub proper and occurring only when other trees are present (which is often the case): the mycorrhiza union of broad-leaved trees (*Quercus, Betula*) and that of conifers (*Pinus, Picea*). The remaining 13 unions are saprophytic and live either in the humus (7 unions) or on special substrata, viz. (i) animal excrements, (ii) rotten fungi, (iii) dead herbaceous plants, (iv) fallen juniper needles, (v) fallen dead twigs and branches, and (vi) buried pine cones. The humus unions are generally linked with special moss (and herb) unions, the mycorrhiza unions and those of special substrata are found scattered throughout the phytocoenose. The same applies to moss unions of special substrata (buried wood, rabbit holes). The alga union of *Prasiola crispa* and *Hormidium flaccidum* (sandy places in Sn) never occurs in combination with other unions, except the shrub union. Four epiphytic unions (bryophytes, algae and lichens) of the junipers are arranged vertically from base to crown; the fifth occurs on bare hard wood. The 4 saprophytic fungus unions on the junipers (dead bark and wood) are mainly arranged horizontally (upper and lower side of slanting trunks respectively). There are also special synusiae on rotten stumps of deciduous trees and of (other) conifers occurring in the juniper scrub. Finally, there are five unions of parasitic fungi, viz. (i) on the base of juniper boles, (ii) on living twigs, (iii) on living needles, (iv) on living herbs and dwarf-shrubs, and (v) on pupae of arthropods in the soil. None of these epiphytic, wood-inhabiting saprophytic, and parasitic synusiae are in any way linked topographically with the terrestrial unions.

Summarizing, we may state that the more than 600 plant species of the Dicrano-Juniperetum can be grouped into 52 unions, 36 of which are terrestrial or soil-occupying. These can be grouped into 12 microcommunities, which belong to 5 microbiotopes. The epiphytic unions are floristically very distinct from the terrestrial, except for the moss and saprophytic fungus synusiae on the extreme bases of the juniper stems. The moss union in question consists of terrestrial species that climb the trees from the ground; the fungus union develops on the bark and then colonises the adjacent soil.

16.4.5.3 *Relations of Synusiae to Associations*

Some synusiae are restricted to one microbiotope or microcommunity, or form a microcommunity of their own, others occur throughout the phytocoenose, either as a continuous separate layer or scattered on special substrata within a certain layer ('azonal' synusiae). The synusiae (unions) are separated either vertically or horizontally. There may also be a (partial) separation in time (herbs in summer, fungi in autumn). Some unions are constantly present, others may be absent from certain stands. The former might be called obligate unions of the association, the latter facultative. The example of the Dicrano-Juniperetum can serve to clarify some notions, for instance the difference between obligate and specific and between facultative and non-specific unions. The moss union of *Hypnum* and *Lophozia* (dm) is constantly present (obligate), but it is not specific (occurs also in other plant associations). The parasitic union of *Gymnosporangium juniperinum* and *G. clavariaeforme* on juniper twigs is specific of the association, but not always present. All unions mentioned here are designated by their name-giving species only, but they contain many more species.

There is also a difference between topographical and ecological dependence. The moss union of *Pseudoscleropodium purum* etc., the herb union of *Sieglingia* and *Galium hercynicum* etc., and the fungus union of *Galerina heterocystis* are topographically linked, because they have similar habitat requirements (north side of shrubs on not too poor sand), but they are not ecologically interdependent. Ecologically dependent unions, for instance the parasitic synusiae of *Juniperus* may be either specific (e.g. the *Lophodermium juniperinum* union on juniper needles) or non-specific (e.g. the *Fomes annosus-Armillaria mellea* - union on the base of the trunks). The latter occurs also in pinewoods and is therefore not topographically dependent on the Dicrano-Juniperetum although it is ecologically dependent. Ecologically dependent unions may even have their optimum elsewhere, e.g. the mycorrhizal synusiae linked with *Pinus* and *Quercus*. A distinction must also be made between topographical linkage of societies (synusiae) to one another within a phytocoenose (syntaxon), and linkage of a particular synusia to the syntaxon. The three moss, herb and fungus unions mentioned above always occur together in the stands of the association, but they also occur outside the Dicrano-Juniperetum, for instance in Violion caninae associations. On the other hand, the *Rhodobryum* moss union and the saprophytic fungus union of *Peniophora laevigata* a.o. are both specific of the association, but they are not necessarily found together within its stands.

From the point of view of classification, the synusiae (unions) that are independent of certain associations, offer the greatest problems. Their widespread occurrence was one of the main reasons to treat them (or all synusiae!) as separate associations. I have called them 'syntaxonomically independent' (BARKMAN 1968) as distinct from ecologically independent synusiae. It is, however, better to speak of syntaxonomic non-specificity (BARKMAN 1970). Besides I made a distinction between syntaxonomic non-specificity and lack of affinity: a certain union may be specific for a certain association, but its nearest floristic affinities may be with unions of quite different associations. It is fundamentally the same problem on a higher hierarchial level, for it means that higher synusial units (federion, ordulus or classicula) into which unions are grouped by floristic affinities may be non-specific in relation to (phytocoenotic) syntaxa even if the unions themselves may be syntaxonomically specific. Both phenomena demonstrate that synusiae are being classified 'unnaturally' when only phytocoenoses are recognized and classified.

16.5 Evaluation

Divergent views have been held among phytosociologists and ecologists on the justification and value of the synusial approach (see also WHITTAKER 1962: 34). A number of the arguments have been met with in the foregoing sections, but it seems useful to give here a more coherent discussion. The following is the author's synthesis of comments from the lectures and discussions of the symposia on 'Biosoziologie' and 'Pflanzensoziologische Systematik' (TÜXEN 1965, 1968), special literature on the problem, arguments used implicity in published vegetation systems, and inquiry among and discussions with Dutch vegetationists. The arguments apply to choice among five positions different phytosociologists have taken on synusial treatment of vegetation: (i) opposition to the recognition of synusial groupings, (ii) acceptance of study of societies as an approach strictly subordinate to study and classification of phytocoenoses-i.e. societies might be described within a given phytocoenose, but they would not receive any classification separate from that of the phytocoenoses into syntaxa, (iii) recognition of some synusiae of special substrates as 'associations' while all other societies are left out of consideration, (iv) acceptance of the synusial approach as a complement to the study of whole phytocoenoses, with both societies and phytocoenoses to be classified, separately, and (v) the view that the synusiae should be the principal basis of vegetation study, to which the study of phytocoenoses is secondary or subordinate.

16.5.1 ARGUMENTS AGAINST THE SYNUSIAL APPROACH

(i) Critics of the synusial approach are often concerned over the multiplication of community-units in addition to an already extensive system of syntaxa for whole-communities. The relative independence is supposed to imply such a multitude of combinations and recombinations of synusiae as to render their treatment as significant units pointless (CAIN, 1936). To these arguments it may be responded (a) that if a particular research does not call for the description of synusiae, it is not necessary to do so when dealing with phytocoenoses, (b) the number of synusiae can be reduced by using a rather broadly conceived 'union' concept, (c) the independence makes it probable that the number of unions in a given area might be even less than that of all their possible combinations (the associations). Hence this argument rather works against its advocates.

(ii) A seemingly more fundamental argument, formulated by GLEASON (1936) and others, would be the existence of many and manysided ecological interactions between synusiae of apparently disparate habitats: trees, by their shade, moisture, and chemistry of leaf litter, affect the growth of herbs, mosses and fungi, and the latter in turn may affect the germination and growth of tree seedlings, etc. The phytocoenose is thus regarded as a closed functional system, the synusiae are not. If this were an argument, the study of the ecology of individuals and populations of one species would not be justified either. Besides, it is not true. Phytocoenoses are only a part of ecosystems and even ecosystems are open systems, interacting with each other.

(iii) More important is the observation that synusiae that are distinct in one situation, seem in others to merge together or to have a different composition. Floristic and ecological differences between unions may gradually disappear from one region to another, as occurs with the difference between hollows and hummocks in oligotrophic bogs from north Finland to west Ireland (Palsa bog - raised bog - flat bog - blanket bog ecocline). Even in a restricted area herb unions that are clearly separable by either a mosaic relation (intracommunity pattern) or by height, in one association, may form a single union in another association or even another stand of the same association. It has also been observed that species may interchange their position from one layer synusia to another, for instance in bog communities. Besides, consistently distinct layer unions above the soil need not correspond with root layers, so that one species may belong to different unions above and in the soil, e.g. in the Rosmarino-Lithospermetum (BARKMAN, 1958a).

To the critic this may mean that synusiae should not be distinguished even where the distinction is clear. To the student of synusiae the occasional blurring of boundaries no more invalidates classification on the synusial level than parallel phenomena do on the level of syntaxa.

(iv) Finally, where layers are distinct, all species of higher layers participate in all lower layers above the soil and in all shallower root layers in the soil during a part of their life-cycle, in the case of natural, self regenerating phytocoenoses. This is the main reason for the ZM school not to regard layer synusiae as distinct. Incidentally, this fact would plead for the distinction of lifeform synusiae instead of layer synusiae (cf. 16.3.1).

16.5.2 ARGUMENTS FOR THE SYNUSIAL APPROACH

These are partly of a practical, partly of a theoretical nature. The practical arguments can be summarized as follows:

(i) The necessity of taxonomic specialization by investigators nearly always compels separate study of vascular plants and fungi; demands on investigators' time may require that really intensive analysis of lichen and bryophyte synusiae be carried out by someone other than the student of vascular plants.

(ii) Different research methods, including different kinds of samples accommodated to different sizes, positions and seasonal timings of organisms, must be applied to phytocoenoses and to different synusiae. In a forest the minumum area of the moss layer is much smaller than that of the tree layer, the minimum area of the terrestrial macrofungi considerably larger. A small plot therefore is by no means representative for the tree layer and the fungi, a plot large enough for a complete species' list of the latter is not suitable for the moss layer, as it will often comprise different moss societies or — to put it another way — it is not suitable because the moss layer is too heterogeneous.

The theoretical arguments are partly typological (iii-iv), partly ecological in nature (v-viii).

Typological arguments. (iii) Ignorance of synusiae or subordination to syntaxa seems illogical as the two may be equivalent regarding (a) size of the concrete stands, (b) sharpness of the boundaries between the stands, (c) floristic differences between adjacent stands (societies, phytocoenoses) or between abstract types (unions versus associations), (d) structural differences between stands or types. In fact, the boundaries between societies are often sharper than between phytocoenoses, for the simpler and more strongly dominated by one species is a community or fraction thereof, the more likely is its boundary to appear definite. In the ZM school, however, boundaries between clones or aggregates (families) of different species are not considered of any phytosociological importance, unless correlated with other floristic or site differences (cf. point vii). Difference in structure between adjacent societies may be as great as that between phytocoenoses accepted as different 'formations'. It is not surprising to find that in the ZM school such societies (as opposed for instance to alternating herb societies) have been treated as syntaxa of the rank of alliances or higher, for instance the Calluno-Genistion pilosae and the Violion caninae (Nardo-Galion) which may form a mosaic in dry heath and which belong to different formations (IV and V) of SCHMITHÜSEN (1961) respectively (cf. however, BARKMAN, 1968, p. 32!).

(iv) Many synusiae are distributed in ways that do not conform to those of the syntaxa in which they occur. To recognize in phytosociology only syntaxa is thus to fail to recognize types and floristic affinities of existing vegetation units on a lower organisational level. This syntaxonomic non-specificity of synusiae has been frequently referred to above. Even specific unions, limited to certain associations, cause difficulties if we try to classify the latter into alliances etc., for the unions in question may be more closely allied with unions of quite different alliances. This syntaxonomic non-affinity (BARKMAN, 1968) is in fact only a matter of non-specificity on a higher level (non-specific federation). An extreme case is that of mobile synusiae (cf. point Vb). These problems can only be solved by classifying synusiae and syntaxa separately.

Ecological arguments.

(v) Some synusiae do not belong to the ecosystem proper of the phytocoenose in which they are found, either (a) because they depend on energy, nutrient or diaspore supply from outside the biocoenose or (b) because the plants themselves may freely enter and leave the biocoenose.

Typical examples of (a) are moss societies near forest edges with inblowing dust and in forest brooks. Both need a constant supply with nutrients from outside the forest. They are usually not considered part of the forest biocoenose (association) and are often considered to represent separate associations. This is open to criticism, for all biocoenoses receive and many are dependent on nutrient supply from outside, either partially (for instance sloping mire associations) or wholly (cave and deep sea biocoenoses). More often than not it is difficult to say whether this supply is indispensable or not. It seems practical, however, not to consider those groupings of species as part of the phytocoenose that live only near the edge, do not perform their life-cycle and are wholly dependent on diaspore supply from other phytocoenoses (vicinism). This is certainly true also of errant (floating) societies (b), like air and water plankton communities and floating water plants (Lemnion), that can be driven in and out of a given phytocoenose by wind and water currents. Such 'dysaptic complexes' (WOLAK, 1970) can be no more considered part of the biocoenose (not even as synusiae) than birds or insects that happen to fly over a vegetation stand or rest there temporarily. They are ecologically independent.

(vi) Many synusiae appear both as phytocoenoses in their own rights and as components of another phytocoenose. In the latter case they do belong to the phytocoenose, but the relation is not obligate, for instance the Rosmarino-Lithospermetum ericetosum occurs in the same region (southern France) in almost the same composition with ('R.-L. pinetosum') and without a canopy of *Pinus halepensis*. Another example are the treeless and tree savannahs in Surinam where the trees are the only difference (VAN DONSELAAR 1965). More often there is an obligate dependence in one region, independence in another region. This holds true for herb and dwarf-shrub synusiae that need the shade of a tree layer in a dry climate, but can be self-supporting in a moist climate or on a north facing slope. Thus the *Rhododendron ferrugineum* union forms an understory in subalpine *Larix-Pinus cembra* woods and a community of its own above the timber line. The same applies to the Swedish *Salix glauca-S. lanata-S. lapponum* thickets occurring both above the timber line ('forest without trees' of the Swedish botanists) and in the subalpine *Betula tortuosa*-woods, and to the treeless Querco-Blechnetum ('Q.-B. extrasilvaticum') in Ireland.

In all examples mentioned it seems illogical to consider such unions as independent associations in one case and to ignore them as vegetation units of whatever kind, when being part of other associations, as is often done.

(vii) Even wholly dependent synusiae may differ in microhabitat. Vertical habitat differences (between layers) are obvious, but the individuality of the layers may be obscured by discrepancy between corresponding layers above and in the soil, and by different life stages of the plants (16.5.1, (iii) and (iv)). Horizontal variation in habitat may be either due to primary or to secondary differences in site factors, the latter as a result of biotic heterogeneity (BARKMAN 1968). Societies based on primary differences (e.g. rocky outcrops in the forest floor) are nearly always regarded as distinct associations, those based on secondary site differences (for instance variation in light climate on the soil as a result of variation in canopy density;

special conditions around tree bases) are often not regarded as such. Yet epiphytic societies and hummocks and hollows in bogs, which belong to the second category, are always separated and rightly so. If we realize that the biocoenose also includes animals, even game tracks, rabbit holes, mole hills and ant heaps belong to the secondary environmental differences. Besides, it is often difficult to know whether a case of biotic heterogenity is not after all due to minor initial (primary) variations in abiotic site factors. It seems therefore more practical to consider all microhabitats as separate societies of the same phytocoenose. However, there is one restriction to be made: the microhabitats should be sufficiently distinct in floristic respect. Otherwise one might consider every gregarious plant species to have its own society (union), for every species has its specific influence on the environment.

(viii) Even in the same microhabitat plants may differ considerably as to function and performance. It seems reasonable to separate such groups as autotrophic, parasitic and saprophytic synusiae, but not as syntaxa. The differences are not sharp. Parasitic fungi may continue their life as saprophytes after the death of their host tree. Among flowering plants the existence of hemiparasites is well-known and many so-called autotrophic plants also absorb organic substances. Difference in habitat exploitation is often expressed in different periodicity. It seems practical to separate the well marked cases, like therophytes (annuals) and perennials, especially if the former appear only once in a number of years (semi-deserts).

The phenomena discussed in points (vii) and (viii) are all part of the broader phenomenon of the evolution of species toward difference in niche (position and manner of function in relation to other species within the community, see article 3). Synusiae comprise groups of species that have evolved toward similar positions, and uses of resources and time, in their communities. Synusiae can thus like the guilds of zoologists (ROOT 1967) be regarded as niche-groupings. Synusiae are of interest not only for descriptive analysis of communities but also for interpretation of community organization.

16.5.3 CONCLUSION

The author finds these arguments in favor of the synusial approach convincing. It is not the case that synusiae are so bound to one another and particular syntaxa, that study of the latter gives sufficient knowledge of the former. It would be convenient to phytosociologists if such were the case; but all the respects in which we have observed that synusiae differ from phytocoenoses and one another–distributional and temporal relations, structure and composition, environmental relations and function, and scale and suitable research methods–imply the appropriateness of accepting synusiae as research objects in their own rights.

The fact that synusiae are both parts of phytocoenoses, interacting with and in many cases dependent on other synusiae, and yet in distribution are relatively independent of particular phytocoenoses and other synusiae, is paradoxical. The paradox is the same as that which applies to species, as these are both interrelated in communities, and individualistic in distribution (WHITTAKER 1962, article 3). As species, despite their interactions, are distributed differently, each according to its own characteristics and ways of relating to environment, so are synusiae, as groups of species responding to different environmental factors, differently distributed. Both sides of the paradox are true, but there is no real contradiction; for the kinds of interactions that occur among species (or synusiae) do not (with some exceptions) imply that they should have the same distributions (WHITTAKER 1962). The significance of both interactions and relative distributional independence implies that both phytocoenoses and societies are appropriate objects of study. We thus decide for viewpoint (iv) among those above (16.5)– acceptance of both approaches to communities as useful and productive, with societies and phytocoenoses to be separately classified into independent hierarchies of synusiae and syntaxa (DU RIETZ 1930, 1932; BARKMAN 1958b, 1970; WILMANNS 1962, 1970).

Parts of communities that might be subject to synusial treatment form a continuous size spectrum from the canopy trees of the forest, to the bacteria of decay in dead microorganisms. They form a second spectrum of relative independence from the fully autotrophic and dominant trees, to the wholly dependent fungal parasites and pathogens. There is no evident way to draw a boundary limiting synusiae to larger fractions of communities or to species to some degree independent of other species. Only judgment and practicality can determine what synusiae are to be accepted as useful for research. Such judgment must exert some limit on the number of synusiae given formal recognition as taxonomic units, if overwhelming numbers of such units are not to be described. Apart from formal designation of units, however, there is no reason why a student should not divide the societies of a community as finely as his particular research makes appropriate. The problems of synusial taxonomy–of distinguishing among multidirectionally related and often intergrading objects and avoiding an excessive number of classes–are not different in character from the problems of the taxonomy of phytocoenoses.

Though the taxonomic problems of societies and communities are similar, their ecological relations are different. Much may be learned from study of synusiae, as phytosociology seeks means for more penetrating investigation of some aspects of communities than that based on syntaxa. As a spectrum permits us to analyse much about composition of light and wave lengths of absorption that is not evident in the color of the light, so synusiology permits us to analyse much about the composition of a community and the character of its species relationships that is not evident in the phytosociology of the community as a whole.

16.6 SUMMARY

An alternative to the study and classification of plant communities as wholes is the approach through societies and synusiae. A society is defined here as a part of a plant community with a specific floristic composition, consisting of plants that belong to the same layer (if layers can be distinguished), occupy the same microhabitat (if these are present) and have essentially the same periodicity and way of habitat exploitation. A synusia is an abstract unit grouping together similar societies that occur in a number of plant communities (phytocoenoses). Different societies of the same microhabitat are grouped together as microcommunities or microcoenoses (abstract type: microcoenon).

The study of synusiae was originated by GAMS (1918), and the influence of the idea spread through different schools of phytosociology with a proliferation of terms. The problems of criteria for defining synusiae and related units and of the appropriate concepts and terms, are discussed; and terms are summarized in Tables I and II. The principal unit for hierarchial classification of societies is the 'union,' defined by floristic composition and analogous to the association. Subordinate units may be termed 'subunion,' 'variant,' 'vicariant' and 'socion,' higher units 'federation,' 'ordulus,' and 'classicula.' The system of synusiae is thus broadly parallel to, but independent of, the system of units (syntaxa) for classification of phytocoenoses. It is proposed here to form the scientific names of the synusiae according to the rules generally accepted for syntaxa, i.e. -etosum for subunions, -etum for unions, -ion for federations, -etalia for orduli, -etea for classiculae.

Applications of synusial approaches are reviewed for (i) vascular plants on land, (ii) societies of epiphytes, (iii) aquatic communities and (iv) soil-inhabiting communities, and (v) as an example, a particular community, a juniper scrub association of the Netherlands and Northwest Germany that the author has studied in detail (unpublished). Though synusiae have had little application to vascular plants in land communities since the work of LIPPMAA, they have been of increasing importance in the study of other kinds of organisms and communities.

Synusiae that do not correspond in distribution to the syntaxa in which they occur may be termed 'syntaxonomically non-specific.' Such independence of distribution of synusiae, compared with syntaxa and with one another, is a general phenomenon of communities and a justification of the synusial approach. Some synusiae, however, are syntaxonomically specific; thus a particular synusia may be a character-union for a given association. Other justifications for the synusial approach include the relative ecological independence of many synusiae and the wide differences among synusiae of a given community in composition and structure, in ecological and functional relationships, and in the appropriate scales for samples to study them.

It is true both that synusiae may be related to one another by interactions among their species when they occur together in the same phytocoenoses, and that they are different or 'individualistic' in their responses to environment within the community and in their distributions. The study of synusiae is consequently a well-justified complement to the study of phytocoenoses and syntaxa. The analysis of a community through its synusiae permits study of relationships among species and kinds of communities that cannot generally be obtained through study limited to observation and classification of phytocoenoses as wholes.

REFERENCES

ALEKSANDROVA, V. D., – 1969 – Classification of Vegetation: Principles of Classification and Classification Systems of Various Phytocoenological Schools. (In Russian). Naukov, Leningrad. 275 pp.

ALLORGE, P., – 1922 – Les associations végétales du Vexin francais. *Revue gén. Bot.* 33: 481—544 etc.; 34: 71—79 etc.

ALMBORN, O. – 1948 – Distribution and ecology of some South Scandinavian lichens. *Bot. Notiser*, Suppl. 1 (2): 1—354.

ALMQUIST, E., – 1929 – Upplands vegetation och flora. *Acta phytogeogr. suec.* 1: 1—622.

ANDERSSON, O. – 1950 – Larger fungi on sandy grass heath and sand dunes in Scandinavia. *Bot. Notiser*, Suppl. 2 (2): 1—89.

ANDREAS, C. H., – 1950 – De Hygrophorusweide een associatie. *Fungus* 20. 66—68.

APINIS, A. E., – 1970 – Das Verhalten der Pilze in bestimmten Graslandgesellschaften. (Engl. summ.) In 'Gesellschaftsmorphologie (Strukturforschung),' ed. R. TÜXEN, *Ber. Symp. Int. Ver. Vegetationskunde*, Rinteln 1966: 172—186.

BALOGH, J., – 1938 – Biosoziologische Studien über die Spinnenfauna des Sashegy (Adler-Berg) bei Budapest. *Festschr. Strand* 4: 464—499.

BARKMAN, J. J., – 1958a – La structure du Rosmarineto-Lithospermetum helianthemetosum en Bas-Languedoc. *Blumea*, Suppl. 6: 113—136.
BARKMAN, J. J., – 1958b – Phytosociology and Ecology of Cryptogamic Epiphytes. Van Gorcum, Assen. 628 pp.
BARKMAN, J. J., – 1968 – Das synsystematische Problem der Mikrogesellschaften innerhalb der Biozönosen. (Engl. summ.) In 'Pflanzensoziologische Systematik,' ed. R. TÜXEN, *Ber. Int. Synp. Vegetationskunde*, Stolzenau/Weser 1964: 21—53.
BARKMAN, J. J., – 1969 – Epifytengemeenschappen. Ch. XIV in 'Plantengemeenschappen in Nederland,' by V. WESTHOFF & A. J. DEN HELD, pp. 272—286. Thieme, Zutphen.
BARKMAN, J. J., – 1970 – Enige nieuwe aspecten inzake het probleem van synusiae en microgezelschappen. (Germ. summ.) *Meded. Landbouwhogeschool Wageningen* 5: 85—116.
BOLLETER, R. – 1921 – Vegetationsstudien aus dem Weisstannental. *Jb. St. Gall. naturw. Ges.* 57 (2): 1—140.
BRAUN-BLANQUET, J., – 1928 – Pflanzensoziologie: Grundzüge der Vegetationskunde. 1st ed. Springer, Berlin. 330 pp. 2nd ed., Springer, Wien, 1951, 631 pp. 3rd ed., 1964, 865 pp.
CAIN, S. A., – 1936 – Synusiae as a basis for plant sociological field work. *Am. Midl. Nat.* 17: 665—672.
CAIN, S. A., & G. M. DE OLIVEIRA CASTRO – 1959 – Manual of Vegetation Analysis. Harper, New York. 325 pp.
CAIN, S. A. & A. J. SHARP – 1938 – Bryophytic unions of certain forest types of the Great Smoky Mountains. *Am. Midl. Nat.* 20: 249—301.
CLEMENTS, F. E., – 1905 – Research Methods in Ecology. Univ. Publ. Co., Lincoln. 334 pp.
CLEMENTS, F. E., – 1907 – Plant Physiology and Ecology. Holt, New York. 315 pp.
CLEMENTS, F. E., – 1916 – Plant succession: an analysis of the development of vegetation. *Publs Carnegie Inst.* Washington 242: 1—512.
CLEMENTS, F. E., – 1928 – Plant Succession and Indicators. Wilson, New York. 453 pp.
CLEMENTS, F. E., – 1936 – Nature and structure of the climax. *J. Ecol.* 24: 252—284.
COOKE, W. B., – 1955 – Fungi, lichens, and mosses in relation to vascular plant communities in eastern Washington. *Ecol. Monogr.* 25: 119—180.
DAUBENMIRE, R., – 1968 – Plant Communities. Harper & Row, New York. 300 pp.
DONSELAAR, J. VAN, – 1965 – An ecological and phytogeographic study of northern Surinam savannas. *Wentia* 14: 1—163.
DRUDE, O., – 1919 – Die Elementar-Assoziation im Formationsbilde. *Bot. Jb.* 55 (Beibl. 122): 45—82.
DU RIETZ, G. E. – 1930 – Classification and nomenclature of vegetation. *Svensk Bot. Tidskr.* 24: 489—503.
DU RIETZ, G. E., – 1932 – Vegetationsforschung auf soziationsanalytischer Grundlage. *Handb. biol. ArbMeth.* 11, 5: 293—480.
DU RIETZ, G. E., – 1936 – Classification and nomenclature of vegetation units 1930—1935. *Svensk Bot. Tidskr.* 30: 580—589.
DU RIETZ, G. E., – 1954 – Die Mineralbodenwasserzeigergrenze als Grundlage einer natürlichen Zweigliederung der nord- und mitteleuropäischen Moore. *Vegetatio* 5/6: 571—585.
DU RIETZ, G. E., – 1957 – Vegetation analysis in relation to homogeneousness and size of sample areas. *8me Congr. int. bot.*, Paris 1954, *C.-R. Séances* Sect. 7 et 8

(Suppl.): 24—35.
DU RIETZ, G. E., – 1965 – Biozönosen und Synusien in der Pflanzensoziologie. (Engl. summ.) In 'Biosoziologie,' ed. R. TÜXEN, *Ber. Int. Symp. Vegetationskunde*, Stolzenau/Weser 1960: 23—42.
ELLENBERG, H., – 1956 – Aufgaben und Methoden der Vegetationskunde. In 'Einführung in die Phytologie' by H. WALTER, Vol. 4, Pt. 1. Ulmer, Stuttgart, 136 pp.
FAVRE, J., – 1948 – Les associations fongiques des hauts-marais jurassiens et de quelques régions voisines. *Beitr. Kryptog. Flora Schweiz* 10 (3): 1—228.
FELFÖLDY, L. – 1941 – A debreceni Nagyerdö epiphyta vegetációja. (Germ. summ.: Die Epiphytenvegetation des Waldes 'Nagyerdö' bei Debrecen). *Acta geobot. hung.* 4 (1): 35—73.
FUKAREK, F., – 1961 – Die Vegetation des Darss und ihre Geschichte. *Pflanzensoziologie* 12: 1—321. Fischer, Jena.
GAMS, H. – 1918 – Prinzipienfragen der Vegetationsforschung: Ein Beitrag zur Begriffsklärung und Methodik der Biocoenologie. *Vjschr. naturf. Ges. Zürich* 63: 293—493.
GAMS, H., – 1927 – Von den Follatères zur Dent de Morcles: Vegetationsmonographie aus dem Wallis. *Beitr. geobot. Landesaufn. Schweiz* 15: 1—760.
GAMS, H., – 1936 – Beiträge zur pflanzengeographische Karte Österreichs. 1. Die Vegetation des Grossglocknergebietes. *Abh. zool.- bot. Ges. Wien* 16 (2): 1—79.
GLEASON, H. A. – 1936 – Is the synusia an association? *Ecology* 17: 444—451.
GRAHAM, V. O., – 1927 – Ecology of fungi in the Chicago region. *Bot. Gaz.* 83: 267—287.
GROSSHEIM, A. A. – 1930 – Zur Frage nach dem Zustandekommen der Pflanzendecke. *Beitr. Biol. Pfl.* 18: 225—268.
GROSSHEIM, A. A. & L. I. PRILIPKO, – 1929 – Geobotanical survey of the Karabakh Steppe, Azerbaijan (In Russian with Engl. summ.). *Trudy geobot. Obsled. Pastb. SSR Azerbaidzh.* 4: 3—130.
HALE, M. E. JR., – 1955 – Phytosociology of corticolous cryptogams in the upland forests of southern Wisconsin. *Ecology* 36: 45—63.
HARTOG, C. DEN, – 1959 – The epilithic algal communities occurring along the coast of the Netherlands. *Wentia* 1: 1—241.
HARTOG, C. DEN, – 1963 – Enige waterplantengemeenschappen in Zeeland. (Engl. summ.) *Gorteria* 1: 155—164.
HARTOG, C. DEN, & S. SEGAL, – 1964 – A new classification of the water-plant communities. *Acta bot. neerl.* 13: 367—393.
HOEK, C. VAN DEN, – 1960 – Groupements d'algues des étangs saumâtres méditerranéens de la côte française. *Vie et Milieu* 11 (3): 390—412.
HÖFLER, K., – 1937 – Pilzsoziologie. *Ber. dt. bot. Ges.* 55: 606—622.
HÖFLER, K., – 1959 – Ueber die Gollinger Kalkmoosvereine. *Sber. öst. Akad. Wiss.*, Math.-nat. Kl. 168: 541—582.
HOSOKAWA, T., – 1951 – On the nomenclature of Aerosynusia. *Bot. Mag.*, Tokyo 64: 107—111.
HOSOKAWA, T., – 1954 – On the *Campnosperma* forests of Kusaie in Micronesia, with special reference to the community units of epiphytes. *Vegetatio* 5/6:351—360.
HÜBSCHMANN, A. VON, – 1957 – Zur Systematik der Wassermoosgesellschaften. *Mitt. flor.-soz. ArbGemein*, Stolzenau, N.F. 6/7: 147—151.
HÜBSCHMANN, A. VON, – 1960 – Einige Ackermoos-Gesellschaften des nordwestdeutschen Gebietes und angrenzender Landesteile und ihre Stellung im pflanzensoziologischen System. *Mitt. flor.-soz. ArbGemein*, Stolzenau, N.F. 8: 118—123.
HUECK, H. J. – 1953 – Myco-sociological methods of investigation, *Vegetatio* 4 (2): 84—101.

HULT, R., – 1881 – Försök till analytisk behandling af växtformationerna. *Meddn Soc. Fauna Flora fenn.* 8: 1—155.

IVERSEN, J., – 1936 – Biologische Pflanzentypen als Hilfsmittel in der Vegetationsforschung — Ein Beitrag zur ökologischen Charakterisierung und Anordnung der Pflanzengesellschaften. *Medd. Skalling-Lab.*, København 4: 1—224.

IWATSUKI, Z. & S. HATTORI, – 1957 – Studies of the epiphytic moss flora of Japan. 8. The bryophytic communities in the *Pinus pumila* association of central Japan. *J. Hattori bot. Lab.* 18: 70—77.

JOVET, P., – 1949 – Le Valois, phytosociologie et phytogéographie. Soc. d'Ed. d'Enseign. Sup., Paris. 389 pp.

KALAMEES, K. – 1965 – On problems and methods in mycosociology (In Russian with Engl. summ.). *Akad. Nauk, Eston. SSR*, Tartu: 14—22.

KALAMEES, K., – 1968 – Mycocoenological methods based on investigations in the Estonian forests. *Acta Mycol.* 4 (2): 327—335.

KATZ, N., – 1929 – Die Zwillingsassoziationen und die homologen Reihen in der Phytosoziologie. *Ber. dt. bot. Ges.* 47: 154—164.

KELLER, B., – 1932 – Die Methoden zur Erforschung der Ökologie der Steppen- und Wüstenpflanzen. *Handb. biol. ArbMeth.* 11, 6: 1—128.

KERNER VON MARILAUN, A., – 1863 – Das Pflanzenleben der Donauländer. Wagner, Innsbruck. 348 pp.

KLEMENT, O., – 1931 – Zur Flechtenflora des Erzgebirges. Die Umgebung von Komotau. *Beih. bot. Zbl.* 48 (2): 52—96.

KLEMENT, O., – 1955 – Prodromus der mitteleuropäischen Flechtengesellschaften *Beih. Repert. Spec. nov. Regni veg.* 135: 5—194.

KOCH, W., – 1925 – Die Vegetationseinheiten der Linthebene unter Berücksichtigung der Verhältnisse in der Nordostschweiz. *Jb. St. Gall. naturw. Ges.* 61 (2): 1—146.

KOPPE, F., – 1955 – Moosvegetation und Moosgesellschaften von Altötting in Oberbayern. *Reprium nov. Spec. Regni veg.* 58 (1/3): 92—144.

KORCHAGIN, A. A., – 1964 – Synusial structure of forest communities. *Proc. X Int. Bot. Congr.*, Edinburgh 1964, Abstracts, p. 403.

KORJINSKY, S. J., – 1888 – La limite nord de la région du tschernosjom dans la partie orientale de la Russie Européenne (Russian). *Trudy Obshch. Estest. imp. kazan. Univ.* 18 (5): 1—253, 22 (6): 1—201.

KORNAŚ, J., – 1959 – Sea bottom vegetation of the Bay of Gdańsk off Rewa. *Bull. Acad. pol. Sci.*, Cl. II, Sér. Sci. biol. 7: 5—10.

KORNAŚ, J. & A. MEDWECKA-KORNAŚ, – 1950 – Associations végétales sousmarines dans le Golfe de Gdańsk. *Vegetatio* 2: 120—127.

KOSKINEN, A. – 1955 – Ueber die Kryptogamen der Bäume, besonders die Flechten, im Gewässergebiet des Päijänne sowie an den Flüssen Kalajoki, Lestijoki und Pyhäjoki. Floristische, soziologische und ökologische Studie I. Helsinki. 176 pp.

KRAJINA, V., – 1933 – Die Pflanzengesellschaften des Mlynica-Tales in den Vysoke Tatry (Hohe Tatra). *Beih. bot. Zbl.*, Abt. 2, 50: 774—957, 51: 1—224.

KRUSENSTJERNA, E. VON, – 1945 – Bladmossvegetation och bladmossflora i Uppsalatrakten. (Engl. summ.) *Acta phytogeogr. suec.* 19: 1—250.

LANGE, M., – 1948 – The agarics of Maglemoose: study in the ecology of the agarics. *Dansk bot. Ark.* 13 (1): 1—141.

LIPPMAA, T., – 1931 – Pflanzensoziologische Betrachtungen. *Acta Inst. Horti bot. tartu.* 2 (3/4): 1—32, and *Loodusuur. Seltsi Aruanded* 38 (1/2): 1—32.

LIPPMAA, T., – 1933a – Aperçu général sur la végétation autochtone du Lautaret (Hautes-Alpes) avec des remarques critiques sur quelques notions phytosocio-

logiques. (Eston. summ.) *Acta Inst. Horti bot. tartu.* 3 (3): 1—108.
LIPPMAA, T., – 1933b – Taimeühingute uurimise metoodika ja eesti taimeühingute klassifikatsiooni põhijooni (Eston. with Germ. summ.: Grundzüge der pflanzensoziologischen Methodik nebst einer Klassifikation der Pflanzenassoziationen Estlands). *Acta Inst. Horti bot. tartu.* 3 (4): 1—169.
LIPPMAA, T., – 1935a – La méthode des associations unistrates et le système écologique des associations. *Acta Inst. Horti bot. tartu.* 4 (1/2, art. 3): 1—7.
LIPPMAA, T., – 1935b – Une analyse des forêts de l'île estonienne d'Abruka (Abro) sur la base des associations unistrates. *Acta Inst. Horti bot. tartu.* 4 (1/2, art. 5): 1—97.
LIPPMAA, T., – 1935c – La méthode des associations unistrates et le système écologique des associations. *Proc. Zesde Int. Bot. Congr.*, Amsterdam 1935, 2: 109—110.
LIPPMAA, T., – 1938 – Areal und Altersbestimmung einer Union (*Galeobdolon-Asperula-Asarum-U.*) sowie das Problem der Charakterarten und der Konstanten. *Acta Inst. Horti bot. tartu.* 6 (2): 1—152.
LIPPMAA, T., – 1939 – The unistratal concept of plant communities (the unions). *Am. Midl. Nat.* 21: 111—145.
LOHMEYER, W. c.s., – 1962 – Contribution à l'unification du système phytosociologique pour l'Europe moyenne et nord-occidentale. *Melhoramento* 15: 137—151.
LORENZ, J. R., – 1858 – Allgemeine Resultate aus der pflanzengeographischen und genetischen Untersuchung der Moore im präalpinen Hügellande Salzburgs. *Flora* 41: 209—221, 225—237, 241—253, 273—286, 289—302, 344—355, 360—376.
LUTHER, H., – 1949 – Vorschlag zu einer ökologischen Grundeinteilung der Hydrophyten. *Acta bot. fenn.* 44: 1—15.
MAAREL, E. VAN DER, – 1965 – Beziehungen zwischen Pflanzengesellschaften und Molluskenfauna. (Engl. summ.) In 'Biosoziologie,' ed. R. TÜXEN, *Ber. Symp. Int. Ver. Vegetationskunde*, Stolzenau/Weser 1960: 184—198.
MARGALEF, R., – 1947 – Los methodos para la investigacion de las comunidades acuaticas adnadas y especialmente las formadas por organismos microscopicos (perifiton, pecton). *Collnea bot.*, Barcinone 1 (3): 247—259.
MARGALEF, R. – 1949 – Las asociaciones de algas en las aguas dulces de pequeño volumen del Noreste de España. (Engl. summ.) *Vegetatio* 1 (4/5): 258—284.
MARGALEF, R., – 1955 – Comunidades bióticas de las aguas dulces del noroeste de España. (Engl. summ.) *Publnes. Inst. Biol. apl.*, Barcelona 21: 5—86.
MEISEL-JAHN, S. & W. PIRK, – 1955 – Über das soziologische Verhalten von Pilzen in Fichten-Forstgesellschaften. *Mitt. flor.-soz. ArbGemein.*, Stolzenau, N.F. 5: 59—63.
MÖRZER BRUINS, M. F., – 1947 – Over levensgemeenschappen. (Engl. summ.) Diss. Kluwer, Deventer. 195 pp.
MOTYKA, J., – 1925 – Die Pflanzenassoziationen des Tatra-Gebirges. II. Die epilithischen Assoziationen der nitrophilen Flechten im polnischen Teile der Westtatra. *Bull. int. Acad. Pol. Sci. Lett.*, Cl. Sci. Math. Nat., Sér. B, 1924: 835—850.
NAKANISHI, S., – 1970 – The bryophyte-lichen vegetation of Mt. Hakusan area. *Scient. Stud. Hakusan Nation. Park* Japan, pp. 174—200. Ishikawa-ken, Japan.
OBERDORFER, E., – 1954 – Ueber Unkrautgesellschaften der Balkanhalbinsel. *Vegetatio* 4 (6): 379—411.
OBERDORFER, E., – 1957 – Süddeutsche Pflanzengesellschaften. *Pflanzensoziologie* 10: 1—564. Fischer, Jena.
OCHSNER, F., – 1928 – Studien über die Epiphytenvegetation der Schweiz. *Jb.*

St. Gall. naturw. Ges. 63 (2): 1—106.
OCHSNER, F., – 1952 – Moose in den Pflanzengesellschaften des Languedoc.*Ber. schweiz. bot. Ges.* 62: 106—122.
OMURA, M., – 1952 – A synecological study on the epiphyte communities of lichens in the subalpine forests of Titibu mountain region in Japan. *Bull. Soc. Pl. Ecol.*, Tokyo 1 (4): 176—181.
OMURA, M., – 1953 – Synecological studies on the epiphyte communities of lichens in the temperate forests of the Iide and Titibu mountain regions in Japan. *Mem. Fac. Sci., Kyushu Univ.*, Ser. E, 1 (3): 147—157.
PASTAK, E., – 1935 – Harilaiu Taimkate (Engl. summ.: The vegetation of the peninsula of Harilaid, Estonia). *Acta Inst. Horti bot. tartu.* 5 (1/2, art. 3): 1—44.
PHILIPPI, G., – 1956 – Einige Moosgesellschaften des Südschwarzwaldes und der angrenzenden Rheinebene. *Beitr. naturk. Forsch. Südw. Dtl.* 15 (2): 91—124.
PHILIPPI, G., – 1963 – Zur Kenntnis der Moosgesellschaften saurer Erdraine des Weserberglandes, des Harzes und der Rhön. *Mitt. flor.-soz. ArbGemein.* Stolzenau, N.F. 10: 92—108.
PHILLIPS, E. A., – 1951 – The associations of bark-inhabiting bryophytes in Michigan. *Ecol. Monogr.* 21: 301—316.
PIRK, W., – 1950 – Pilze in Moosgesellschaften auf Brandflächen. *Mitt. flor.-soz. ArbGemein.*, Stolzenau, N.F. 2: 3—5.
PIRK, W., – 1952 – Die Pilzgesellschaft der Baumweiden im mittleren Wesertal. *Mitt. flor.-soz. ArbGemein.*, Stolzenau, N.F. 3: 93—96.
PIRK, W. & R. TÜXEN, – 1949 – Das Coprinetum ephemeroides, eine Pilzgesellschaft auf frischem Mist der Weiden im mittleren Wesertal. *Mitt. flor.-soz. ArbGemein.*, Stolzenau, N.F. 1: 1—7.
PIRK, W. & R. TÜXEN, – 1957 – Das Trametetum gibbosae, eine Pilzgesellschaft modernder Buchenstümpfe. *Mitt. flor.-soz. ArbGemein.*, Stolzenau, N.F. 6/7: 120—126.
POORE, M. E. D., – 1955 – The use of phytosociological methods in ecological investigations. I. The Braun-Blanquet system. *J. Ecol.* 43: 226—244.
RAUNKIAER, C., – 1934 – The Life Forms of Plants and Statistical Plant Geography. Clarendon, Oxford. 632 pp.
REGEL, C., – 1923 – Assoziationen und Assoziationskomplexe der Kola Lappmark. *Bot. Jb.* 58: 607—635.
REGEL, C., – 1927 – Zur Klassifikation der Assoziationen der Sandböden. *Bot. Jb.* 61: 263—284.
RICHARDS, P. W., – 1939 – The bryophyte communities of a Killarney oakwood. *Annls. bryol.* 9: 108—130.
RICHARDS, P. W. – 1952 – The Tropical Rain Forest: An Ecological Study. Cambridge Univ., New York, 423 pp.
ROOT, R. B., – 1967 – The niche exploitation pattern of the blue-gray gnatcatcher. *Ecol. Monogr.* 37: 317—350.
RÜBEL, E., – 1925 – Vorschläge zur Untersuchung von Buchenwälder. *Veröff. geobot. Inst. Rübel*, Zürich, Beibl. 3: 1—35.
RÜBEL, E., – 1933 – Versuch einer Uebersicht über die Pflanzengesellschaften der Schweiz. *Ber. geobot. Inst. Rübel*, Zürich 1932: 19—30.
RYDZAK, J., – 1961 – Tree lichens in the forest communities of the Białowieża National Park. (Polish & Russ. summ.s) *Annls Univ. Mariae Curie-Skłodowska*, Lublin, C 16 Sect. C. Biol., 16 (2): 17—47.
SCAMONI, A., – 1963 – Einführung in die praktische Vegetationskunde. 2nd ed. Fischer, Jena. 236 pp.
SCHEYGROND, A., – 1931 – Het plantendek van de Krimpenerwaard. IV. Sociographie van het hoofd-associatie-complex Arundinetum-Sphagnetum. *Ned.*

kruidk. Archf. 1932 (1): 1—184.

SCHMID, E., – 1941 – Vegetationsgürtel und Biocoenose. *Ber. schweiz. bot. Ges.* 51: 461—474.

SCHMITHÜSEN, J., – 1959 – Allgemeine Vegetationsgeographie. In 'Lehrbuch der allgemeinen Geographie,' E. Obst, vol. 4: 1—261. de Gruyter, Berlin.

SCHMITZ, W. – 1965 – Die Soziologie aquatischer Mikrophyten. (Engl. summ.) In 'Biosoziologie,' ed. R. TÜXEN, *Ber. Symp. Int. Ver. Vegetationskunde*, Stolzenau /Weser 1960: 120—139.

SCHUMACHER, A. – 1944 – Ueber Calypogeia arguta im Bergischen Lande. Beil. z. 14. *Rundbr. Zentralstelle für Vegetationskartierung*, Stolzenau.

SCHWEERS, A. C. A. – 1949 – De Hygrophorusweide, een associatie. *Fungus* 19: 17—18.

SCOTT, G. A. M. & J. M. ARMSTRONG, – 1966 – The altitudinal sequence of climax vegetation on Mt. Anglem, Stewart Island. *N.Z. J. Bot.* 4 (3): 283—299.

SIRGO, V. – 1936 – Emajõe alamjooksul Peipsiäärsel madalikul asuvaist taimeühinguist (Engl. summ.: Plant unions of the swamps at the mouth of the River Emajogi). *Acta Inst. Horti bot. tartu.* 5 (1/2, art. 4): 1—64.

ȘTEFUREAC, T. I., – 1941 – Cercetări sinecologice şi sociologice asupra Bryophytelor din codrul secular Slatioara (Bucovina). (French summ.) *Anal. Acad. rom.*, București, ser. III, 16 (Mem. 27): 1—197.

SULMA, T., – 1935 – Beiträge zur Ökologie und Verbreitung der Flechten auf dem Lubliner Hügelland. *Bull. Int. Acad. pol. Sci. Lett.*, Cl. Sc. Math. Nat., Sér. B (1), 1935 (1—3): 77—100.

SZCZAWINSKY, A., – 1953 – Corticolous and lignicolous plant communities in the forest associations of the Douglas Fir forest on Vancouver Island. Thesis, Univ. of Brit. Columbia, Vancouver. 283 pp.

TOMILIN, B. A. – 1964 – Environmental factors affecting the distribution of fungi among plant communities (In Russian with Engl. summ.) *Bot. Zh. SSSR* 49 (2): 230—239.

TOMSON, A., – 1937 – Sõrve taimkate (French summ.: La végétation de la presqu' île de Sõrve, Estonie). *Acta Inst. Horti bot. tartu.* 6 (1, art. 1): 1—87.

TÜXEN, R., – 1937 – Die Pflanzengesellschaften Nordwestdeutschlands. *Mitt. flor.-soz. ArbGemein. Niedersachsen* 3: 1—170. *Jber. naturh. Ges. Hannover*, 1929/30-1935/36, 81—87: 1—170.

TÜXEN, R., – 1950 – Grundriss einer Systematik der nitrophilen Unkrautgesellschaften in der Eurosiberischen Region Europas. *Mitt. flor.-soz. ArbGemein.*, Stolzenau, N.F. 2: 94—175.

TÜXEN, R., – 1964-6 – Bibliographia Phytosociologica Cryptogamica. Pars II: Bibliographia mycosociologica I, II. *Excerpta bot.* B, 6 (2): 135—160; 6 (3): 161—178, 7 (3): 220—224.

TÜXEN, R. (ed.), – 1965 – Biosoziologie. *Ber. Symp. Int. Ver. Vegetationskunde*, Stolzenau/Weser, 1960, 350 pp. Junk, The Hague.

TÜXEN, R. (ed.), – 1968 – Pflanzensoziologische Systematik. *Ber. Symp. Int. Ver. Vegetationskunde*, Stolzenau/Weser, 1964, 346 pp. Junk, The Hague.

TÜXEN, R., A. VON HÜBSCHMANN, & W. PIRK, – 1957 – Kryptogamen- und Phanerogamen-Gesellschaften. *Mitt. flor. – Soz. Arb. Gemein.*, Stolzenau/Weser, N.F. 6/7: 114—118.

UBRISZY, G. DE, – 1940 – Adatok a Nyírség gombavegetációjának ismeretehéz. (French summ.: Contribution à la connaissance de la végétation mycologique de Nijirseg). *Acta geobot. hung.* 3: 65—78.

UBRISZY, G. DE, – 1956 – Ujabb vizsgálatok az erdötipusok talajlakó nagygombáinak társulasi viszonyairól. (Germ. summ.: Neuere Untersuchungen über die Cönologie bodenbewohnender Grosspilze der Waldtypen). *Acta bot. hung.*

2: 391—424.
VAGA, A., - 1940 - Fütotsönoloogia põhiküsimusi (Engl. summ.: On some fundamental problems in phytocoenology). *Acta Comment Univ. tartu*, Ser. A, 35 (6): 1—152.
VILLERET, S., - 1954 - Contribution à la biologie des Algues des tourbières a Sphaignes. *Bull. Soc. scient. Bretagne* 29: 1—246.
VILLERET, S., - 1965 - Quelques aspects biologiques de la répartition des Algues d'eau douce. (Engl. summ.) In 'Biosoziologie,' ed. R. TÜXEN, *Symp. Int. Ver. Vegetationskunde* Stolzenau/Weser 1960: 140—149.
VRIES, D. M. DE, - 1926 - Het plantendek van de Krimpenerwaard I. Phytosociologische beschouwingen: Begrippen, wetten, bouwbeschrijvende methodiek. *Ned. Kruidk. Archf.* 1925: 215—275.
VRIES, D. M. DE, - 1932 - Grondslag van een Nederlandse plantensociographische naamgeving. *Ned. kruidk. Archf.* 1931: 517—527.
WALDHEIM, S., - 1944 - Mossvegetationen i Dalby-Söderskogs Nationalpark. *K. Svenska VetenskAkad. Avh. Naturskydd.* 4: 1—142.
WALDHEIM, S., - 1947 - Kleinmoosgesellschaften und Bodenverhältnisse in Schonen. *Bot. Notiser*, Suppl. 1 (1): 1—203.
WARMING, E., - 1909 - Oecology of Plants: An Introduction to the Study of Plant Communities. Oxford Univ., Oxford. 422 pp.
WESTERDIJK, J., - 1949 - The concept 'association' in mycologie. *Antonie van Leeuwenhoek* (J. Microbiol. Serol.) 15: 187—189.
WESTHOFF, V., - 1968 - Einige Bemerkungen zur syntaxonomischen Terminologie und Methodik, insbesondere zu der Struktur als diagnostisches Merkmal. (Engl. summ.) In 'Pflanzensoziologische Systematik,' ed. R. TÜXEN, *Ber. Symp. Int. Ver. Vegetationskunde*, Stolzenau/Weser 1964: 54—70.
WESTHOFF, V. & A. J. DEN HELD, - 1969 - Plantengemeenschappen in Nederland. Thieme, Zutphen. 324 pp.
WHITTAKER, R. H., - 1962 - Classification of natural communities. *Bot. Rev.* 28: 1—239.
WILMANNS, O., - 1958 - Zur standörtlichen Parallelisierung von Epiphyten- und Waldgesellschaften. *Beitr. naturk. Forsch. SüdwDtl.* 17 (1): 11—19.
WILMANNS, O., - 1959 - Epiphytengesellschaften Nordgriechenlands im Vergleich mit denen Mitteleuropas. *Phyton* 8 (1/2): 175—182.
WILMANNS, O., - 1962 - Rindenbewohnende Epiphytengemeinschaften in Südwestdeutschland. *Beitr. naturk. Forsch. SüdwDtl.* 21 (2): 87—164.
WILMANNS, O., - 1970 - Kryptogamen-Gesellschaften oder Kryptogamen-Synusien? In 'Gesellschaftsmorphologie (Strukturforschung),' ed. R. TÜXEN, *Ber. Symp. Int. Ver. Vegetationskunde*, Rinteln 1966: 1—7.
WILMANNS, O. & H. BIBINGER, - 1966 - Methoden der Kartierung kleinflächiger Kryptogamengemeinschaften. *Bot. Jb.* 85 (3): 509—521.
WOLAK, J., - 1970 - Complexe horizontal et complexe vertical. (Germ. & Engl. summs.) In 'Gesellschaftsmorphologie (Strukturforschung),' ed. R. TÜXEN, *Ber. Symp. Int. Ver. Vegetationskunde*, Rinteln 1966: 136—141.

17 RUSSIAN APPROACHES TO CLASSIFICATION OF VEGETATION

Vera D. Aleksandrova

Contents

17.1	Introduction: Brief Historical Survey	169
17.1.1	The Early Period	169
17.1.2	Development of Modern Viewpoints	170
17.1.3	More Recent Trends	174
17.2	Objects of Classification	175
17.2.1	Phytocoenoses	175
17.2.2	Synusiae	177
17.2.3	Complexes and Continua	179
17.3	Units of Classification	180
17.3.1	Associations	180
17.3.2	Formations	184
17.3.3	Higher Units	185
17.4	Summary-Conclusion	189

17 RUSSIAN APPROACHES TO CLASSIFICATION OF VEGETATION

17.1 Introduction: Brief Historical survey

17.1.1 THE EARLY PERIOD

The first suggestions of classifying vegetation in Russia can be found in works of early naturalists (TATISHCHEV 1740, see TATISHCHEV 1950, KRASHENINNIKOV 1755, ENGELMANN 1810, and others) and in papers by travelers–geographers and floristic botanists of the middle of the XIX century (RUPRECHT 1845, TRAUTVETTER 1849-1851, BORSHCHEV 1865, MIDDENDORF 1867, and others). Typological characteristics of vegetation were considered at that time in only a fragmentary manner, and usually in connection with floristic and general description of particular regions. Appearance of investigations which focused special attention upon geobotanical classification dates back to the 1880's when these problems were recognized primarily in two areas–forest typology and studies of steppe vegetation.

The peculiarities of forest typology in Russia were due to the demands of forestry activities (GENKO 1889, see GENKO 1902, GUTOROVICH 1897, MOROZOV 1912, SEREBRENNIKOV 1913, KRÜDENER 1916, and others). The result was an essentially ecological classification: a 'forest type' comprised forest stands grouped on the basis of similarities in the envoronment and resemblance in stand quality. A certain forest type consequently might include forest stands with different dominant tree species. At the same time Russian classification of forests was being influenced by foreign scientists, thus GORDYAGIN (1900) employed a method similar to that of the Uppsala school in his investigations of the West Siberia forests. Along with Russian terms he used Latin terminology in defining forest types and proposed for every association 'a principal group of species' on the basis of species constancy.

Classification of steppe vegetation (KRASNOV 1886 et al. KORZHINSKI 1888 et al. TANFILYEV 1897 et al., KOSTYCHEV 1900, GORDYAGIN 1900-1, and others) was developing at this time principally in connection with the zonal division of the vast steppe areas of Russia. Terms like 'formation', 'association', and 'vegetation type' were used; but no real classification was expressed thereby, for these terms were used to designate any plant communities.

In 1902 FLEROV & FEDCHENKO published their classification, comprising all vegetation types of the European part of Russia. The major unit of classification was 'a group of plant communities'; and several 'groups of communities' were suggested: 1) water group, 2) mire group, 3) forest group, 4) steppe group, 5) meadow group, 6) bare slopes group, and 7) cultured (anthropogenic) group. Each 'group' was further subdivided into 'communities' or 'formations' (these terms were employed as synonyms). Classification was based upon environmental relations of communities and on the dominant species; dynamics of vegetation was also considered. However, no theoretical justifications for use of taxonomic categories and definition of units of vegetation were proposed.

17.1.2 DEVELOPMENT OF MODERN VIEWPOINTS

The fundamentals for developing a theoretical background for Russian classification of vegetation were established by SUKACHEV and RAMENSKI. They both expressed their views (at the same time) in December 1909 in reports in Moscow (SUKACHEV 1910, RAMENSKI 1910); SUKACHEV proceeded from the idea of discrete units of vegetation, whereas RAMENSKI conceived vegetation as continuous. RAMENSKI (1915-1953) developed further his theory of vegetative cover taken as a continuum and delineated a variety of quantitative approaches to studies of vegetation, including an ingenious method of ordination (more detailed description thereof can be found in article 5 in this volume).

SUKACHEV repeated in his report (1910) the definitions of community and formation offered in his earlier paper 'Forest formations and their relationships in the Bryansk woods' (1908): 'The plant community ('Einzelbestand' of SCHRÖTER, 'association locale' of FLAHAULT, forest stand in forestry) is taken to mean an association of organisms in which mutual influence is exerted by plants on one another and between plants and the environment. The plant formation ('Bestandtypus' of SCHRÖTER, forest stand type in forestry) comprises those plant communities of a given geographic region which are characterized by essentially the same ecology, i.e. displaying essentially similar interrelations between plants as well as between the latter and the environment' (SUKACHEV 1908: 5-6). In his report in 1909 SUKACHEV (1910: 150) repeated almost literally this definition, already referred to as characteristic of the 'Leningrad school', and implying simultaneously in the single term 'community' (in the middle 1930's and later, 'phytocoenose') both the chorological unit of the plant cover (stand, 'Einzelbestand') and the phytosociological concept of associated growth of plants (community, 'Pflanzengemeinschaft'). After the appearance of the proceedings of the Brussels Congress (FLAHAULT & SCHRÖTER 1910), SUKACHEV began to use in the latter sense the term 'association' instead of 'formation.'

In 1915 SUKACHEV wrote a paper containing a first suggestion of a coordinate scheme of plant associations in relation to environmental gradients, which was further developed in 1927 and in other works (Fig. 1). This method gained much support from Russian scientists and is widely employed in classification of forests and other vegetation types (cf. Fig. 2). SUKACHEV (1917, 1928, et al.) proposed a hierarchic system of taxonomic categories for communities the lowest unit of which was the association, the main unit of middle rank the formation, while the broadest unit was the vegetation type. The terminology proposed by SUKACHEV, and the meanings he implied thereby, have become most widely accepted in Soviet geobotany.

The classification of vegetation has been further enriched with the introduction of the notion of 'placor' by VYSOTSKI (1909, 1930, et al.). The word is derived from Greek ($πλάξ$ or $πλάκα$ and $ορος$) – upland habitats which are adequately drained under given macro-

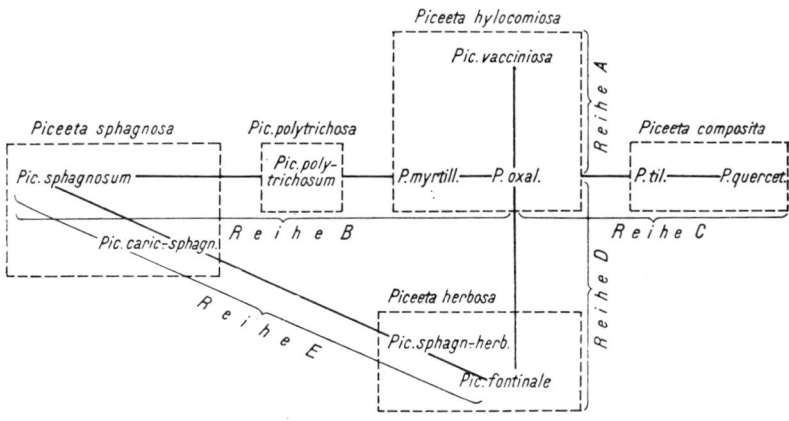

Fig. 1. A compound ecological series for spruce forest (Piceeta) according to SUKACHEV (1928: 218, fig. 48, cf. 1932).P.oxal – Piceetum oxalidosum, P.til. – Piceetum tiliosum, P. quercet. – Piceetum quercetosum, P. sphagn.- herb.– Piceetum sphagnoso-herbosum, P. myrtill. – Piceetum myrtillosum, Pic. caric.-sphagn. – Piceetum caricoso-spagnosum. Spruce forest with oxalis (Piceetum oxalidosum) is the central or climax type, from this radiate series ('Reihe'): A – toward drier, nutrient-poor soils; B – toward wetter soils with stagnant water; C – toward soils with more favorable nutrient content; D – toward wet soils with moving water; E – connects the latter with the bog forests.

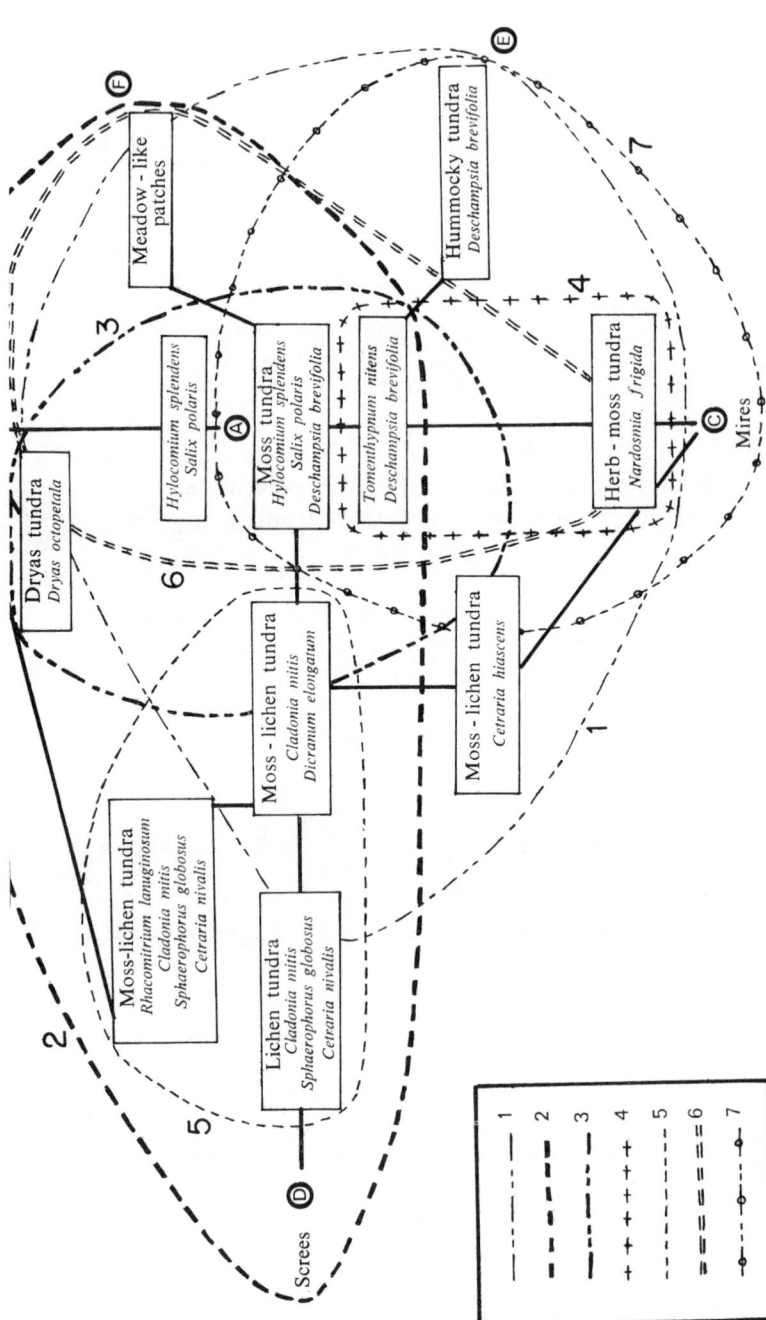

Fig. 2. A compound ecological series for tundra communities, and the distributional areas occupied by some plant species on the Southern Islands of Novaya Zemlya (ALEKSANDROVA 1956: 231, Fig. 10). The moss tundra community at *A* is central to the whole pattern, from this radiate ecological series: *AB* – toward drier sites with a thinner snow cover; *AC* – toward wet, poorly drained sites with heavier snow cover; *AD* – toward screes and rockslopes, showing the effect of increase in

climatic conditions. Making use of this definition and referring to ideas expressed by SIBIRTSEV (1900) on 'zonal' soils, PACZOSKI (1915) introduced notions of zonal, azonal and extrazonal vegetation; ALEKHIN (1915: 409) independently of the latter proposed the terms zonal and intrazonal. 'Zonal' vegetation, occupying placors, is analogous to the climax concept in the sense of CLEMENTS. 'Azonal' vegetation (which can be seen in nonplacoric habitats of any zone, e.g. flood plain meadows) and 'intrazonal' vegetation (which occurs only within certain zones, e.g. sphagnum bogs) are together loosely analogous to the subclimax concept of CLEMENTS. 'Extrazonal' vegetation meets its analogue in the terms pre- and postclimax (CLEMENTS 1936); this is vegetation 'spreading outside its zone' (ALEKHIN 1936: 366), e.g. forest stands on north-facing slopes in the steppe zone. However, in contrast to CLEMENTS, Russian authors used these concepts without requiring successional relationships between placoric and non-placoric formations.

ALEKHIN, who was in controversy with SUKACHEV on a number of points, formed the 'Moscow school' in Russian geobotany including DOKHMAN, KATS, URANOV and others. Disagreements with SUKACHEV involved both concepts and terminology; ALEKHIN argued in particular against the double meaning of the term 'community' as defined by SUKACHEV. Scientists of the Moscow school extensively studied the methodology of investigation and treatment of data. The significance of the means of detecting analytical characteristics of plant communities was emphasized (ALEKHIN 1932, et al.); and a critical evaluation in practice was applied to the methodological recommendations of the Uppsala school concerning the constancy concept, minimal area and others. KATS (1933, et al.) contributed to the studies of the association (= sociation) as a small, homogeneous unit defined on the basis of the dominants of layers. DOKHMAN (1930, 1936, et al.) developed principles for detecting and classifying of community complexes.

In the 1920's in the Ukraine POGREBNYAK (1929, et al.) investigated classification in his own way, basing his classification on ordinating communities along gradients of two factors: moisture and nutrient condition of soil. The forest type, as conceived by representatives of this school (cf. the Finnish school), comprises forest sites with the same climax-type and all successional stands, including areas without any trees (ploughed fields and the like).

The system of concepts and ideas for classification was developing in Russia under the primary influence of Russian ecological and genetical traditions, derived especially from ideas of the great Russian naturalist DOKUCHAEV which influenced MOROZOV, SUKACHEV, and their contemporaries. On the other hand, there was marked influence of the phytocoenologists of northern Europe, particularly those of the Uppsala school (HULT, SERNANDER, DU RIETZ, and others), and WARMING, as well as CAJANDER, SCHRÖTER, DRUDE, BROCKMANN-JEROSCH, and RÜBEL. The American trend of classifying of vegetation by dynamics as developed by COWLES and CLEMENTS, and the approaches to classification offered by BRAUN-BLANQUET in the southern Europe, were not popular among Russian geobotanists at that time. The development of principles of classification of vegetation in Russia was influenced also by the vast territories of the country and the great variety of vegetation found there.

In 1934 a discussion took place in Leningrad on fundamental problems in phytocoenology. Leading Soviet phytocoenologists participated in this: SUKACHEV, ALEKHIN, SHENNIKOV, OVCHINNIKOV, and others. Reports of this discussion were published in *Soviet Botany*, 5, 1935. The principal attention was given to discussion of concepts and terms pertaining to the main purposes of phytocoenology, and the meaning of the lower units of classification of vegetation; the different approaches of the 'Leningrad school' (SUKACHEV) and 'Moscow school' (ALEKHIN) to these problems were defined and expressed. In these years a series of works appeared concerning major units of vegetation classification (KOROVIN 1934, VASILYEV 1935, GORODKOV 1935, ILYNSKI 1935, PROZOROVSKI 1935, SHENNIKOV 1935, ALEKHIN 1936, 1938, and others). In these same years (the 1930's) a system of ideas emerged as characteristic of vegetation clasification in Soviet geobotany. Despite some still existing controversy, there are common traditional ideas shared by all Soviet geobotanists to this day that differ from those of foreign schools.

17.1.3 MORE RECENT TRENDS

In the 1940's [1]) LAVRENKO used principles of classification based upon dominant ecological variants of the growth-forms (ecobiomorphs) for classification of steppes (1940), compilation of small-

---

[1]) In these years SUKACHEV began to write on the biogeocoenose (SUKACHEV 1942, 1945, et al.). This concept cannot be directly employed in classification of vegetation, since the biogeocoenose is not the subject of phytocoenology but of a different discipline – biogeocoenology – that concerns the biogeocoenotic geosphere (SUKACHEV 1964), i.e. the ecosphere in the sense of COLE (1958). A biogeocoenose comprises along with the phytocoenose the animals thereof, and the soil, atmospheric gases and water in a state of mutual exchange with the living organisms. 'Biogeocoenose' can be considered either as a synonym of the holocoen in the sense of FRIEDERICHS (1930), or as an ecosystem within the boundaries of one phytocoenose (of one stand) (LAVRENKO & DYLIS 1968).

scale maps of vegetation (1941, et al.), and regional geobotanic characterization of the USSR (1947). These years saw also the development of genetical classification (LESKOV 1943, SOCHAVA 1944, 1945, OVCHINNIKOV 1947, et al.). Of particular importance was the introduction of the phratria concept (SOCHAVA 1944, et al.), extensively used by SOCHAVA and his school in classification of vegetation and vegetation mapping.

Towards the end of the 1950's publications appeared resuming discussion of the vegetation continuum (PONYATOVSKAYA 1959, 1961, VASILEVICH 1960, 1966a, 1968a, 1969, RABOTNOV 1963, ALEKSANDROVA 1965, 1969, TRASS 1966, FREY 1967, DOLAKHANOV 1970, and others). In the 1960's there has been a tendency toward revision of traditional ideas on classification and the concepts of association and formation, and toward exploring new methods of classification (VASILEVICH 1963b, 1966b, 1968c, 1969, 1971, NITSENKO 1963, 1966, MIRKIN 1965, TRASS 1965, 1971, IPATOV 1966, 1969, ISACHENKO 1967, 1969, NORIN 1966, 1970, FREY 1966b, 1967, 1971, KARAMYSHEVA 1967, SOCHAVA 1968, 1970, KARAMYSHEVA & RACHKOVSKAYA 1968, RYSIN & KOVALENKO 1968, APALIA 1968, 1971, LUKICHEVA & SABUROV 1971, and others). Further publications deal with quantitative geobotany and application of mathematical methods to classification of vegetation (VASILEVICH 1960, 1963a, 1967a, 1967b, 1969, 1971, FREY 1963, 1967, 1968, 1971, NESHATAEV 1968, 1971a, VOLKOVA 1969, SAMOYLOV 1970, NORIN 1971, MIRKIN 1970a, and others).

In order to describe Russian approaches to classification of vegetation further, the discussion will divided as follows: the objects of classification, the association, the formation, and the higher units of vegetation.

17.2 Objects of Classification

17.2.1 PHYTOCOENOSES

The main object of classification of vegetation is considered to be a phytocoenose, as understood by SUKACHEV (1957: 15) 'The 'phytocoenose, or plant community, is taken to mean any plot of vegetation on the given area uniform in its composition, synusial structure, and the pattern of interactions among plants and between these and the environment.' This definition, repeatedly expressed by SUKACHEV with but minor modifications (1954: 316, et al.), and similar definitions given by SHENNIKOV (1964: 12, et al.) are most widely accepted in the USSR. As already been mentioned, this definition of 'phytocoenose', or 'community' implies double meaning: the chorological unit (stand, 'Bestand') and phytosociological concept (plant community, 'Pflanzengemeinschaft') are called by the same term. In a discussion with SUKACHEV lasting over 20 years, ALEKHIN proposed to divide these two concepts – the second to be referred as a 'plant community' (or 'phytocoenose'), the first to be termed a 'parcel of association' (other terms have been suggested, e.g. 'phytochora' – NORIN, 1970). ALEKSANDROVA (1969: 22-36) considers it reasonable to associate with the term 'phytocoenose' its original meaning of a chorological ('topographical') unit, or stand, for which this term had been suggested by GAMS (1918: 421), [1]) with the word 'community' to serve as a general, rankless term for definition of any vegetation unit both chorological and synecological. The above given definition by SUKACHEV is, however, the most widely accepted in Russian geobotany.

Since the phytocoenose as a chorological unit is defined on the basis of similarity or homogeneity of vegetation within the limits of a given parcel of vegetation cover, the meaning of phytocoenose depends upon the way the concept of 'similarity' is defined. YAROSHENKO (1961: 50) writes: 'As introduced by SUKACHEV into the phytocoenose definition, the character of similarity of composition and interaction of structural parts is not so readily discernible as might appear at first glance. The trouble is that the so-called mosaic phytocoenoses are too often met in nature; these display a pattern of patches differing from each other in species composition and structure. In such a case similarity of the phytocoenose as a whole must be understood to mean only a similar mosaic'. The different elements of a mosaic in the phytocoenose were called 'microgroups' or 'microaggregations' (YAROSHENKO 1961, et al.) or 'microphytocoenoses' (LAVRENKO 1959). 'Each microgroup includes only a small part of the whole community, but it includes all the community's layers at a given place, and in this respect the microgroup or microphytocoenose differs from the 'synusia' (YAROSHENKO 1961: 57). A question then arises should any 'microcoenoses' be considered as separate phytocoenoses, or as fragments from different phytocoenoses installed within the given phytocoenose? LAVRENKO (1959:31) uses the following criterion: If a plant cover contains a patch which differs from the surrounding vegetation then, if the roots of plants surrounding this patch join beneath it, we have a microgroup

---

[1]) The first who used the term 'phytocoenose' was PACZOSKI (1915, 1921), however he implied another meaning by this term, namely, he proposed to call 'phytocoenose' a stand with one species (e.g. stand of *Phragmites communis*), while the 'community' was a stand comprising different species. Note also that PACZOSKI (1896, 1921) was the first to use the term 'phytosociology.'

(microphytocoenose) that is part of a mosaic phytocoenose. If, on the other hand, the roots of plants surrounding the patch in question contact it only along its boundary, the patch is to be regarded as a different phytocoenose or a fragment from another phytocoenose; and in such a case we see not a mosaic phytocoenose, but a phytocoenose-complex or its fragments. (A fragment of a phytocoenose is a plot smaller than minimal or representative area.) This criterion is considered crucial in discriminating mosaic-structure and complex-structure: The term 'mosaic structure' denotes non-uniformity of composition within a phytocoenose, 'complex-structure' non-uniformity of a higher rank, the elements of which represent different phytocoenoses or their fragments. Searches for objective criteria for detecting phytocoenoses and their boundaries in the field are being continued with the aid of statistical methods (NITSENKO 1948, 1971, BYKOV 1957: 50, VASILEVICH 1967a, 1971, LOBANOVA 1971, TROTSENKO 1971, and others).

17.2.2 SYNUSIAE

Within a phytocoenose, as its distinctive parts, synusiae are recognized. The term 'synusia' was defined so broadly by GAMS (1918) that confusion in use of the concept resulted (see also article 16). In the USSR the term is used with two different meanings.

ALEKHIN (1936: 325-327, et al.) applied 'synusia' to an assemblage of species of similar environmental relations and life-forms within a given phytocoenose. This concept is close to the 'ecological groups' of ELLENBERG, (1950, 1952) the 'bioecogroups' of APALIA (1968), and the like. But the term 'synusia' has been more widely accepted in the USSR in the sense of LIPPMAA (1933, 1939, et al.), for the concept initially termed by him the 'unistratal association.' Synusiae in the sense of LIPPMAA are communities of plants growing together which, in the course of long-lasting natural selection, have become adapted to co-habitation under the conditions of certain environmental relations. This gives deep phytocoenological meaning to the synusia and justifies use of synusiae as natural units of classification. The relative independence of synusiae is readily illustrated by such phenomenon as SOCHAVA (1930: 28) described in the Ural mountains and called the 'incumbation' (overlaying) and 'decumbation' (segregation) of layers and formation of 'incumbational (overlay) series'. Fruitful investigations in the classification of synusiae in the sense of LIPPMAA are being carried out in the USSR by TRASS (1964, et al.) and his disciples (MARTIN, 1967, et al.). TRASS offered a system of units for classification of synusiae (Table I) employing the units of LIPPMAA and, in part, the terminology proposed by Du Rietz (1930: 497). These taxonomic units were used by Mirkin (1970b, et al.) in classification of forests with regard to their ground layer.

Table I

System of taxonomic units for classification

of synusiae (monosynusial units) and phytocoenoses

(polysynusial units) by TRASS (1964:104).

(See also article 16, Table I.)

Monosynusial units	Polysynusial units
Society	Sociation
Union	Association
Federation	Association-group
Formion	Formation
Panformion	Formation-group
Type of monosynusial vegetation	Type of polysynusial vegetation

The smallest part of the plant cover that can be distinguished as a phytocoenological unit is the 'coenocell' that Ipatov (1966) described as 'an elementary unit of social life of plants.' The coenocell comprises all plants found in the area of immediate influence by a single plant; it can also be defined as an assemblage of all plants growing within the 'phytogenic field' of biotic interactions with a single individual, as described by Uranov (1965). Recently Vasilevich (1971) provided statistical evidence for an elementary pattern phenomenon in plant cover called by him a 'coenoquant.'

17.2.3 COMPLEXES AND CONTINUA

The importance of units of the plant cover broader than the phytocoenose for classification is acknowledged by Soviet geobotanists. Such units are complexes (DIMO & KELLER 1907, DU RIETZ 1921, 1930, DOKHMAN 1930, 1936, et al.) and compound complexes (DOKHMAN l.c., et al.). All these units are termed by ISACHENKO (1967, 1969, et al.) 'combinations': microcombinations, mesocombinations, macrocombinations. Macrocombinations are similar to the concept of 'Fliesen' of SCHMITHÜSEN (1968: 126, 236). Recently significant contributions have been made in the USSR to classification based on units of this kind (ISACHENKO 1967, 1969, GURICHEVA et al. 1967, KARAMYSHEVA & RACHKOVSKAYA 1968, 1969 et al.), especially for large-scale mapping of vegetation.

In Soviet geobotany renewed attention has been given to the problem of the degree of objectivity of these units, as affected by the vegetation continuum conception (VASILEVICH 1960, 1966a, 1968a et al., IPATOV 1966, TRASS 1966, FREY 1967, ALEKSANDROVA 1965, 1969, DOLUKHANOV 1970, et al.). It has been concluded that, although continuity is an inherent and crucial feature of vegetation, the continuum is not structureless, not amorphous, not homogeneous. Vegetation forms a structured, patterned continuum, heterogeneous and unequal in its various parts, with regular repetition under similar conditions of similar patterns of plant populations. There are, moreover, cases both of continuous transitions from one population pattern to another, and of more or less 'steep' transition from one 'isotropic' phytocoenose to another (VASILEVICH 1967a: 92, see also NITSENKO 1948, 1971). Relative discontinuity in a vegetation continuum is related to the scale of investigation (MASING & TRASS 1963, SOCHAVA 1965a: 4, VASILEVICH 1968a, et al.). There is no conflict between the conception that phytocoenoses exist as parts of partly or wholly continuous patterns of vegetation, and treatment of those phytocoenoses as objects of classification (cf. RAMENSKI 1924, 1938, WHITTAKER 1956, 1967, and articles 3 and 5 in this volume).

Thus, for classification of vegetation Russian geobotanists use a number of units, each of which has its significance and its own area of application. These units are listed above. The principal object of classification is considered to be the phytocoenose or plant community; the abstract classes or community-types include the association, formation, and other units of the hierarchic system into which phytocoenoses are classified.

17.3 Units of Classification

17.3.1 ASSOCIATIONS

Russian geobotanists in the 19th century used both the terms 'association' and 'formation' without any real taxonomic meaning. The definition of the association concept and the theoretical foundation for distinguishing associations were originally presented by SUKACHEV. In his initial papers SUKACHEV called the small typological unit of vegetation, distinguished on the base of layer dominants, a 'formation' (SUKACHEV 1908: 56, 1910: 150, cf. HULT 1881). Following the appearance of the proceedings of the Brussels Congress (FLAHAULT & SCHRÖTER 1910) he changed this term in favour of 'association'. At the later Amsterdam Congress the term 'sociation' was accepted for this unit (DU RIETZ 1936). Throughout the following it should be kept in mind that the Russian 'association' corresponds in general to the 'sociation' of Western authors.

In the 1930's much attention was given to the theoretical and methodological foundation of the association concept as a small, homogeneous unit, and to the technique for distinguishing of such associations, in the works of KATS (1929, 1933, et al.); the same attitude was shared by LESKOV (1943) and more recently by NITSENKO (1963, et al.). The majority of the Soviet geobotanists, however, are inclined to give the term 'association' not to the smallest homogeneous unit, but to a broader one, to a group of phytocoenoses 'with a given system of variation' (SHENNIKOV, 1964: 395). The association as a such a unit was used by SUKACHEV (1928, 1931, et al.), ALEKHIN (1935, 1936, 1938); BYKOV (1957), SOCHAVA (1961, 1968), MASING & TRASS (1963), MIRKIN (1968), APALIA (1968, 1971), SAMOYLOV (1970), SHELYAG-SOSONKO (1971), and others. It is considered possible to distinguish subdivisions of this broader association: subassociations (SUKACHEV 1928: 119, SHENNIKOV 1964: 392, et al.), association variants (VASILEVICH 1968b, APALIA 1968, 1971, MATVEEVA 1968, et al.), narrowly defined sociations (ALEKHIN 1935, 1938; MASING & TRASS 1963, TRASS 1964, SHENNIKOV 1964, et al.), plant combinations (VASILEVICH 1963b, 1968b), dominant modifications (MIRKIN 1968), etc.

An effort toward substantiating the meaning of the association as a natural unit was offered by VASILEVICH (1968b: 872). VASILEVICH defines the plant association as the first (the smallest) natural group of communities, the taxonomic distinctiveness of which is shown statistically: a group, forming a more or less isolated cluster of standsamples in an abstract multi-dimensional model (Fig. 3). Units lower than the association (association variants, etc.) are under-

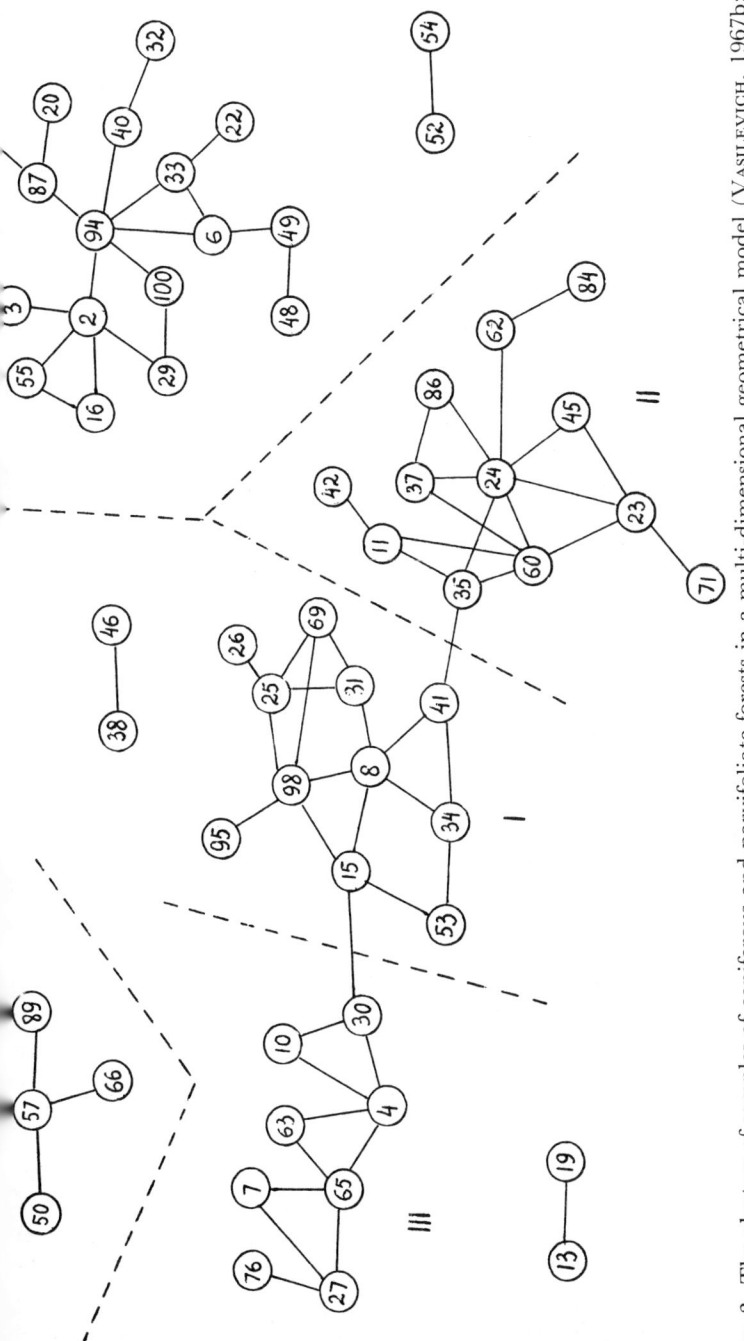

Fig. 3. The clusters of samples of coniferous and parvifoliate forests in a multi-dimensional geometrical model (VASILEVICH, 1967b: 48, fig. 1). Samples (phytocoenoses) differing less than by 25 R^2 value are joined by a line. I – V – group numbers. Dashed lines separate different groups. The clusters might be considered as typological units or noda of the forests studied. $R^2 = (\sqrt{a_1} - \sqrt{a_2})^2 + (\sqrt{b_1} - \sqrt{b_2})^2 + (\sqrt{c_1} - \sqrt{c_2})^2$ .... where $a_1, b_1, c_1, \ldots$ are coverages (per cents) of species in the first sample, $a_2, b_2, c_2, \ldots$ coverages of the same species in the second sample.

181

stood as parts (to be considered as artificial) of this natural group. Evidently a similar but intuitive attitude was shared by SUKACHEV, who called the association 'a lower taxonomic unit' but described subassociations within it.

The manner of distinguishing associations according to the layer dominants is traditional for Russian classification of vegetation, and is most widely accepted. A number of Soviet geobotanists, however, do not share this view. RAMENSKI (1938: 333, 1950, 1952, et al., article 5) most energetically argued against classification by dominants, emphasizing use of the complete list of species and basing the systematization of phytocoenoses on ordination. DOKHMAN (1954, 1960) also used for distinguishing associations not the dominants (at least not only the dominants), but a 'character-group of associated species'; a similar approach was used by DOLUKHANOV (1957) and ISACHENKO & RACHKOVSKAYA (1961). The inadequacy of using dominants for vegetation classification was pointed out also by TRASS (1965). A 'dominant-determinant' approach to classification was elaborated by MIRKIN (1965, et al.), and a similar principle was successfully employed by SAMOYLOV (1970) in classification of meadow communities. Distinction of associations on the basis of bioecogroups, rather than dominants, was developed by APALIA (1968, 1971), etc. For the classification of meadows and other herbaceous phytocoenoses, where the relative abundance of species may vary considerably with annual fluctuations, the concepts of 'ecological replacement groups' and 'change dominant-complex' are proposed and used for classification of phytocoenoses (RABOTNOV 1963, 1965, NITSENKO 1963, MIRKIN 1968), In his recent papers SOCHAVA (1968, et al.) gives the term 'epiassociation' to a unit still broader including the climax community with successional stages (SOCHAVA 1968: 17, 21).

The technique for distinguishing plant associations, characteristic of the Russian approach to classification of vegetation, is described by NESHATAEV (1971b). A collection of stand-samples is taken through geobotanical analysis of either sample plots (quadrats) or within natural boundaries of phytocoenoses as estimated visually. The choice of stand-samples is the same as that suggested by BRAUN-BLANQUET (1964: 24): samples are chosen at typical stands, taking the 'typical' to mean the observed repeated correspondence of similar patterns of species with similar biotopes. In this way the very choice of stand-samples predestines the distinction of certain associations. The second step is tabulation of the samples, which have already been referred to one or several related associations. Examination of such a table (no formal procedure is implied) makes possible clarifying the associations and their distinctions, and permits their definitions to be stated. The third step is the description of association characteristics according to dominants and subdominants, position in ecological and ecologo-phytocoenological series, etc; also analysis of the composition of life-forms and ecological and geographical species groups is performed, and the analytic and synthetic characters of the association are determined (PONYATOVSKAYA 1952, SHENNIKOV 1964: 395-404, etc.).

The names of associations are formed in Soviet geobotany in two ways. The first implies a binary name: the noun, first, is the genus of the dominant with the suffix '-etum'; while the adjective following the genus is the dominant in a subordinate layer with the suffix '-osum'. In case two dominants exist in one layer both are included in the name of the association, so that the name of one (in the dative) is united with the other through a dash, for example: Tilieto-Quercetum aegopodiosum, Piceetum hylocomioso-myrtillosum. When further specification of the dominant is needed, species names may be incorporated as follows: Spahgneto fusci-Sphagnetum magellanicum eriophorosum vaginati, and the like. For subassociations the name is expanded thus: Nardetum sphagnosum subass. agrostosum caninae, and so on. Sometimes the adjective can be derived from common features of the association in question, e.g.: Quercetum robori inundatum, Nardetum alpinum, Nardetum purum, Nardetum pastoralis, etc. (SHENNIKOV 1964).

The second method, rather widely employed in the USSR, was advocated by ALEKHIN. The name of an association is composed of the dominants and subdominants of each layer, with the dash separating the dominants of different layers while the plus sign joins codominants of one layer,e.g.: ass. Pinus silvestris+Picea abies− Vaccinium myrtillus− Pleurozium schreberi+Hylocomium splendens. This system permits quite effective characterization of community structure by the major species of all the vegetation layers; but (a) it is applicable only to the smallest homogeneous units (sociations), and (b) it becomes quite artificial when the plant cover has no clearly defined layers. MIRKIN and SAMOYLOV, to distinguish associations according to dominants and ecological determinants, employ a system of underlining. Thus SAMOYLOV (1970) in the names of associations underlines dominants with a solid line, ecological determinants (indicator species) with a dashed line; species belonging to one associated group (statistically determined) are united with a dash, those belonging to different groups with a plus.

The traditional method as outlined above is used most widely in the USSR, but an alternative that is developing uses random sample choice and statistical analysis of the set of stand-samples collected (VASILEVICH 1960, 1963a, b, 1967a, b, 1969, 1971, FREY, 1963, 1966, 1967, 1971, KRAUKLIS & MEDVEDEV 1966, NESHATAEV 1967, 1971a, MARTIN 1967, 1969, VOLKOVA 1969, IPATOV 1969, MEDVEDEV 1969, MIRKIN 1970a, SAMOYLOV 1970, BUŠS 1971, LOBANOVA 1971, NORIN 1971, TROTSENKO 1971, et al.).

17.3.2 FORMATIONS

The formation is an important concept in the system of units characteristic of Russian approaches to classification of vegetation. This term was used at the end of the 19th and beginning of the 20th centuries for different kinds of categories. Thus, for example, GORDYAGIN (1901) like KORZHINSKI distinguished 'four steppe formations, namely the meadow-steppe, the shrub-steppe, the Stipa-steppe and the stony-steppe,' understanding the formation as a large physiognomic division of vegetation; SUKACHEV, on the other hand, in his papers of 1908 and 1910 considered a formation to be the smallest typological unit. After the latter unit had been given the name 'association,' the term 'formation' found use in Russian geobotanical literature for the principal unit of intermediate rank, comprising all communities with the same species dominant. This definition of the formation concept has been firmly established in Russian geobotany: 'Associations with the major layer composed of the same dominant species are united into a formation. Thus, to the formation Pineta silvestris belong all pine forest associations with *Pinus silvestris* dominant, however different the lower layers may be. Various meadows with *Alopecurus pratensis* dominant constitute the formation Alopecureta pratensis. Existence of two dominants in the major layer distinguishes certain formations, e.g. Pineto-Betuleta, and so on' (SHENNIKOV 1964: 410-411). Similar definitions of 'formation' were given by ALEKHIN (1936: 350, 1944: 165), LAVRENKO (1959: 61, 1968: 56), VORONOV (1963: 299), and others. Thus the Russian use of 'formation', like that of 'association,' differs from usage standard in the West, where the formation is a physiognomic unit (see article 13). The Russian formations are dominance-types in the sense of WHITTAKER (1962, article 14), and correspond to the concepts of association and consociation of many American and British authors. Despite the rather artificial means of definition, the formation as a dominance-type is convenient because of the simplicity of its definition and its apparent objectivity (any researcher may classify stand-samples into the same units). These practical advantages explain the fact that this unit contributed so much to generalizing the whole body of evidence gained by Russian geobotanists over the period of 1930's to 1950's. Particularly extensive geobotanical investigations were made in the 1930's, after the war toward the end of the 1940's and in the 1950's when aeromethods were used. As a result an inventory or registration of natural meadow and pasture lands of all zones was accomplished (LARIN 1965, et al.); registration of the forests of the country has been done (TSEPLYAEV 1965, et al.); and a registration of bogs comprising detailed data of their areas and distribution was possible (NIKONOV 1955, et al.). A great many vegetation maps of middle and small scale have been compiled (LAVRENKO 1941, 1950, LAVRENKO & SOCHAVA 1956, et al.); numerous regional monographs and reviews have been published (KOROVIN 1934, GORODKOV 1935, SHENNIKOV 1938, ZINSERLING 1938, LAVRENKO 1940, PROZOROVSKI 1940, SOCHAVA 1948, LAVRENKO & SOCHAVA 1956, et al.); and geobotanical characterization of regions of the USSR has been accomplished (LAVRENKO 1947). Primarily because of the system of classification based on dominants it has been possible to synthesize in a short period of time the results of registration of vegetation over the whole vast and most variable territory of the USSR.

In the 1960's however, along with the development of more detailed investigations the problem of the formation concept in the USSR was exposed to a critical reconsideration, particularly in connection with large-scale mapping of vegetation (SOCHAVA 1962, 1965a, ISACHENKO 1962, 1969, et al.). Some geobotanists in Russia refuse to accept the definition of formations by dominants and find more useful ecological and floristic criteria, structural characteristics of communities, and their successional relationships (DOKHMAN 1954, 1960, KOLESNIKOV 1956, DOLUKHANOV 1957, SOCHAVA 1961, YAROSHENKO 1961, et al.). Other scientists suggest discarding the formation as a category in the hierarchy of vegetation units (MIRKIN 1965, 1968, SAMOYLOV 1970). Instead of the traditional hierarchy–vegetation type, formation, association group, association–, MIRKIN employs the following scheme–vegetation type, vegetation subtype, section of the subtype (association class), association group, association, dominant modification. Along with this he considers it possible to retain the concept of 'formation' as a useful rankless unit.

17.3.3 HIGHER UNITS

The term 'vegetation type' designates the principal unit of major rank in Russian geobotany. Acceptance of this term in the sense of BROCKMANN-JEROSH & RÜBEL (1912) is due essentially to the works

Table II

Classification scheme for pine forests as made by URANOV

(ALEKHIN 1938)

Taxonomic categories	Phytocoena, community-types
Vegetation type	Lignosa
Formation class	Aciculilignosa
Formation group	Aciculisilvae
Formation	Pineta silvestris
Association group	Pineta silvestris hylocomiosa
Association	Pinus silvestris - Vaccinium vitis - idaea - Hylocomia
Sociation	Pinus silvestris - Vaccinium vitis - idaea - Pleurozium schreberi + Hylocomium splendens

of ALEKHIN (1936, 1938, 1944, et al.). The hierarchy of subordinate units in the classification scheme suggested by ALEKHIN is exemplified in Table II. Such an approach to major units was justified in the initial stages of geobotany, as a means for preliminary systematization of the rather incomplete data on plant cover of great regions of the Earth. Considering this type of classification SOCHAVA (1944: 4) very reasonably wrote: 'Doubtless this approach employs very superficial physiognomical features... There is no doubt of the artificial nature of the major physiognomic units of such a classification.'

Most widely used in the USSR as major units of classification are the vegetation types in the sense of LAVRENKO (1940, 1947, 1959, et al.). According to LAVRENKO (1947: 10), 'a vegetation type is determined essentially on the basis of ecologic-biological characteristics of the dominant species in the coenosis, or édificateurs.' This approach, called by LAVRENKO (1959: 62) 'physiognomic-ecological,' allows distinction of types of communitics by structure and environmental relations and as units of landscape significance, such as steppe, meadow, savanna etc, instead of such abstract concepts as Lignosa, Herbosa, and the like. The steppe vegetation type, for instance, is defined by LAVRENKO (1954: 174) as follows: 'Les steppes, en tant que type de végétation, comprennent les associations d'Herbacées vivaces microthermiques et xérophiles (resistantes au froid et à la sécheresse), parmi lesquelles les Graminées cespiteuses tiennent la première place.' Consequently, the types of vegetation of LAVRENKO are similar to the formations of American and West Europaean authors, as defined by physiognomic-ecological criteria. The weak point of this approach is that rather seldom do we observe in nature communities with marked dominance of a single growth-form or ecobiomorph, frequently two or more ecobiomorphs share dominance. That is especially true for the vegetation of boreal bogs, tundras, savannas etc. ALEKSANDROVA (1971) proposed as a complementary criterion the 'character synusia,' i.e. a synusia that is not always a dominant one, but the presence of which is particularly characteristic of a given vegetation type. Thus, for the tundra vegetation type observed on automorphic soils of tundra-zone placors the character synusiae are those of hekistotherm (sensu CANDOLLE 1874), hypoarctic (sensu TOLMACHEV 1932, YURTSEV 1966) and arctoalpine dwarf shrubs and half-prostrate (sensu SEREBRYAKOV 1964) shrubs, with more or less abundant mosses.

Along with the physiognomic-ecological approach most widely accepted in the USSR a historic-genetical approach has gained popularity; in this the principal criterion is the historical (florogenetic and phylocoenogenetic) relationship between communities. The idea of classification of vegetation based on floristic or florogenetic relationships originated in Russia very long ago (KRASNOV 1888, PACZOSKI 1891, 1915, KORZHINSKI 1888, SUKACHEV 1915, 1928, SINSKAYA 1933, LESKOV 1943, et al.). The most fruitful developments of this genetic approach to classification have been those of OVCHINNIKOV (1947, 1957, et al.) and SOCHAVA (1944, 1945, 1961, 1965b, 1967, et al.). OVCHINNIKOV proposed the concept of florocoenotype; in his definition a 'florocoenotype is an assembly of plant formations, dominants of which have developed through the common adaptive evolution under certain long-lasting physiographic conditions' (OVICHINNIKOV 1947: 18-19). In case of the Central Asia vegetation 24 florocoenotypes have been distinguished by this author.

SOCHAVA has developed a system of classification the central concept of which is the 'phratria' of formations (Greek φρατρια), defined as follows: 'Formation classes are united into phylogenetic series, or complexes, which can be given the name of phratriae of plant forma-

Table III

The system of vegetation classification proposed

by SOCHAVA (1961:15)

Taxonomic categories	The Olenek-Vilyui plateau	The Amur-Zeya region	The Kazakhstan melkosopochnik
Association	Moss (Tomenthypnum nitens, Aulacomnium turgidum) dryas (Dryas crenulata) larch light forests	Herbs (Carex sutschanensis, Scabiosa fischeri, Tanacetum sibiricum) lespedeza oak forests	Grass-forbs (Stipa rubens, Stipa lessingiana) steppe with mesophytic element in flora
Association group	Moss-lichen-dwarf shrub larch light forests	Herb-lespedeza oak forests	Forbs-Stipa rubens steppe
Subformation	Central Siberia subformation	Upper Amur subformation	Northern Kazakhstan subformation
Formation	Northern taiga larch light forests (Larix dahurica)	Quercus mongolica forests	Stipa lessingiana steppe
Formation group	Northern taiga larch forests and light forests	–	Stipa-steppe
Formation class	Forests and light forests of Larix dahurica	Manchurian broad-leaved forests	Transvolga-Kazakhstan caespitose grass steppe
Phratria of formations	Angara phratria	Manchurian phratria	Transvolga-Kasakhstan phratria
Vegetation type	Boreal	Nemoral	Steppe type
Suite of vegetation types	Arcto-herb-forest suite		
System of vegetation types	Northern extratropical		
Phytosphere			

tions. Phratriae include formation classes that are related phylocoenogenetically, but frequently are diverse physiognomically. One phratria comprises, as a rule, forest, shrub, and herbaceous formation classes related phylogenetically' (SOCHAVA 1944: 11). The hierarchy of categories employed by SOCHAVA is summarized in Table III. Under the term 'vegetation type' SOCHAVA implies meaning different from that understood in the physiognomic-ecological approach: 'vegetation types' in SOCHAVA's sense are formed as a result of grouping not phytocoenoses but the vegetation complexes (combinations) of landscapes and regions; it is thus related to the 'Pflanzendivisionen' of SCHMITHÜSEN (1968: 315). Thus, for example, the 'boreal vegetation type' according to SOCHAVA is formed through grouping vegetation combinations composed of taiga forests, bogs, meadows and their anthropogenic successional communities.

The system offered by SOCHAVA reveals in the course of its classification by phylocoenogenetical considerations, that different types of communities display different regional relationships. For example, moss-lichen-dwarf shrub light forests of *Larix dahurica* are found only within the boundaries of the Angara phratria. SOCHAVA (1964a, b, 1967) has developed a system of classification of the world vegetation including 98 phratriae. The broadest unit of vegetation defined by BRAUN-BLANQUET, the 'Gesellschaftskreis' (community-circle), in many respects reminds one of the phratria of SOCHAVA, although the Kreis is broader than the phratria. Thus, for example, the vegetation of Europe (tundra not included) is described by BRAUN-BLANQUET as three Kreisen, while SOCHAVA describes in the same area 17 phratriae; LEBRUN (1947) distinguished 6 Kreisen in Africa, SOCHAVA (1964a: 13) 22 phratriae. Physiognomical convergence revealed in ecologically related conditions (even in the case of complete floristic non-identity) permits a different classification according to physiognomic-ecological principles, as described above in LAVRENKO's system and discussed in article 13 in this volume.

17.4   SUMMARY-CONCLUSION

1) The physiognomic-ecological approach to classification of vegetation is the most widely used in Russian geobotany. The greatest support is given to the system used by LAVRENKO, which is based upon the dominants of the major stratum: on the ecobiomorph (growth-form) of the dominant for distinguishing higher units and on dominant species for distinguishing middle and lower units of vegetation.

2) In classifying vegetation for regional description and mapping, the concepts of zonal, intrazonal and extrazonal vegetation are widely accepted. The concept of zonal vegetation relates to that of placor habitats and is analogous to CLEMENTS' climax-formation (cf. WHITTAKER 1953: 42, WALTER 1954 et al.); however no successional relationship of zonal with non-placoric intrazonal (subclimax) and extrazonal (preclimax and postclimax) formations need be implied.

3) The principal lower unit of classification in Soviet geobotany is the 'association,' distinguished according to stratal dominants. It is closely related to the sociation of Scandinavian phytocoenologists, although it is, as a rule, a somewhat broader unit.

4) Usage of the term 'formation,' significantly differs from that by the West European and American scientists. A majority of Soviet geobotanists employ 'formation' for a major category of middle rank, distinguished by the dominant species of the upper layer, and consequently close to the association and consociation of Anglo-American ecologists (CLEMENTS, TANSLEY).

5) It is characteristic of the Russian approach to use the term 'vegetation type' to designate higher units of vegetation. A majority of Soviet geobotanists use it for broad ecologic-physiognomical units such as meadows, steppe, savannas, and the like, distinguished according to the dominant growth-form. The vegetation type in this sense is similar to the formation of British and American authors. SOCHAVA (1944) has developed a distinctive system including as higher units one termed 'vegetation type' (which is, however, understood not in the physiognomic-ecological but in a regional-historical sense and comprises communities with different dominant ecobiomorphs), and as major divisions of vegetation types the 'formation phratriae.'

6) Methods for collecting vegetation samples in the field and classifying them have, in the traditional Russian approach, been based on the idea of discontinuity of both the objects of classification (phytocoenoses, as stands) and taxonomic units (associations, etc.), as well as upon choice in the field of sample plots considered 'typical' of taxonomic units. In the 1960's a critical revision of traditional concepts has led to increasing acceptance of two alternative approaches: (i) Vegetation has been treated as a continuum (as originally proposed by RAMENSKI many years ago), and conceived as a heterogeneous, structured pattern of community intergradation often including relative discontinuity. (ii) Random choice of the stand samples in the field is combined in many works with statistical treatment of the data obtained.

REFERENCES

ALEKHIN, V. V., – 1915 – Types of Russian steppe (In Russian). *Izv. Bot. sada Petra Velikogo* 15(3-4): 405-432.
ALECHIN W. W., – 1932 – Die vegetationsanalytischen Methoden der Moskauer Steppenforscher. *Handb. Biol. ArbMeth.* 11, 6: 335-373.
ALEKHIN, V. V., – 1935 – Main concepts and main units in phytocoenology (In Russian). *Sov. Bot.* 1935(5): 21-34.
ALEKHIN, V. V., – 1936 – Vegetation of the USSR in its main zones (In Russian). In: 'Fundamentals of Botanical Geography,' H. WALTER & V. ALEKHIN, pp. 306-715. Biomedgiz, Moscow & Leningrad.
ALEKHIN, V. V., – 1938 – Schemas de classification de la végétation (In Russian).

Sov. Bot. 1938(3): 96-98.
ALEKHIN, V. V., – 1944 – Geography of plants (In Russian). Sovetskaya Nauka, Moscow. 455 pp.
ALEKSANDROVA, V. D. (ALEXANDROVA, V. D.), – 1956 – Vegetation of the southern Island of Novaya Zemlya between 70° 56' and 72° 12' n.l. (In Russian). *Rastit. krain. Sev. SSSR i ee Osv.* 2: 187-306.
ALEKSANDROVA, V. D., – 1965 – On the problem of distinguishing phytocoenoses in a vegetative continuum. (In Russian with Engl. summ.) *Bot. Zh. SSSR* 50(9): 1248-1259.
ALEKSANDROVA, V. D., – 1969 – Classification of Vegetation: Principles of Classification and Classification Systems of Various Phytocoenological Schools (In Russian). Nauka, Leningrad 275 pp.
ALEKSANDROVA, V. D., – 1971 – Principles of zonal subdivision of Arctic vegetation (In Russian with Engl. summ). *Bot. Zh. SSSR* 56(1): 3-21.
APALIA, D. K., – 1968 – Distinguishing of plant associations on the base of bioecogroups (In Russian). Abstract of a report. See: NORIN, 1968.
APALIA, D. K., – 1971 – Distinguishing of plant associations on the base of the indicator bioecogroups. (In Russian). In, 'Metody vydeleniya rastit. assotsiatsi,' pp. 38-51. Nauka, Leningrad.
BORSHCHOV, I. G., – 1865 – Materials for botanical geography of the Aralo-Caspian region (In Russian). *Zap. imp. Akad. Nauk* 7: 1-90.
BRAUN-BLANQUET, J., – 1964 – Pflanzensoziologie: Grundzüge der Vegetationskunde. 3 Aufl. Springer, Wien. 865 pp.
BROCKMANN-JEROSCH, H. & E. RÜBEL., – 1912 – Die Einteilung der Pflanzengesellschaften nach ökologisch-physiognomischen Geschichtspunkten. Engelmann, Leipzig. 72 pp.
BUŠS, K. K., – 1971 – Some problems of biometric analysis of anthropogenic successions on drained forest areas (In Russian with Engl summ..) *Bot. Zh. SSSR* 56(1): 22-30.
BYKOV, B. A., – 1957 – Geobotany (In Russian). 2nd ed. Izd. Akad. Nauk Kazak SSR, Alma-Ata. 381 pp.
CANDOLLE, ALPH DE, – 1874 – Constitution le règne végétal de groupes physiologiques. *Archives Sci. phys. nat.*, Geneva 50: 1-38.
CLEMENTS, F. E., – 1936 – Nature and structure of the climax. *J. Ecol.* 24: 252-284.
COLE, L., – 1958 – The ecosphere. *Scient. Amer.* 198(4): 83-96.
DIMO, N. A. & B. A. KELLER, 1907 – In the semi-desert region (In Russian). Izd. Saratovsk. gubern. Zemstva, Saratov. 215 pp.
DOKHMAN, G. I., – 1930 – Phytosociological analysis of vegetation of the Starobyelskie steppe (In Russian). *Isv. ass. Nauchno-issled. inst. physico-math. fak. I Moskow Univ.* 3(2-A): 212-232.
DOKHMAN, G. I., – 1936 – On some classification units of the vegetation complexes (In Russian with Germ. summ.) *Zemlevêdênie* 38(3): 294-326.
DOKHMAN, G. I., – 1954 – Vegetation of Mugodzhary (In Russian). Geografgiz, Moscow. 235 pp.
DOKHMAN, G. I., – 1960 – On the system of diagnostic characters of vegetation units (In Russian with Engl. summ.) *Bot. Zh. SSSR* 45(5):637-648.
DOLUKHANOV, A. G., – 1957 – Problems of the coenological forest classification in connection with convergence features in vegetation (In Russian). *Tez. dokl. deleg. s'ezda VBO*, Leningrad, 4(2): 18-23.
DOLUKHANOV, A. G., – 1970 – The questions of typology of mountain forests in connection with the nature of relative continuity of their plant cover (In Russian with Engl. summ.) *Trudý mosk. Obshch. ispýt Prir*, old. biol. 38: 24-33.

Du Rietz, G. E., – 1921 – Zur methodologischen Grundlage der modernen Pflanzensoziologie. Holzhausen, Wien 267 pp.
Du Rietz, G. E., – 1930 (1932) – Vegetationsforschung auf soziationsanalytischer Grundlage. *Handb. biol. ArbMeth.*, 11, 5: 293-480.
Du Rietz, G. E., – 1936 – Classification and nomenclature of vegetation units 1930-1935. *Svensk Bot. Tidskr.* 30: 580-589.
Engelmann, G. I., – 1810 – The Theoretical and Practical Manual for Drainage of Lands (In Russian). St. Petersburg. 291 pp.
Flerov, A. F. & B. A. Fedchenko, – 1902 – Handbook for Studying Plant Communities of Middle Russia (In Russian). Izd. Sabashnikovykh, Moscow. 184 pp.
Frey, T., – 1966 – On the significance of Czekanowski's index of similarity. *Zastosow. Mat.* 9(1): 1-7
Frey, T. E. A., – 1963 – On the use of the correlation coefficient of interspecific relations (In Russian). *Bot. Zh. SSSR* 48(2): 235-239.
Frey T. & L. Võhandu, – 1966 – Uus meetod klassifikatioomühikute püstitaniseks (In Estonian with Engl. summ.) *Eesti NSV Tead. Akad. Toim.* biol seer. 4: 565-576.
Frey, T. E.-A., – 1967 – On the mathematical-phytocoenological methods of classification of vegetation (In Russian). Tartu State University, Tartu. 18 pp.
Frey, T. E., – 1968 – Cluster analysis of the sample plots descriptions with 'K' criterion (In Russian). Abstract of a report. See: Norin, 1968.
Frey, T. E., – 1971 – Cluster analysis of the sample plots descriptions with 'K' criterion (In Russian). In, 'Metody vydeleniya rastit. assotsiatsi,' pp. 226-241. Nauka, Leningrad.
Friederichs, K., – 1930 – Die Grundfragen und Gesetzmässigkeiten der land- und forstwirtschaftlichen Zoologie insbesondere der Entomologie. Parey, Berlin. 417 pp.
Gams, H., – 1918 – Prinzipienfragen der Vegetationsforschung: Ein Beitrag zur Begriffsklärung und Methodik der Biocoenologie. *Vjschr. naturf. Ges. Zürich* 63: 293-493.
Genko, N. K., – 1902 – Description of the Białowieza (In Russian). *Lêsnoi Zh.* S-Peterb. 5: 1012-1056, 6: 1269-1302.
Gordyagin, A. Ya., – 1900-1901 – Materials for studying soils and vegetation of the Western Siberia (In Russian). *Trudy Obshch. Estest. imp. kazansk. Univ.*, Kazan 34(3):3-222, 35(2): 223-528 (+36).
Gorodkov, B. N., – 1935 – Vegetation of the Tundra Zone of the USSR (In Russian). Izd. Akad. Nauk SSSR, Moscow & Leningrad. 141 pp.
Guricheva, N. P., Z. V. Karamysheva & E. I. Rachkovskaia, – 1967 – On the compilation of a legend to a large-scale vegetation map in the desert-steppe subzone of the Kazakhstan (In Russian). *Geobot. Kartogr.*, Leningrad 1967: 57-67.
Gutorovich, I. I., – 1897 – Notes of a northern forester (In Russian). *Lêsnoi Zh.* S-Peterb. 2: 216-228, 5: 789-799.
Hult, R., – 1881 – Försök till analytisk behandling af växtformationerna. *Meddn Soc. Fauna Flora fenn.* 8: 1-155.
Ilynski, A. P., – 1935 – The higher taxonomic units in geobotany (In Russian). *Sov. Bot.* 1935(5): 49-65
Ipatov, V. S., – 1966 – On the concept of phytocoenosis and an elementary cell of the social life of plants (In Russian). *Vest. leningr gos. Univ.* 15, ser, biol., 3: 56-62
Ipatov, V. S., – 1969 – On the choice of typical objects in geobotanical investi-

gations (In Russian). *Problemij Bot.* 11: 13-21.

ISACHENKO, T. I., – 1962 – Principles and methods of generalisation in large, middle and small-scale geobotanical mapping (In Russian). In: 'Printsipy i metody geob. kartogr', pp. 28-46. Moscow Leningrad.

ISACHENKO, T. I., – 1967 – On the cartography of serial rows and microbelts in valleys and lake-basins (In Russian). *Geobot. kartogr.*, Leningrad 1967:42-56.

ISACHENKO, T. I., – 1969 – The structure of the vegetation cover and mapping (In Russian). *Geobot. kartogr.*, Leningrad 1969:20-33.

ISACHENKO, T. I. & E. I. RACHKOVSKAYA, – 1961 – Principal zonal types of the Northern Kazakhstan steppe Vegetation (In Russian). *Trudij bot. Inst. Akad. Nauk SSSR*, ser. 3 (Geobotanika) 13:133-397

KARAMYSHEVA, Z. V., – 1967 – The experience in the treatment of the descriptions of sample areas of steppe communities by the method of Braun-Blanquet (In Russian). *Bot. Zh. SSSR*, 52(8): 1132-1144.

KARAMYSHEVA, Z. V. & E. I. RACHKOVSKAJA, – 1968 – On the complication of a small-scale map of a steppe territory of Kazakhstan (In Russian). *Geobot. kartogr.* Leningrad, 1968: 5-21.

KARAMYSHEVA, Z. V. & E. I. RACHKOVSKAYA, – 1969 – A vegetation map of the middle part of the Tersakkan river basin (In Russian). In: 'Biologicheskie issledovaniya v Kazakhstane,' I: 481-485. Nauka, Leningrad.

KATZ, N. J. (KATS), – 1929 – Die Zwillingsassoziationen und die homologen Reihen in der Phytosoziologie. *Ber. dt. bot. Ges.* 47: 154-164.

KATZ, N. J., – 1933 – Die Grundprobleme und die neue Richtung der Phytosoziologie. *Beitr. Biol. Pfl.* 21:133-166. (In Russian, *Byull. mosk. Obshch. Ispýt. Prir.*, otd biol. 43(2): 188-221, 1934).

KOLESNIKOV, B. P., – 1956 – A review of the forest formation of Primorye and Priamurye (In Russian). In, 'Akad. V. N. Sukachevu k 75-let. so dnya rozhdeniya,' pp. 286-305. Izd. Akad Nauk SSSR, Moscow & Leningrad.

KORZHINSKI, S. I., – 1888-91 – Northern boundary of the chernozem region of the East part of the European Russia in botanic-geographical and pedological points of view (In Russian). *Trudij Obshch. Estest. kazan. Univ.* Kazan, 18(5): 1-253, 22(6): 1-201.

KOROVIN, E. P., – 1934 – Vegetation of the Central Asia and Southern Kazakhstan (In Russian). Moscow & Tashkent. 480 pp. 2nd ed., 1961-2, Tashkent, 2 parts, 452 & 547 pp.

KOSTYCHEV, P. A., – 1900 – Relations between soils and some plant formations (In Russian). *Dnev. S'êzda russk. Estestvoisp. Vrach.* 8(9): 6-7.

KRASHENINNIKOV, S. P., – 1755 – Description of the Kamchatka Land (In Russian). St. Petersburg. 2 vols., 438 & 319 pp.

KRASNOV, A. N., – 1886 – Geobotanical investigations in the Kalmyk steppe (In Russian). *Izv. Russk. Geogr. Obshch.* 22(1): 1-22.

KRASNOV, A. N., – 1888 – An outline of development of flora of the southern part of the Eastern Tian Shan (In Russian). *Zap. Russk. Geogr. Obshch.* 19: 1-413.

KRAUKLIS, A. A. & YU. O. MEDVEDEV., – 1966 – The principles of showing vegetation in series of large-scale maps in connection with mapping dynamics of natural environment (In Russian). *Geobot. kartogr.* Leningrad 1966: 26-35.

KRÜDENER, A. A., – 1916 – Fundamentals of the forest types classification. I and II (In Russian). In, 'Materily po izuch. russk. lesa,' pp. 1-318. Petrograd.

LAVRENKO, E. M., – 1940 – Steppe of the USSR (In Russian). In, *Rastitelnost SSSR* 2: 1-206 Moscow & Leningrad.

LAVRENKO, E. M., – 1941 – An Explanatory Text to the Vegetation Map of the USSR, scale 1: 5 000 000 (In Russian). Moscow & Leningrad. 132 pp.

LAVRENKO, E. M. (ed.), – 1947 – Geobotanical Regionalization of the USSR (In Russian). Izd Akad. Nauk SSSR, Moscow & Leningrad. 152 pp.
LAVRENKO, E. M. (ed.), – 1950 – Vegetation map of the European part of the USSR, scale 1: 2 500 000. Explanatory text (In Russian). Izd. Akad. Nauk SSSR, Moscow & Leningrad 288 pp.
LAVRENKO, E. M., – 1954 – Les steppes de la région stéppique Eurasienne (Géographie, dynamisme, histoire). *Essais de Bot.* 1: 157-194. Éd. l'Acad. Sci. URSS, Moscou & Léningrad.
LAVRENKO, E. M., – 1959 – Main regularities of plant communities and the ways of their investigation (In Russian). In, *Polevaya geobotanika* 1: 13-78. Izd. Akad. Nauk SSSR, Moscow & Leningrad
LAVRENKO, E. M., – 1968 – On the next task of the exploration of the plant cover in connection with botanico-geographical regionalization of the USSR (In Russian). In, 'Osnovn. Probl. Sovrem. Geobot,' pp. 45-68. Nauka, Moscow & Leningrad.
LAVRENKO, E. M. & N. V. DYLIS, – 1968 – Achievements and current tasks in the investigation of terrestrial biogeocoenoses in the USSR. (In Russian). *Bot. Zh. SSSR*, 53(2): 155-167.
LAVRENKO, E. M. & V. B. SOCHAVA, – 1956 – Plant cover of the USSR. Explanatory text to 'Geobotanical Map of the USSR' scale 1: 4 000 000. (In Russian). Izd. Akad. Nauk SSSR, Moscow & Leningrad. 2 parts, 460 & 971 pp.
LARIN, I. V., – 1965 – Natural meadows and pastures and how became they reestablished. (In Russian). In, *Probl. Sovrem. Bot.* 1: 172-205. Izd. Akad. Nauk SSSR, Moscow & Leningrad.
LEBRUN, J., – 1947 – La végétation de la plaine alluviale au sud du lac Édouard. *Inst. Parcs Natl. Congo Belge, Explor. Parc Natl. Albert*, 4 Mission Lebrun 1: 1-800, Bruxelles.
LESKOV, A. I., – 1943 – Principles of the natural system of plant associations (In Russian). *Bot. Zh. SSSR* 28(2): 37-52.
LIPPMAA, T., – 1933 – Taimeühingute uurimise metoodika ja eesti taimeühingute klassifikatsiooni põhijooni (Eston. with Germ. summ.: Grundzüge der pflanzensoziologischen Methodik nebst einer Klassification der Pflanzenassoziationen Estland). *Acta Inst. Horti bot. tartu.* 3, 4: 1-169.
LIPPMAA, T., – 1939 – The unistratal concept of plant communities (the unions). *Am. Midl. Nat.* 21: 111-145.
LOBANOVA, V. F., – 1971 – The quantative analysis of the boundaries of some meadow phytocoenoses of Karelia. (In Russian with Engl. summ.). In, *Kolichestv. metody analiza rastit.*, 2: 178-184. Riga.
LUKICHEVA, A. N. & D. N. SABUROV, – 1971 – Methods of grouping samples into plant associations in connection with the landscape structure (In Russian). In, 'Metody vydel. rastit. assotsiatsi'. Nauka, Leningrad.
MARTIN, YU. L., – 1967 – Development of lichen synusia on the glacier moraines of the Polar Ural (In Russian). Izd. Akad. Nauk SSSR, Sverdlovsk. 22 pp.
MARTIN, YU. L., – 1969 – Ordination method in lichenocoenology (Russian with Engl. summ.). In, 'Kolichestv. metody analiza rastit.', pp. 83-86. Tartu.
MASING, V. V. & H. H. TRASS, – 1963 – Elaboration of some theoretical problems in the works of Estonian geobotanists. (In Russian with Engl. summ.). *Bot. Zh. SSSR* 48(4): 473-485.
MATVEYEVA, N. V., – 1968 – The structure of the vegetation of the main types of tundras in the middle course of the river Piasina (the Western part of the Taimyr Peninsula). (In Russian with Engl. summ.) *Bot. Zh. SSSR* 53(11): 1588-1603.

MEDVEDEV, YU. O., – 1969 – Some phytocoenological and dynamic peculiarities of the dark-coniferous forest (In Russian with French summ.). In, 'Yuzhnaya tayga Priangarya,' pp. 219-262. Nauka, Leningrad.
MIDDENDORF, A. E., – 1867 – Siberische Reise. Bd 4, pp. 525-783, Die Gewächse Sibiriens. Akad. Wissensch., St. Petersburg.
Mirkin, B. M., – 1965 – On the ecological classification of the meadow vegetation of flood-plains (In Russian with Eng. summ.). *Bot. Zh. SSSR* 50(3): 324-334.
MIRKIN, B. M., – 1968 – The criteria of dominants and determinants in the classification of phytocoenoses (In Russian with Engl. summ.). *Bot. Zh.SSSR* 53(6): 767-778.
MIRKIN, B. M., – 1970a – Introduction to Quantitative Methods of Vegetation Analysis (In Russian). Izd. Bashkirsk, gos. Univ., Ufa. 87 pp.
MIRKIN, B. M., – 1970b – Main Problems of Geobotany (In Russian). Izd. Bashkirsk, gos. Univ., Ufa. 88 pp.
MOROZOV, G. F., – 1912 – Forest Science. I. Introduction to Forest Biology (In Russian). St. Petersburg. 83 pp.
NESHATAEV, YU. N., – 1967 – An attempt of the large-scale geobotanical mapping of the forest-steppe oak forest (In Russian). *Uchenije Zap. leningr. gos. Univ.* 331: 87-117.
NESHATAEV, YU. N., – 1968 – A selective-statistical method for distinguishing plant associations (In Russian). Abstract of a report. See: Norin, 1968.
NESHATAEV, YU. N., – 1971a – A selective-statistical method for distinguishing plant associations (In Russian). In, 'Metody vydel. rastit. assotsiatsiy,' pp. 181-205. Nauka, Leningrad.
NESHATAEV, YU. N., – 1971b – Method for grouping of vegetation samples in the curriculum of the Geobotany Department of the Leningrad State University (In Russian). In, 'Metody vydel. rastit. assotsiatsiy,' pp. 23-37. Nauka, Leningrad.
NIKONOV, M. N., – 1955 – Regionalisation of the peatbogs in conncetion with its utilization in national economy (In Russian). *Trudij Inst. Lesa, Akad, Nauk SSSR Mosk.* 31: 49-63.
NITSENKO, A. A., – 1948 – The boundaries of plant associations in nature (In Russian). *Bot. Zh. SSSR* 33(5): 478-495.
NITSENKO, A. A., – 1963 – On some controversial theoretical problems in geobotany (In Russian with Engl. summ.). *Bot. Zh. SSSR* 48(4): 486-501.
NITSENKO, A. A., – 1966 – On the criteria for distinguishing plant associations (In Russian with Engl. summ.). *Bot. Zh. SSSR* 51(8): 1085-1098.
NITSENKO, A. A., – 1971 – An attempt to apply quantiative analysis to the problem of causality of border-lines in vegetation cover (In Russian with Engl. summ.). In, 'Kolichstvennye metody analiza rastitelnosti,' pp. 216-220. Riga.
NORIN B. N., – 1966 – On the zonal types of vegetational cover in the Arctic and Subarctic zones (In Russian with Engl. summ.). *Bot. Zh. SSSR.* 51(11): 1547-1563.
NORIN, B. N., – 1968 – A conference on the methods of distinguishing of plant associations held in Leningrad on May 15-17, 1967 (In Russian). *Bot. Zh. SSSR* 53(6): 872-878.
NORIN, B. N., – 1970 – The functional structure of plant aggregations in the forest-tundra zone (In Russian with Engl. summ). *Bot. Zh. SSSR* 55(2): 170-183.
NORIN, B. N., – 1971 – Application of the similarity coefficient to classification of the forest-tundra microaggregations (In Russian) In, 'Metody vydeleniya rastit, assotsiatsi,' pp. 206-225. Nauka, Leningrad.
OVCHINNIKOV, P. N., – 1947 – About principles of classification of vegetation (In

Russian). *Soobshch. Tadzhiksk. fil. Akad. Nauk SSSR*, Stalinabad 2: 18-23.
OVCHINNIKOV, P. N., – 1957 – The florocoenotypes and their value for classification of vegetation in the Middle Asia. (In Russian). In: *Tezisy dokl. S'ezda VBO*, Leningrad, 7: 28-36.
PACZOSKI, J. K., – 1891 – Stages of floral development (In Russian). *Vestnik estetstvozn.*, Kiev, 8: 261-270.
PACZOSKI, J. K., – 1896 – Zycie gromadne roślin. *Wszechświat* 15. Warszawa. (Reprinted with Engl. transl., 'Social life of plants,' *Bibljot. Bot.*, Krakow, 2: 1-40, 1930).
PACZOSKI, J. K., – 1915 – Vegetation of the Kherson district. I. Forests (In Russian). Izd. Olkhovnikova, Kherson. 202 pp.
PACZOSKI, J. K., – 1921 – Fundamentals of Phytosociology (In Russian). Izd. Stud. komit. S-Kh. technikuma, Kherson. 346 pp.
POGREBNYAK, P. S. (POGREBNJAK P. S.), – 1929 – Über die Methodik der Standortsuntersuchungen in Verbreitung mit Waldtypen. *Verh. II Int. Kongr. forstl. Versuchanstalten*
PONYATOVSKAYA, V. M., – 1952 – Of several years' investigation of the herb communities of the Southern Kirghiz forest belt (In Russian). *Trudy bot. Inst. Akad. Nauk SSSR*, ser. 3 (Geobotanika) 8: 156-240.
PONYATOVSKAYA, V. M., – 1959 – On two trends in phytosociology (In Russian). *Bot. Zh. SSSR* 44(3): 402-407. Engl. transl. in *Vegetatio*, 10:373-385, 1961).
PROZOROVSKI, A. V., – 1935 – Zonal types of the Soviet Central Asia deserts (In Russian). *Isv. Russk. Geograf. Obshch.* 67(3): 318-331.
PROZOROVSKI, A. V., – 1940 – Semideserts and deserts of the USSR (In Russian). In, *Rastitelnost SSSR* 2:207-480. Izd. Akad Nauk SSSR, Moscow & Leningrad.
RABOTNOV, T. A., – 1963 – An attempt to use of the idea of vegetation continuum in study of the plant cover of Wisconsin (In Russian). *Byull. mosk. Obshch. Ispyt. Prir.*, otd biol. 68(4): 147-151.
RABOTNOV, T. A., – 1965 – On the character of the structure of polydominant meadow coenoses (In Russian with Engl. summ.). *Bot. Zh. SSSR* 50(10): 1396-1408.
RAMENSKI, L. G., – 1910 – The comparative method for ecological study of plant communities (In Russian). *Dnev. S' ězda russk. Estestvoisp. Vrach.*, Moscow 12(4): 389-390.
RAMENSKI, L. G., – 1915 – About the quantative study of plant cover (in Russian). In, *Materialy po organiz. i kulture kormov. ploshchadi* 12: 105-140. St. Petersburg.
RAMENSKI, L. G., – 1924 – Main regularities of the plant cover (In Russian). *Věstnik opijtnogo děla Stredne-Chernoz. Obl.*, Voronezh pp. 37-73.
RAMENSKI, L. G. (RAMENSKY, L. G.), – 1929 – On the method of comparative treatment and systematization of plant lists and other objects determined by several factors with unlike actions (In Russian). *Trudij Soveshch. geobot.-lugov. sozvan., (Gos.) lugovoi Inst.* (Transl. in *Beitr. Biol. Pfl.* 18: 269-304, 1930).
RAMENSKI, L. G., – 1932 – Die Projektionsaufnahme und Beschreibung der Pflanzendecke. *Handb. Biol. ArbMeth.* 11, 6: 137-190.
RAMENSKI, L. G., – 1938 – Introduction to the Complex Soil-Geobotanical Investigation of Lands (In Russian). Selkhozgiz, Moscow. 620 pp.
RAMENSKI, L. G., – 1950 – Classification of lands according to their plant cover (In Russian). *Problemÿ Bot.* 1: 484-512.
RAMENSKI, L. G., – 1952 – About some principles of the current geobotany (In Russian). *Bot. Zh. SSSR* 37(2): 202-211.
RAMENSKI, L. G., – 1953 – On ecological study and systematization of plant communities. (In Russian). *Byull. mosk. Obshch. Ispÿt. Prir.*, otd. biol. 58(1):

35-54.
RUPRECHT, F., – 1845 – Flora Samoedorum cisuralensis. St .Petersburg. 67 pp.
RYSIN, L. P. & Z. M. KOVALENKO, 1968 – On the possible use of the methods of the BRAUN-BLANQUET school in our geobotanical investigations. (In Russian with Engl. summ.). *Byull. mosk. Obshch. lspÿt. Prir.*, otd. biol. 73(1) 93-114.:
SAMOYLOV, YU. I., – 1970 – A tentative classification of meadows in the floodplains of the river Msta. (In Russian with Engl. summ.) *Bot. Zh. SSSR* 55(10): 1419-1431.
SCHMITHÜSEN, J., – 1968 – Allgemeine Vegetations Geographie. 3 Aufl. de Gruyter, Berlin. 463 pp.
SEREBRYAKOV, I. G., – 1964 – Life-forms of the higher plants and their study (In Russian) In, *Polevaya Geobot. (Field geobotany)* 3: 146-208. Izd. Akad. Nauk SSSR, Moscow & Leningrad.
SEREBRENNIKOV, P. P., – 1913 – The forest-types and their value in forestry (In Russian). *Lêsnoi Zh.* S-Peterb. 1-2: 39-72.
SHELYAG-SOSONKO, YU. R., – 1971 – The oak-woods of Podolia (In Russian). *Bot. Zh. SSSR* 56(4): 512-516.
SHENNIKOV, A. P., – 1935 – Principles of the botanical classification of meadows (In Russian). *Sov. Bot.* 1935(5): 35-49
SHENNIKOV, A. P., – 1938 – Meadows vegetation of the USSR (In Russian). In, *Rastitelnost SSSR* 1:429-647. Izd. Akad Nauk SSSR, Moscow & Leningrad.
SHENNIKOV, A. P., – 1964 – Introduction to Geobotany (In Russian). Izd. Leningr. gos. Univ., Leningrad. 447 pp.
SIBIRTSEV, N. M., – 1900 – Pedology. I. (In Russian). Izd. Skorokhodova, St. Petersburg. 212 pp.
SINSKAYA, E. N., – 1933 – The principal features of the forest evolution of the Caucasus, in connection with the history of species (In Russian with Engl. summ.). *Bot. Zh. SSSR* 18(5): 370-406, (6): 487-515.
SOCHAVA, V. B., – 1930 – A timber line in the Lyapin Ural maintains (In Russian). *Trav. Mus. bot. Acad. Sci. Russ.* 22: 1-47.
SOCHAVA, V. B., – 1944 – An attempt of the phylogenetical systematics of plant associations. (In Russian). *Sov. Bot.* 1944(1): 3-18.
SOCHAVA, V. B., – 1945 – Phratriae of plant formations of the USSR and their phylocoenogenesis (In Russian). *Dokl. Akad. Nauk SSSR.* 47: 60-64.
SOCHAVA, V. B., – 1948 – Geographical connections of the plant cover on the territory of the USSR (In Russian). *Uchenÿe Zap. Leningr. pedagog. Inst.* 73: 3-89.
SOCHAVA, V. B., – 1961 – Problems of classification of vegetation, typology of physico-geographical facies and biogeocoenoses (In Russian). *Trudÿ Inst. Biol.* Sverdlovsk (Uralsk. fil. Akad. Nauk SSSR) 27: 5-22.
SOCHAVA, V. B., – 1962 – Mapping problems in geobotany (In Russian). In, 'Printsipy i metody geobot. kartograf,' pp. 5-27. Izd. Akad. Nauk SSSR, Moscow & Leningrad.
SOCHAVA, V. B., – 1964a – A new world vegetation map (In Russian). *Geobot. kartogr.*, Leningrad 1964: 3-15.
SOCHAVA, V. B., – 1964b – Vegetation (of world and continents). Explanatory text (In Russian). In, 'Fiziko-geogr. atlas mira,' pp. 280-283. Izd. Akad. Nauk SSSR, Moscow.
SOCHAVA, V. B., – 1965a – Current problems of large-scale vegetation cartography (In Russian). *Geobot. kartogr.*, Leningrad 1965: 3-10.
SOCHAVA, V. B., – 1965b – Vegetation. *Soviet Geogr.* (Rev. Transl.) 6(5-6): 348-359.
SOCHAVA, V. B. (SOTCHAVA, VICTOR), – 1967 – Carte universelle de la végétation

du monde. *Rev. roumaine biol., Sér. bot.* 12(2-3): 239-242.
SOCHAVA, V. B., – 1968 – Plant communities and dynamics of natural systems. (In Russian). *Doklady Instit. geogr. Sibiri i Dalnego Vostoka, Irkutsk* 20: 12-22.
SOCHAVA, V. B., – 1970 – Theoretical principles of steppe geosystems topology. (In Russian). In, 'Topologiya stepnykh geosistem,' pp. 3-12. Nauka, Leningrad.
SOUKATCHEV (see SUKACHEV).
SUKACHEV, V. N., – 1908 – Forest formations and their interrelations in the Bryansk woods (In Russian). *Trudy po lesn. opytn. Delu Ross.* 9: 1-61.
SUKACHEV, V. N., – 1910 – On the plant formation (In Russian). *Dnev. S'êzda russk. Estestvoisp. Vrach.*, Moscow 12(2): 150, *Pochvovêdênîe* 12(4): 393, 1910.
SUKACHEV, V. N., – 1915 – Introduction to Plant Community Science (In Russian). Izd. Panafidinoi, Petrograd. 127 pp.
SUKACHEV, V. N., – 1917 – About terminology in the plant community science (In Russian). *Zh. russk. bot.Obshch.* 2(1-2): 1-19.
SUKACHEV, V. N., – 1927 – Manual for Study of Forest Types (In Russian). Novaya derevnya, Moscow. 166 pp. 2nd ed., 1930, Moscow & Leningrad, 318 pp. 3rd ed., 1931, Selkhozgiz, Moscow & Leningrad, 328 pp.
SUKACHEV, V. N., – 1928 – Plant Communities (Introduction to Phytosociology) (In Russian) 4 ed. Kniga, Leningrad & Moscow. 232 pp.
SUKATSCHEW, W. N., – 1932 – Die Untersuchung der Waldtypen der osteuropäischen Flachlandes. *Handb. biol. ArbMeth.* 11, 6: 191-250.
SUKACHEV, V. N., – 1942 – Idea of development in phytocoenology (In Russian). *Sov. Bot.* 1942(1-2): 5-17.
SUKACHEV, V. N., – 1945 – Biogeocoenology and phytocoenology. *Dokl. Akad. Nauk SSSR* 47(6): 447-449.
SOUKATCHEV, V. N., – 1954 – Quelques problèmes théoriques de la phytocénologie. In, *Essais de botanique*, 1: 310-330. Ed. l'Acad. Sci, URSS, Moscou & Léningrad
SUKACHEV, V. N., – 1957 – General principles and a program for study of forest types (In Russian). In,'Method. ukazan. k izuchen. tipov lesa,'V.N. SUKACHEV, S. V. ZONN & G. P. MOTOVILOV, pp. 9-63. Izd. Akad. Nauk SSSR, Moscow.
SUKACHEV, V. N., – 1964 – Main concepts of forest biogeocoenology (In Russian). In, 'Osnovy lesn. biogeotsenologii,' pp. 5-59. Moscow.
TANFILIEV, G. I., – 1897 – Physico-geographical regions of the European Russia (In Russian). *Trudy Voln. econom. obshch.* 1, 33 pp. Izd. Demakova.
TATISHCHEV, V. N., – 1950 – Selected Works on the Russian Geography (In Russian). Izd. Akad. Nauk SSSR, Moscow & Leningrad. 248 pp.
TOLMACHEV, A. I., – 1932 – Flora of the central part of the Eastern Taimyr. I. (In Russian). *Trudy polyarn. Kom.* 8: 1-126.
TRASS, H. H., – 1964 – Application and validity of the synusial method in phytocoenology (In Russian with Engl. summ.) In, 'Izuchenie rastit, ostrova Saaremaa,' pp. 82-111, Izd. Akad. Nauk Est. SSR, Tartu.
TRASS, H. H., – 1965 – Importance of the plant communities dominants to classification of vegetation (In Russian). *Problemy sovremennoi botaniki* 1: 247-250. Nauka, Moscow & Leningard.
Trass. H. H., – 1966 – On vegetation discontinuity and continuity (In Russian with Engl. summ.). *Trudy mosk. Obshch. ispyt. Prir.*, otd. biol. 27: 167-182.
TRASS, H. H., – 1971 – The development of a quantitative approach to the investigation of vegetation (In Russian with Engl. summ.). *Bot. Zh. SSSR* 56(4): 457-464.
TRAUTVETTER, E. R., – 1849-51 – Die Pflanzengeographischen Verhältnisse des Europäischen Russlands. Häcker, Riga. 3 parts ,51, 64, & 82 pp.

TROTSENKO, G. V., – 1971 – The application of the ordination method for analysis of the boundaries among some tundra associations (In Russian with Engl. summ.). In, *Kolichestvennye metody analiza rastit.* 2: 288-292. Riga.
TSEPLYAEV, V. P., – 1965 – Forestry in the USSR (In Russian). Lesnaya promyshlennost, Moscow. 408 pp.
URANOV, A. A., – 1965 – The phytogenic field (In Russian). *Problemy̆ sovremennoi botaniki*, 1: 251-254. Nauka, Moscow & Leningrad.
VASILEVICH, V. I., – 1960 – On application of statistical methods to study of plant associations (In Russian). *Vest. leningr. gos. Univ.* 9, ser. biol. 2: 64-70.
VASILEVICH, V. I., – 1963a – A tentative morphological analysis of the meadow continuum (In Russian). *Bot. Zh. SSSR* 48(11): 1653-1659.
VASILEVICH, V. I., – 1963b – Statistical approach to plant association (In Russian). *Trudy̆ bot. Inst. Akad. Nauk SSSR*, ser. 3 (Geobotanika), 15: 94-105.
VASILEVICH, V. I., – 1966a – A study of vegetational continua (In Russian with Engl. summ.). *Trudy mosk. Obshch, ispy̆t, Prir.*, otd. biol. 27: 59-69.
VASILEVICH, V. I., – 1966b – What is the natural classification (In Russian). In, 'Filosofsk. problemy sovrem. biol,' pp. 177-190, Moscow & Leningrad.
VASILEVICH, V. I., – 1967a – About the methods of analysis of the boundaries of phytocoenoses (In Russian with Engl. summ.). *Byull. mosk. Obshch. Ispy̆t. Prir.*, otd. biol. 72(3): 85-93.
VASILEVICH, V. I., – 1967b – A continuum in the coniferous and parvifoliate forest of the Karelian isthmus (In Russian). *Bot. Zh. SSSR* 52(1): 45-53.
VASILEVICH, V. I., – 1968a – Commentary on the continuum concept of vegetation. *Bot. Rev.* 34(3): 312-314.
VASILEVICH, V. I., – 1968b – Principles for distinguishing of plant associations. Abstract of a report. See: NORIN, 1968.
VASILEVICH, V. I., – 1968c – Natural classification in phytocoenology (In Russian with Engl. summ.). *Trans Tartu State Univ. (Tartu Ülik. Toim.)* 211: 136-147.
VASILEVICH, V. I., – 1969 – Statistical Methods in Geobotany (In Russian). Nauka, Leningrad. 231 pp.
VASILEVICH, V. I., – 1971 – The representative area of the local abundance and coenoquant concept (In Russian with Engl. summ.). In, *Kolichestv. metody analiza rastitelnosti* 2: 33-38. Riga.
VASILYEV, YA. YA., – 1935 – A concept of 'forest type' and a scheme of the forest types classification (In Russian). *Sov. Bot.* 1935(1): 36-63.
VOLKOVA, V. G., – 1969 – Detailed plans of vegetation and the method of complex ordination (In Russian). *Geobot. kartogr.* Leningrad 1969: 42-51.
VORONOV, A. G., – 1963 – Geobotany (In Russian). Vysshaya shkola, Moscow. 273 pp.
VYSOTSKI, G. N. (VYSSOTZKY, G.), – 1909 – On phyto-topological maps, methods of making them and their practical value (In Russian). *Pochvovêdênïe* 11(2): 97-124.
VYSOTSKI, G. N. (WYSSOTZKY, G. N.), – 1930 – Skizzen über die hydrologischen Grundlagen der Bodenkunde. *Pochvovêdênïe* 1930(4): 5-31.
WALTER, H., – 1954 – Klimax und zonal Vegetation. *Angew. Pfl Soziol.*, Wien, Festschrift Aichinger 1: 144-150.
WHITTAKER, R. H., – 1953 – A consideration of climax theory: the climax as a population and pattern. *Ecol. Monogr.* 23: 41-78.
WHITTAKER, R. H., – 1956 – Vegetation of the Great Smoky Mountains. *Ecol. Monogr.* 26: 1-80.
WHITTAKER, R. H., – 1962 – Classification of natural communities. *Bot. Rev.* 28(1): 1-239.
WHITTAKER, R. H., – 1967 – Gradient analysis of vegetation. *Biol. Rev.* 42: 207-264.

Yaroshenko, P. D., – 1961 – Geobotany (In Russian). Izd. Akad. Nauk SSSR, Moscow & Leningrad. 474 pp.
Yurtsev, B. A., – 1966 – Hypoarctic botanico-geographical belt and the origin of its flora (In Russian). *Komarov. Chten.* 19, 93 pp. Nauka, Moscow & Leningrad.
Zinserling, Y. D., – 1938 – Vegetation of mires (In Russian). In, *Rastitolnost SSSR* 1: 355-428. Izd. Akad. Nauk SSSR, Moscow & Leningrad.

18 NORTH EUROPEAN APPROACHES TO CLASSIFICATION

Hans Trass and Nils Malmer

Contents

18.1	Introduction	203
18.2	Historical Background	203
18.2.1	Periods	203
18.2.2	Carl von Linné	204
18.2.3	Hampus von Post	204
18.2.4	Ragnar Hult	205
18.2.5	Rutger Sernander	205
18.3	The Uppsala School	206
18.3.1	Development: Gustav Einar Du Rietz	206
18.3.2	Boundaries between Communities	208
18.3.3	Minimum Area Concept	209
18.3.4	Rule of Constancy	210
18.3.5	Definition by Dominants	212
18.3.6	Classification Units	214
18.3.7	The Outcome for the Uppsala School	216
18.4	Swedish Approaches	217
18.4.1	Sampling Technique	217
18.4.2	Tabular Arrangement	221
18.4.3	Community-Types of Low and Middle Rank	222
18.4.4	Higher Groupings	223
18.5	Other Scandinavian and Baltic Approaches	228
18.5.1	The Finnish School	228
18.5.2	The Estonian School	228
18.5.3	Danish Schools	229
18.5.4	Icelandic Vegetation	230
18.5.5	Norwegian Mountain Vegetation	231
18.6	Conclusion	232
18.7	Summary	233

18 NORTH EUROPEAN APPROACHES TO CLASSIFICATION

18.1 Introduction

In the four Scandinavian countries – Sweden, Norway, Denmark and Iceland – and the four Baltic countries – Finland, Estonia, Latvia, and Lithuania – vegetation study and the development of classifications have proceeded along a number of paths. There are, however, linkages among the several schools of vegetation science in the Scandinavian and Baltic area, because of which WHITTAKER (1962) grouped them together in a 'Northern Tradition.' The leading plant geographers and phytocoenologists of these countries, e.g. WARMING and RAUNKIAER in Denmark, POST, SERNANDER, FRIES, and DU RIETZ in Sweden, NORDHAGEN and DAHL in Norway, CAJANDER in Finland, and LIPPMAA in Estonia have to a considerable extent influenced vegetation study not only in Northern Europe but, in numerous cases, on a broader scale. On the role of these North-European naturalists in vegetation studies WHITTAKER (1962: 38) comments, 'The magnitude of the contribution of the smaller nations at the Scandinavian and Baltic area (and Switzerland) is one of the striking features of the history of ecology as a whole.'

Other articles in this volume review the Finnish school (article 15) and the synusial approach of which Lippmaa's work in Estonia is a part (article 16). In Scandinavia itself the tradition of intensive local study of vegetation is longest, and problems of classification have been subject to most extensive investigation, in Sweden. We shall therefore in this review concern ourselves primarily with the 'school of Uppsala' and other Swedish approaches, secondarily with other Scandinavian and Baltic work.

18.2 Historical Background

18.2.1 PERIODS

In Swedish vegetation study four periods can be distinguished: (i) from the first half of the 18th century to the middle of the 19th century – from LINNÉ to POST – i.e., the time when vegetation groupings first made their appearance in Swedish natural science, (ii)

203

from the middle of the 19th century to the beginning of the 20th century – from POST until the foundation of the Uppsala Institution of Plant Biology (1914) i.e. the formation of the inductive approach in Swedish vegetation science, (iii) from the main work of FRIES (1913) and foundation of the Institution to World War II – i.e. the golden age of the Uppsala school, and (iv) the post-war period – i.e. the time when the Uppsala school merged in other trends and schools and its influence weakened.

18.2.2 CARL VON LINNÉ

As a number of DU RIETZ' studies (1942c, 1957a, 1957b) have indicated, several elements of vegetation study can be found already in LINNÉ's (1707–1778) works. LINNÉ realized the great diversity of plant cover and described a number of its types – fens and bogs, eutrophic and oligotrophic lakes, coniferous forest belts in mountains. He also presented a typology of the habitats of plants ('stationes plantarum'), characterized indicator plants, etc.

18.2.3 HAMPUS VON POST

According to DU RIETZ (1921) Hampus VON POST (1822–1911) was the founder of phytocoenology in Sweden. 'He was the first to distinguish communities of different ranks and to apply an exact method to bounded quadrats for the analysis of these, to introduce a growth-form system useful (in contrast to HUMBOLDT's) for European vegetation, and to oppose consciously the deductive approach to plant sociological research in the field.' (DU RIETZ 1921: 49). POST (1951) was of the opinion that in the study of vegetation one should proceed primarily from its composition. For that purpose he described sample plots of definite size, estimating relative importance of each species according to a 5-degree abundance scale. He applied the term 'locale' ('vegetationslokaler') to the main unit of vegetation, with relative homogeneity as the main feature of this unit. He divided locales into subunits and united them into his highest category, the 'group' ('vegetationsgrupper'). In Central Sweden he distinguished six vegetation groups–coniferous forests, deciduous forests, meadows, cultural vegetation, fresh water vegetation, and salt water vegetation.

18.2.4 RAGNAR HULT

Ragnar HULT (1857–1899), a Swede by nationality who lived in Finland, a lecturer at the University of Helsinki who wrote most of his studies in Swedish, is considered by DU RIETZ (1921: 57) one of the founders and most outstanding students of phytocoenology. His major work was an 'Essay toward analytical treatment of plant formations' (in Swedish, HULT 1881).

HULT was an antagonist of the so-called 'deductive' approach to research, believing that vegetation units distinguished according to habitat conditions are artificial and even pseudo-scientific. HULT writes (1881: 8–9): 'We can thus in work that has the investigation of the physiognomy of vegetation as its goal, distinguish two directions: the deductive and the inductive ... The former assumes as known the external conditions on which the occurrence of each species is dependent. It thus builds from the general principle that, so long as the distributional relations of species are equivalent, to the same complex of local relationships must correspond the same formation... The inductive school proceeds from the observation of the manner of grouping of plant communities as this appears in nature. Through analysis of the composition of formations it seeks the laws for these, and through the analysis of external relationships under which formations occur it seeks the laws for their distribution on the earth's surface. It comes from actualities to principles, from the particular to the general.' When studing plant communities HULT proceeded from rather detailed structural divisions – he distinguished between layers, main-forms and vegetation-forms. He describes his vegetation units as 'formations' (e.g., Pineta cladineta, Cladineta pura, etc.) by strictly observing their growth-form structure, and for illustrative purposes representing their structure in graphs. In his later writings HULT paid more attention to environmental conditions along with his physiognomic methods. In his analysis of the alpine plant communitives of Northern Finland (HULT 1885) he accepted 29 'formations.' Though these are physiognomic units, defined by stratal structure, they are narrow units on the level of current sociations or associations, not formations. Among these 'formations' he recognized two large groups – multistratal and unistratal formations. This was a classification surpassing in detail and effectiveness all preceding treatments of the plant cover.

18.2.5 RUTGER SERNANDER

In 1886 HULT organized a plant geographic field seminar in

Finland where he demonstrated his methods. Among the participants was a 20-year-old Swedish botanist, Rutger SERNANDER (1866–1944), who after his return home began to propagate HULT's views in Sweden. SERNANDER (1894, 1898) emphasized the importance of small sample plots of definite size in studying plant communities, and introduced a percentage scale for estimating the importance values of plant species. His many-sided investigations inspired a lively interest in vegetation studies and classification among botanists in Sweden. During the first decade of the present century dozens of papers were published in Sweden describing vegetation types, proceeding primarily from POST's, HULT's and SERNANDER's methods. The most important of these papers came from the University of Uppsala where, under the influence of SERNANDER, the 'Uppsala school' of vegetation study developed.

18.3 The Uppsala School

18.3.1 DEVELOPMENT: GUSTAV EINAR DU RIETZ

Characteristics of the Uppsala school emerged during a period of years as a result of intensive research and discussions among Swedish botanists. The formation of school's ideas was greatly influenced by the Plant Biological Seminar organized by SERNANDER in 1907 and by the Uppsala Institution of Plant Biology – Växtbiologiska Institution – that won world-wide fame after its foundation by SERNANDER in 1914 (R. E. FRIES 1950). In the years 1913–1921 a number of papers contributing to the development of phytocoenology were published, among them both monographs of field research and analyses of community theory.

In 1913 Thore C. E. FRIES (1892–1936) published his classic monograph on alpine and subalpine vegetation in Lapland (FRIES 1913). Here FRIES took into regular use the term 'association.' The association was regarded as a vegetation type uniform, on the whole, in physiognomy and floristic composition; whereas the formation was characterized by physiognomic correspondence without reference to floristic composition. These small and numerous 'associations' represented the emergence, in the line of evolution from the units of POST and HULT, in essentially its modern form of the unit that was to be known as the 'sociation.' Associations were grouped into 'series,' broader units that had been proposed by NILSSON (1902, see 18.4.4).

FRIES' paper was by 1920 followed by a series of monographs, many of them on mountain vegetation – SAMUELSSON (1917a, 1917b),

MELIN (1917), TENGWALL (1920), and SMITH (1920) – and the following decade produced monographs dealing with diverse communities – Osvald 1923, 1925, MALMSTRÖM 1923, BLOMGREN & NAUMANN 1925, DU RIETZ 1925a, b, ALMQUIST 1929, BOOBERG 1930, and the Norwegian work of NORDHAGEN (1923, 1928). Some of these authors agreed with FRIES in considering floristic-physiognomic classifications as correct and natural (SMITH, TENGWALL), while others prefered ecological (environmental) classification. Thus for example, SAMUELSSON (1917a: 31) writes that the natural distribution of plant communities of a definite area can be obtained only if their ecology has been studied as thoroughly as possible. Otherwise, these aims will be inaccessible to us and we confine ourselves to groupings, to a smaller or greater extent artificial. Ideas of this kind were criticized by DU RIETZ, TENGWALL, OSVALD, and other representatives of the emerging Uppsala school.

From field experience and discussion a *theory* of phytocoenology – a coherent set of principles and concepts linked with one another and field technique – took shape. The leader in the formulation of this theory, the major figure in advocating the distinctive approach to vegetation of the school of Uppsala, was G. E. DU RIETZ (1885–1967), who stated the viewpoint in a series of papers: DU RIETZ (1917) discussing the 'association-complex' and other concepts; DU RIETZ, FRIES, & TENGWALL (1918) criticizing the deductive approach and presenting principles for the determination of minimum-area, with a survey of Scandinavian plant formations; DU RIETZ, FRIES, OSVALD, & TENGWALL (1920) outlining the theory of constant species and the principles for their application in the delimitation of vegetation units; and DU RIETZ (1921), the major work summarizing and generalizing the views of the Uppsala school.

The last two are most important in stating the theory, which we may summarize. Each association is considered to possess a number of species (or at least one species) that occur in all the plots of the association, provided these plots exceed the minimum area. The association may be defined by these species, the 'constants,' which occur in 90 percent or more of samples, either locally or throughout the range of the association. The number of constants normally exceeds the number of species in the intermediate constancy classes; the association is thus made up of a nucleus of constant species, forming the dominant part of the vegetational structure of the community, and a variable group of incidental and accompanying species. The association was thought also to be characterized by a definite minimum area for sampling, related to its richness in species and the size of individual plants. When sample size was increased below the minimum area, the number of constants in-

207

creased rapidly; but the number of constants remained the same with further increase in sample size beyond the minimum area. Associations were thought to contact one another at sharp boundaries, where the group of species characterizing one abruptly replaced the group characterizing the other; the transition between associations was normally extremely small in relation to the area of the associations themselves. From these principles a full definition of the association could be formulated (Du Rietz et al. 1920, Du Rietz 1923): An association is a complex of species-combinations which recur with especial frequency in nature and possess a common nucleus of species (constants) almost never lacking and present in more or less definite quantitative relations; this complex is as a rule sharply bounded in relation to other comparable species-combinations – i.e. through the lack or relative rarity of intermediate species-combinations. We may consider how the component ideas of the theory were treated by further study. We shall discuss only some aspects of the polemic battles of the northern and southern schools, for which the interested reader may see also Du Rietz (1924, 1928), Du Rietz & Gams (1924), Kylin (1926), Nordhagen (1928), Faegri (1937), Braun-Blanquet (1925), Rübel (1925, 1927), Wangerin (1925), Pavillard (1927, 1935), Lüdi (1928), and the summaries of Whittaker (1962), Aleksandrova (1969) and Shimwell (1971).

18.3.2 Boundaries between Communities

One of the most important theoretical assertions of the Uppsala school, and of Du Rietz (1921, 1922, 1924) particularly, was the normal occurrence of sharp boundaries between plant communities. Du Rietz (1921: 195—6) concluded that, '...boundaries are, as a rule, astonishingly sharp and distinct...diffuse boundaries and slowly intergrading, continuous transitions – such as might well be characteristic of associations of sites with continuous change in environmental factors – have not yet been found in any exactly analyzed case.' In the course of time only the wording and the imperative character of this assertion changed (1923, 1924). Having studied the vegetation of other areas (Swiss Alps), Du Rietz admits that narrow transitional belts can exist between stands in the areas greatly disturbed by man, but considered that phytocoenoses are integral in nature and have sharp boundaries. Du Rietz (1923) states that as among plant species there are also among associations 'good' and 'weak' ones, but he regards the former as typical and prevalent.

Since in general Du Rietz preferred compromises and took

into consideration the results obtained by others, his steadfastness to this tenet should be considered exceptional. It may be pointed out that by the time he started his researches a number of ecologists and plant sociologists had already published data showing the obscurity of boundaries between communities and the frequently continuous character of plant cover (RAMENSKY 1915, 1918, 1924, see article 5, NEGRI 1914, GLEASON 1926). DU RIETZ's concept of sharp boundaries was criticized by a number of his contemporaries (FRÖDIN 1922, KYLIN 1923, NORDHAGEN 1928, LÜDI 1928), and was later opposed by the continuum theory as this was worked out and penetrated into several ecological schools (see summaries: ALEKSANDROVA 1965, VASSILEVITCH 1966, GOODALL 1963, MCINTOSH 1967, NITZENKO 1969, TRASS 1966, WHITTAKER 1962, 1967). The attachment to the idea of sharp boundaries may be explained psychologically by the fact that the denial of this tenet would have led to the necessity of renouncing other basic principles of classification closely associated with the problem of boundaries. In this connection it should be pointed out that several of DU RIETZ' colleagues, leaders of great schools, are or were antagonists of the continuum theory (BRAUN-BLANQUET 1928b, PAVILLARD 1928, NICHOLS 1929, see also WHITTAKER 1962: 82, SUKATSCHEV 1964).

18.3.3 MINIMUM AREA CONCEPT

Each association was considered by DU RIETZ to have a characteristic minimum area. Using his own and other Uppsala phytocoenologists' data, DU RIETZ (1921: 174) drew the conclusion that in Scandinavia minimum areas for associations are as follows: in floristically poor forests $1-4$ m², in floristically rich forests slightly larger, in shrubby formations, heaths, bogs, pure lichen formations and presumably on floristically poor meadows they are, as a rule, under 1 m², in floristically rich shrubby formations and on most of the meadows $1-4$ m², and on floristically richest meadows over 4 m². One notes that the ideas of minimum area and constant species are coupled: The minimum area is that in which all the constants occur, and the constants are the species that occur in the minimum area.

The minimum area concept, as based on constants, was criticized by a number of phytocoenologists and ecologists (NORDHAGEN 1922, PEARSALL 1924, LIPPMAA 1931, 1933, CAIN 1934, 1938, 1943, ASHBY 1936, SCHENNIKOV 1937, RAMENSKY 1937, VESTAL 1949, GOODALL 1952 – a good bibliography, 1961, GREIG-SMITH 1964, URANOV 1966, et al.). Several authors reached the conclusion that the sizes of minimum areas presented by Uppsala researchers were unreal, being frequently considerably larger in nature (KONOVALOV & POVARNITZYN 1927, NOSKOVA 1928, KONOVALOV 1935). SCHALYT (1935) showed that in the steppe the minimum area is not determinable; a more or less steady increase in the constants takes place as the analysis area is enlarged. Instead of the minimum area based on the constants, the species-area curve ('Artenzahl-Arealkurve') method has been widely applied (BRAUN-FLANQUET 1928a, 1951, CAIN 1938, TÜXEN 1970). DU RIETZ (1957a: 11) points out that '...minimal area should not be confused with the minimal area (Minimalraum) of BRAUN-BLANQUET, which rests on the species-area-curve, not on the constant-area-curve as our minimal area.' DU RIETZ (1957c: 12) goes on to say that '...the importance of the minimal area concept was obviously somewhat overrated in the vigorous discussions of the twenties. Nevertheless, the investigations mentioned had the important practical result that Scandinavian phytosociologists went over from the use of large and not very homogeneous sample areas to smaller, purer and more homogeneous ones...'

On the other hand, the minimal area concept based not on constants but on the whole floristic composition of the community has also been criticized as not mathematically demonstrable (see GREIG-SMITH 1964). GOODALL (1961: 191) writes that '...the hypothesis of continuous increase in variance is inconsistent with the concept of a homogeneous plant community, but fully in accord with that of vegetational continua.' There have been several attempts, not evidently successful, to express minimal area more objectively (CAIN 1943, GOODALL 1954, VESTAL 1949, ARCHIBALD 1949, et al.), but we shall not dwell upon them here. The term 'representative area' has been widely applied in Soviet phytocoenology (RAMENSKY 1924, 1937, 1938). Thus it is necessary to distinguish among three notions – minimum area (constants), minimal area (floristic composition) and representative area (complex of features in Soviet practice). The representative area is the minimal sample size that embraces all basic features of a given community, including floristic composition, quantitative interrelations between species, and structure (URANOV 1966); but no satisfactory statistical method for the determination of this complex notion has been worked out. Perhaps, for reasons involving fundamental characteristics of communities, it is not possible to do so.

18.3.4 RULE OF CONSTANCY

The notion of constants was introduced by the Swiss naturalist

BROCKMANN-JEROSCH (1907); DU RIETZ made this concept central to his theory of the association. He stressed repeatedly that the constants characteristic of each association are already noticeable in case of relatively small quadrats, while in further enlarging these the constants do not increase in number (DU RIETZ 1921). Illustrating the constant, accessory and accidental species in graphs (op. cit., Fig. 5, 6, 7, 8, 9) DU RIETZ asserted that there is an abrupt division or cleft between the constants and other species. The property stated for constants – that they gain constancy already in small quadrats and are consequently sharply distinguished from the rest of the species – is considered by DU RIETZ as a most essential feature of the association. DU RIETZ (1921: 200) held the view that constancy relations come into being as a result of the competition among plant species. Constants serve as the nucleus from which it is necessary to proceed in the study and grouping of plant communities.

The Uppsala school's belief in constants and their essential role in classifying communities has been much criticized (ARRHENIUS 1921, NORDHAGEN 1924, KYLIN 1926, LIPPMAA 1931, 1933, 1938, ASHBY 1936, VAGA 1940, KONOVALOV & POVARNITZYN 1927, SCHENNIKOV 1937, NOSKOVA 1928, RAMENSKY 1937, URANOV 1925, 1966, SCHALYT 1935, GREIG-SMITH 1964, GOODALL 1952, 1961, VESTAL 1949, et al.). Vegetation *lacks* the clear division between constants and other species groups that the Uppsala phytosociologists thought they had found. NORDHAGEN (1922: 43) wrote: 'The assertion of DU RIETZ, that there is a fundamental distinction between constants and accessory species, is quite untenable.' KYLIN (1926: 158) asserted: 'The constancy concept of the Swiss school is in principle correct, which one cannot say of that of the Uppsala school. If, however, we accept the strong requirements of the constants in the Uppsala school and add these as criteria to the constancy concept of the Swiss, then I think we retain something useful for phytosociology.' In plant sociological studies calculation of the constancy (or presence) degree of each species is one basis of estimating the degree of homogeneity of a set of samples for a community-type. In this use the role of constancy is reduced from a theoretical concept and essential characteristic of vegetation units, to just one of several diagnostic features of such units.

LIPPMAA (1938) subjected the nature of constants to thorough analysis. Having compiled the distribution maps of the Galeobdolon-Asperula-Asarum union on the basis of character species and constants, he found that *'In the areal definition of a Union it is not equivalent whether one proceeds from character-species or from the constants.* Actually, only the character-species permit an areal definition.

The constant species are often widely distributed species, that only because of their abundance are always recurring in stand samples and are in this way constant, though this constancy of course in no way precludes their occurrence as constants in one or more other unions as well.' (LIPPMAA 1938: 83—84).

18.3.5 DEFINITION BY DOMINANTS

The abandonment of the cleft between constants and other species (with the implications of this for the concepts of the minimum area and the association)was a decisive setback for the Uppsala theory. Already, however, associations had been recognized by the dominant species of strata; and concept was now accommodated to practice – the association was to be defined by dominants (DU RIETZ 1930a, b, 1932, 1936). The artillery of criticism pivoted to follow the school into its new position. Meanwhile the Uppsala 'association' was renamed the 'sociation' (DU RIETZ 1930a, b, 1936).

A sociation was defined as a particular combination of stratal units, with each stratal unit in turn being characterized by the dominance of one or more species. (The stratal units in question have been at different times termed 'consocions,' 'socions,' and 'unions,' see DU RIETZ 1930a, b, 1936 and article 16.2.5). In practice the sociation was characterized by the dominant species of the different strata; a different and broader unit, the consociation, was characterized by the dominant species of a single layer (usually the uppermost) while the other layers might be heterogeneous to any degree. Definition of the sociation by stratal dominants did not imply that every possible combination of dominants became a different sociation; for in a given layer a sociation might be dominated by a number of species occurring in more or less variable quantities. (DU RIETZ 1930b: 376). Nevertheless, it is evident that the number of stratal combinations may be very large, particularly in multistratal vegetation. Numerous examples can be presented to illustrate the abundance of sociations: OSVALD (1923) distinguished 164 in a bog five by eight miles in area, KATZ (1928) found 81 on six small fens in the Moscow Region, BOOBERG (1930) 79 on a fen in Northern Sweden, see also MATVEYEVA (1967), etc. We shall give only a modest example of the combinations of dominants into sociations from Estonian work (Table I, MASING 1969). Five 'associations' are each characterized by *Pinus sylvestris* and one or two other socions or stratal dominants as indicated in the heading of the table; and each association groups together sociations differing in stratal structure in other respects. Thus the nine roman nu-

TABLE I

Sociations described in Estonian bogs (MASING 1969)

...tions presented in the columns of the table belong to the following associations: I Pinus ...stris – Calluna vulgaris – Sphagnum Ass. II Pinus sylvestris – Ledum palustre – Sphag... Ass. – III Pinus sylvestris – Ledum palustre – Pleurozium Schreberi Ass. IV Pinus syl... ...is – Calluna vulgaris Ass. V Pinus sylvestris – Cladonia Ass.

		Dominants of moss layer			
inant ...ee ...er, ...sity	Dominants field layer	Sphagna (Sphagnum angusti- folium etc. (excl. S. fuscum)	Forest mosses (Pleurozium Shreberi etc.)	Lichens (Cladonia species, subgenus Cladina)	Absent (Very thin layer)
	Ledum palustre	II	III	—	III
	Chamaedaphne calyculata	II	III	—	III
	Vaccinium uliginosum	II	III	—	III
s sylvestris	Vaccinium myrtillus	—	III	—	III
raised	Vaccinium vitis-idaea	—	III	V	III
orm)	Calluna vulgaris	I	IV	V	—
	Empetrum nigrum	I	IV	—	IV
	Rubus chamaemorus	I	—	—	—
	A very sparse layer	II	—	V	—

meral III's in the table are nine sociations grouped together into the Pinus sylvestris-Ledum palustre-Pleurozium schreberi 'association,' that differ either in dominance of *Pleurozium schreberi* or occurrence of only a sparse moss layer on the one hand, and in dominance of *Ledum palustre* or one of four other shrub species in the field layer on the other hand. The roman numerals in the body of the table thus represent twenty-two different sociations, and these do not exhaust the combinations that could be recognized as such.

From such observations the basing of classification on dominant species has been widely criticized (see also BECKING 1957: 424, WHITTAKER 1962, and article 14). PAVILLARD (1935: 212) wrote 'The practical result of the initiative of the Upsala group was ... to pulverize the vegetation of their respectively countries into an innumerable multitude of minute rudimentary groups ... generally covering a very restricted area ...'. GOODALL (1953: 41) – 'Communities defined on the dominant species of the highest stratum only are apt to be too broad for general use; if they are defined on a combination of dominants in each stratum, the vegetation is often divided into a large number of inconveniently small units.' VAGA (1940: 150) – 'Sociation sensu DU RIETZ is a piece of multistratal vegetation theoretically cut out in such a manner that each of its strata (socion) consists of one or more dominant species. The number of such theoretical units may be very great and therefore 'sociation' can not be considered as the fundamental multistratal unit.' In the Soviet Union as well, where formerly classification by dominants almost completely prevailed (excluding RAMENSKY and Estonian authors), this practice has been opposed and the basing of associations on a complex of features advocated (RAMENSKY 1952, NITZENKO 1961, MASING & TRASS 1963, TRASS 1963a, 1963b, 1965b, MIRKIN 1965, 1968, et al.).

It by no means follows from these criticisms that study of sociations with the Uppsala methods is useless. Sociations express the structural variation of plant communities, in some vegetation they are effective means for the detailed study of those communities. But as observed by VAGA above, these units in their multiplicity are not 'natural,' fundamental,' units in terms of which all vegetation or even northern vegetation must be conceived. Neither in constancy nor in dominance did the Uppsala school find a redoubt in which its units, as expressions of a theory of the structure of vegetation in general, could be defended.

18.3.6 CLASSIFICATION UNITS

There has thus been a progression in the criteria to be used in defining vegetation units from constants through dominants towards diagnostic species resembling those of the School of BRAUN-BLANQUET. Representatives of the Uppsala school in their first papers regarded *constant species* and *physiognomy* as primary features for distinguishing the main unit of vegetation classification (DU RIETZ et al., 1920: 18 – 'An association is a plant community with certain constants and a certain physiognomy'). DU RIETZ (1930b: 307) later referred to *constant dominants* 'A sociation is a stable phytocoenose of essentially homogeneous species composition, that is at least with constant dominants in each layer.' – and to *dominants* as the criteria for distinguishing the basic units. DU RIETZ (1932: 63 – 64) – 'A consocion is a relatively homogeneous plant population of species that all belong to the same layer, dominated by a certain species or a few species together. Such a consocion can, either alone or with one or more other consocions of other layers, form a sociation – that is, the plant community forming the whole vegetation of its site and formed by either a certain consocion or a combination of certain consocions of different layers.' DU RIETZ (1957c: 25) – 'The lowest unit in the series of biocoenoses, the sociation, is a biocoenose with a definite and homogeneous composition in each layer represented, i.e. each layer has the same dominant or dominants in every part of the sociation concerned, and the general list of species is at least not more variable than in the association of which the sociation forms a part.' While emphasizing dominants and sociations, Uppsala phytosociologists were already devoting attention to diagnostic species and associations. Associations may currently be distinguished on the basis of two types of diagnostic species (see 18.4.3 and article 20) – character-species ('Leitarten,' 'Kennarten') and differential-species ('Scheidearten,' 'Trennarten') (ALBERTSON 1946 GJAEREVOLL 1956, SJÖRS 1956, GILLNER 1960, et al.).

The most essential clarification was made possibly by the understanding that there might be different basic vegetation units as defined by dominants or by diagnostic species. At the Sixth Botanical Congress in Amsterdam in 1935 (DU RIETZ 1936, CAIN 1936) three resolutions were accepted: (i) To use the term *sociation* for vegetation units characterized mainly by dominance in the different layers, in the sense of Scandinavian phytosociologists; (ii) To use the term *association* for vegetation units characterized mainly by characteristic- and differential-species in the sense of Zürich-Montpellier (i.e. of BRAUN-BLANQUET), or at least for units of the same order of sociological value; *subassociation* and *facies* can, where appropriate, be used for their subordinate units; and (iii) To unite sociations and associations into *alliances* in the sense of Zürich-Montpellier, and the alliances into higher units. It was thus accepted that the association in the sense of BRAUN-BLANQUET was suitable for the conditions of southern Europe and as the general unit, whereas the appropriateness of the sociation was more limited, applying especially to the relatively poor flora of Scandinavia and other areas where community-types are best defined by dominance.

During the earlier period of the Uppsala school only a few units, mainly 'associations,' 'formations,' and NILSSON's – 'series,' were distinguished. By the end of the nineteen-twenties a complicated system of classification units was worked out (DU RIETZ 1929, 1930a, 1930b, 1932). A certain rapprochement to BRAUN-BLANQUET's school was thus taking place; BRAUN-BLANQUET's fundamental unit, the association, was accepted, but his hierarchy (alliance, order, class) was not yet applied. The units for the classification of *phytocoenoses* were as follows – sociation, consociation, association, federation, subformation, formation, panformation. Units for *synusiae* were – socion, consocion, associon, federion, subformion, formion, panformion. Among the *complexes* of phytocoenoses were distinguished – mosaic complexes, belt-complexes, vegetation regions, vegetation zones, and horizons. In 1932, after presenting in detail his views on vegetation classification (DU RIETZ 1930b), DU RIETZ (1932: 79—80) wrote in an optimistic manner: 'Whereas sociations and consociations are characterized through their dominants, and often also through other constants, associations and federations are given unity through the sociological affinity of the different dominants of their component consociations and through character-species in the sense of BRAUN-BLANQUET (1928a: 52). I give below a first effort toward a synopsis of lichen communities on rocks in the wet-halophyte zone of eastern Sweden, which should above all show in a concrete example, *that it is by no means impossible to combine study of the sociations of Northern phytosociology with that of the associations and alliances of the school of Braun-Blanquet.*'

But DU RIETZ had to go further in his pursuit of compromise. In 1935 at the Sixth International Botanical Congress in Amsterdam (DU RIETZ 1935, 1936) he introduced BRAUN-BLANQUET'S units (alliances in the first place). By the time of the Eight International Botanical Congress in Paris, 1954, he had worked out the following system of units: panformation, formation, sub-formation (order in the sense of BRAUN-BLANQUET), alliance, suballiance, association, subassociation, consociation and sociation (DU RIETZ 1957c). Meanwhile NORDHAGEN (1937, 1943) and others had also found it possible to combine use of sociations as lower units with the higher units of the school of BRAUN-BLANQUET, or to adapt the full hierarchy of that school to northern vegetation (see also DAHL 1957, GILLNER 1960, IVARSSON 1962, 1971, KIELLAND-LUND 1967, and HALLBERG 1971).

18.3.7 THE OUTCOME FOR THE UPPSALA SCHOOL

Thus were many of the distinctive ideas of the Uppsala school abandoned. The theory of the community-unit formulated by DU RIETZ 1921, DU RIETZ et al. 1920) was phytosociology's clearest and most promising statement of what WHITTAKER (1956, 1962) has called the 'association-unit theory' – the assertion that there exist among natural communities clearly-defined, sharply-bounded 'natural' units, inherent in the structure of vegetation and therefore necessarily the basis of phytocoenological method. The 'association' of the school of BRAUN-BLANQUET is a working assumption, more justified by its usefulness than by any systematic, theoretical interpretation (BRAUN-BLANQUET 1932, 1951). If asserted as a theory about a 'natural' unit, as distinguished from a useful convention, the BRAUN-BLANQUET association is as vulnerable to criticism on the basis of the principles of species individuality and community continuity as was the Uppsala sociation (*vide* discussions in, BRAUN-BLANQUET 1925, 1928b, DU RIETZ & GAMS 1924, LENOBLE 1926, 1928, ELLENBERG 1954, POORE 1956, WHITTAKER 1962). It was a signal endeavor of DU RIETZ as a scientist to offer more than a working assumption: a theory, an interpretation of the *meaning* of associations as phenomena. Under scrutiny, however, the theory suffered only misfortune. The assumption of sharp boudaries, the rule of constancy, the concept of minimum area, the interpretation of the association as a 'concrete' unit, and the underlying belief that the interactions among species would organize them into discrete units – all these fell before research that revealed aspects of the complexity and continuity of species and community relationships (see articles 2, 3).

As the theory was abandoned the practice of basing vegetation study on the constants lost its justification, and the school's practice developed from the doctrine of constants through the emphasis of dominants to acceptance of diagnostic species as bases of classification at higher levels, at least. In the process the school of Uppsala has largely become part of the expanding sphere of the school of BRAUN-BLANQUET. There is no longer a school of Uppsala as such; Swedish phytosociology now represents more a tradition than a well-defined method. A persistent feature of this tradition is the stress on small-scale variation in vegetation and the interest in vegetation units of low rank, for example the prominent position of sociations and unions in vegetation description. This emphasis may be ascribed not only to the tradition and its methods, but also to a flora rather poor in number of species and therefore including many vegetation types with few and poorly specialized species. It is often difficult to find species fulfilling the requirements for character-species, at least on the level of associations.

Thus there is no longer antagonism in principle, only difference in emphasis regarding small-scale units, quantitative sampling, stratal structure, recognition of dominance, and application of diagnostic species. We shall consider next the Swedish approach that has thus developed from the school of Uppsala, before considering further the Scandinavian and Baltic context of Swedish work.

18.4 Swedish Approaches

18.4.1 SAMPLING TECHNIQUE

Phytosociological description of vegetation has always been founded on field sampling alternating with studies of the relevés at home. The detailed sample procedures of Du RIETZ (1921: 215—216, 1930b: 370—371, 1942a) have been more or less closely followed by most authors. As in other schools the sampled areas of vegetation have been rather arbitrarily selected. For example SJÖRS (1948: 281) states that the sampling areas 'were selected in such a way that the vegetation within a square should always be as homogeneous as possible and that the different facies of the association should be represented as fully as possible in the squares.' Du RIETZ (1957c: 31) makes the same recommendation for sociations.

There has always been a strong demand for homogeneity not only of individual samples but also of the set of samples representing a sociation. For the individual sample Du RIETZ (1957c: 31) recommends that 'each small quadrat must be more homogeneous than the sociation as a whole.' Relative homogeneity or consistency with one another of the samples in a set representing a community-type is 'homotoneity' in the sense of DAHL (1957). Vegetational analyses presented (e.g., OSVALD 1923, Figs. 21, 26, 31, 43, 44 and corresponding tables) show that not only did analyzed samples always have to be referred to the same community-type, but also that the same dominating species had to be found in each of the layers represented in those samples, i.e., a sort of quantitative homotoneity (IVARSSON 1962). Such a high demand for sample consistency may bring about an important selection in the choice of sampling areas. During the last twenty years, however, the demand must have been considerably less and the representativeness of the material probably correspondingly better. Compare, for example, the samples of OSVALD (1923) and SJÖRS (1948). More stress has been laid on a qualitative consistency based on species presence (IVARSSON 1962: 26, MALMER 1962: 48, TYLER 1969a: 28).

Scandinavian phytosociologists have sometimes worked in a simple way, noting only the species occurring within an area of variable size with homogeneous vegetation, in some cases completing the species list with information about dominants and estimates of quantity and frequency (e.g. MELIN 1917, MALMSTRÖM 1937, WALDHEIM 1944a, PETTERSSON 1958, LINDGREN 1970). Such species lists have been recommended as a first step or means of study when time does not permit further analysis (Du RIETZ 1930b: 383—385, 1957c). More generally, however, small square analysis (Du RIETZ 1942a, 1957c) has always been characteristic of Scandinavian phytosociology. This is a detailed analysis of exactly delimited small sample areas of a definite size and usually (but not necessarily) in the form of a square. Their size should be larger than the minimum area (see 18.3.3), which in this case means the smallest area of a plant community in which all its constant species occur. This defi-

nition thus deviates from that given by the BRAUN-BLANQUET school
(e.g. BRAUN-BLANQUET 1951: 88, ELLENBERG 1956: 18). The size
of sampling areas most commonly used has been 1 m². However,
larger sampling areas have sometimes been used, for example,
4 m² in species-rich meadows and 16 m² in forest phytocoenoses,
but hardly ever larger ones than this. Smaller sampling areas are
frequently used, for example 0.25 m² in epilithic lichen vegetation
and in bog vegetation since the 1940's. The epiphytic vegetation on
trunks has been investigated with still smaller sampling areas varying from 0.2 to 0.04 m².

In earlier work the sampling areas were arbitrarily placed
wherever the community under investigation was found. They
could be placed either near each other on the same site or on several
separate localities (DU RIETZ 1930b: 431). Such a procedure is still
common, but today often the small square analysis is carried out
within larger areas (usually 50—1000 m²) representing a homogeneous stand or 'segment' of the investigated community or, for
example, on bogs often a mosaic complex of different communities
(DU RIETZ 1957c). Within such a stand a number of small squares
are placed in such a way that will they give a picture of the variation
of the vegetation within the stand. Only a few authors have used a
random or systematic distribution of the quadrats. If a sufficient
number of quadrats are laid out in a given stand, species occurrence
in the stand is effectively measured as *frequency*, the percentage of
samples from the stand in which a species is present. In early work
of the Uppsala school both frequency in this sense and the percentage of samples each representing a different stand in which a
species occurs (which percentage is a *constancy*) were termed
'constancy,' and some of the statements about the 'rule of constancy'
are actually statements about frequencies.

TABLE II
Coverage Scales

Degree of cover	HULT-SERNANDER-DU RIETZ		BRAUN-BLANQUET
	Part of the area covered	Middle of cover class	Part of the area covered
1	at most 1/16	1/32	less than 1/20
2	1/16–1/8	3/32	1/20–1/4
3	1/8–1/4	6/32	1/4–1/2
4	1/4–1/2	12/32	1/2–3/4
5	more than 1/2	24/32	more than 3/4

In small square analysis the coverage is estimated for each species or group of species present. Most authors have used the HULT-SERNANDER-DU RIETZ five-degree scale for this purpose (DU RIETZ 1921: 225), as given together with the BRAUN-BLANQUET (1951) scale in Table II. Earlier this scale was less distinctly defined (cf. MELIN 1917: 5, OSVALD 1923: 43), but since about 1920 this scale has been used to estimate just coverage. One notes that this scale does not involve number of shoots, or amount of plant biomass, or sociability, and in that way deviates from many other scales used in phytosociology.

Several authors have made minor additions to this HULT-SERNANDER-DU RIETZ scale. OSVALD (1923 etc.), for example, used + or − after the figure for the degree of cover in order to indicate a higher or lower cover than the middle of the cover class. Some authors have extended the scale by dividing either or both of the classes 1 and 5. Thus, for example, GILLNER (1960), IVARSSON (1962) and TYLER (1969a, 1969b) indicate when there are only one or two shoots of a species and the cover much less than 1/16 of the sampling area. Class 5 has by several authors (e.g. STEEN 1954, IVARSSON 1962, SJÖGREN 1961, 1964, TYLER 1969a) been divided in two in somewhat different ways. GILLNER (1960) and ANDERSSON (1970) have used the BRAUN-BLANQUET cover scale (BRAUN-BLANQUET 1951: 59). Several Finnish authors (e.g. KALELA 1939, TUOMIKOSKI 1942, HAVAS 1961) have directly estimated the coverage in per cent.

Also other methods for a quantitative estimation of the plant species in the quadrat analysis have been used. (See also Danish work of RAUNKIAER, 1909-10, 1912, 1918, and BÖCHER, 1936). LAGERBERG (1916) and later MALMSTRÖM (1923) used another method with a systematic distribution of 25 or 50 small (0.1 m²) subquadrats within a larger sample quadrat of 25 m². The point quadrat method published by LEVY & MADDEN (1933) and used by LINDQUIST (1931) and JULIN (1948) is rather similar. In this case the cover of a species is estimated from its percentage occurrence at 100 systematically arranged points within an exactly delimited area of 1 m². Even quite exact measurements of the cover for phytosociological purposes have been performed (THUNMARK 1931).

Most authors have treated the whole phytocoenoses even if DU RIETZ (1936, 1957c) following GAMS (1918) and LIPPMAA (1935, 1939) has strongly emphasized the different layers as independent units. NORDHAGEN (1943: 32 ff.), however, takes an opposite position. A synusial approach has been adopted mainly for lake-vegetation (e.g. DU RIETZ et al. 1939 and KAARET 1953) and for bryophyte, lichen and algal communities (see article 16). Irrespective of the treatment of the stratification in the vegetation there has always been a strong demand for the completeness in the species lists viz., at least all macrophytes (vascular plants, bryophytes, lichens, and some large algae) occurring within the sampling area should be listed.

18.4.2 Tabular Arrangement

From the beginning tables worked out by Scandinavian phytosociologists have had a very characteristic form from which the individual authors usually have deviated only slightly. The species occurring in each layer of the vegetation studied are grouped together. In terrestrial phytocoenoses rarely more than four layers are distinguished, viz., the tree layer, the shrub layer, the field (herb and dwarf-shrub) layer, and the bottom (ground surface) layer (Du Rietz 1921: 133—134). Within each layer the species are arranged according to their growth-forms in alphabetic order. In terrestrial vegetation the most commonly distinguished growth-forms are trees, shrubs, dwarf-shrubs, herbs, graminoids, bryophytes, and lichens. This means that within a vegetation table of this type a species has a fixed position in relation to the other species depending on its growth-form and layer. In such a table no consideration is given as to whether species characterize the community, nor to constancy, frequency, coverage, or any other quantitative aspect. Separate systems have been worked out for the stratification and growth-forms of aquatic vegetation (Du Rietz 1921: 134, 1930b: 390—392, 1932, 1940, Thunmark 1931, Wassén 1966: 24—27; cf. also Luther 1949 and article 16).

In a vegetation table the samples from the same locality or site usually are brought together. This is especially important when a number of small square analyses have been performed within the same stand. In that way it is possible to compare with one another at the same time the quadrats from a stand and the different stands. During the last decades many authors have presented survey tables of plant communities in which only the frequencies of the species in the community (Waldheim 1947), or the frequencies together with some indication of coverages (Sjörs 1954, Gjaerevoll 1956, Persson 1961, Malmer 1962 etc.) are given. Such survey tables usually exclude species which occur only with low frequency in all communities. Often the species and the plant community-types are arranged in such a way that the table illustrates vegetational gradients distinguished during the investigation.

18.4.3 COMMUNITY-TYPES OF LOW AND MIDDLE RANK

In Scandinavian phytosociology the greatest interest has always been concentrated on plant communities of low rank, perhaps due to the field technique adopted. As indicated above, the sociation (until 1930 designated 'association,' cf. DU RIETZ 1930b: 304 and 307, or 'soziotypus' according to NORDHAGEN 1922 and MALMSTRÖM 1923) was regarded as the fundamental unit, at that time characterized by fixed constant species and physiognomy (DU RIETZ et al. 1920, DU RIETZ 1921, OSVALD 1923). More than one constant species could dominate an association (DU RIETZ 1921: 186), and variants were sometimes distinguished by differences among the constant species. Later authors have given stronger importance to dominance when distinguishing sociations (18.3.6); and sociations were designated by naming the dominants in each layer, for example, the 'Calluna vulgaris-Cladonia rangiferina-silvatica-Soziation' (DU RIETZ 1930b: 434).

Groups of sociations ('Assoziationen') with close floristic and physiognomic relationship have been designated 'Assoziationsgruppen' (DU RIETZ 1921, OSVALD 1923). Sociations with a common dominant in one layer but not in the others may be brought together in a consociation (DU RIETZ 1930b: 311, 1932, THUNMARK 1931, ALBERTSSON 1946; cf. WARÉN 1926), but this unit has not been much used.

From about 1935 onwards (DU RIETZ 1936, 1942a, 1942b) classifications have given less emphasis to just the dominant and constant species. More stress is put upon the whole community composition, manifested as a regularly repeated combination of species. As especially important however, have been regarded the two kinds of indicator species, viz., species confined to only one plant community-type (character-species) and species confined to one of two community-types compared with each other (differential-species) (cf. 'ledart' and 'skiljeart,' DU RIETZ 1942a, 1942b). Within each of these types it is possible to distinguish species which are exclusive, i.e. entirely confined to one community only or to one of two communities being compared with each other, and species that are only preferential, i.e. occurring as a dominant or with high frequency in one community-type but not confined to that community-type (SJÖRS 1948, 1954). This system of diagnostic species is not the same as that of BRAUN-BLANQUET, and generally differential-species have had greater importance than character-species, at least on lower levels. Community-types delimited in this way are, however, more or less equivalent to associations in the sense of the BRAUN-BLANQUET school. Most authors have used this term for designating them, but some others have used the neutral term 'plant community,' e.g. MÖRNSJÖ (1969).

As units of lower rank within an association several authors have used 'variants,' which have been treated separately because of minor differences in species composition (PERSSON 1961, MALMER 1962, SONESSON 1970). Less attention has been paid by them to phytosociological units corresponding to sociations, even if the latter have been frequently used as subordinate units. DU RIETZ (1942a: 125—126) and ALBERTSSON (1946) distinguish sociations within associations, whereas NORDHAGEN (1943) and GJAEREVOLL (1949, 1950, 1956) try to find character- and differential-species making it possible to group sociations into associations in the sense of Middle-European plant sociologists. The associations are designated by using the names of one or a few of the species characteristic or of importance for the association in any other way. Generally the species names are used directly, for example the Pinus-Carex globularis-Spaghnum parvifolium-association (SJÖRS 1948: 140). It is less common to use a suffix to species names such as the common -etum and -etosum.

The authors using a synusial approach have followed many of the same ideas. Thus, for example, LINDQUIST (1931, 1938, 1954) dealing with the field and bottom layers separately in South Scandinavian broad-leaved forests distinguishes societies (in 1931 called 'socions'), and unions (only 1938) with reference to dominant species, constants and physiognomy. The 'isozionen' (DAHLBECK 1945) from seashore meadows are similar in character. JULIN (1948), ARNBORG (1940, 1943) and IVARSSON (1962, 1971) have instead used indicator species for separating the unions of the different layers in forest vegetation or areas where forest vegetation is colonizing a formerly open landscape. WALDHEIM (1944b, 1947), KRUSENSTJERNA (1945), and SJÖGREN (1961, 1964) have treated bryophyte and lichen communities of different phytocoenoses according to these general principles, SJÖGREN (1964: 16—23) also the field layer of broad-leaved forests. Article 16 further reviews synusial approaches.

18.4.4 HIGHER GROUPINGS

There has been weaker interest in the systematic arrangement of the basic vegetational units into higher ones in Scandinavian phytosociology. In the earlier Scandinavian phytosociological works the lower units were grouped by their physiognomy, stratification and dominating growth-forms. DU RIETZ offered a more elaborate system of formations (Du RIETZ 1921: 136—140, 1925b, OSVALD 1923, MALMSTRÖM 1923) and formal hierarchies based on dominance (Du RIETZ 1930b, 1936, see also 18.3.6), but these were not successful. Several modern authors have grouped associations (e.g. Du RIETZ 1942b, ALBERTSSON 1946, 1950, GJAEREVOLL 1956, PERSSON 1961, SJÖGREN 1961, 1964, MALMER 1962, MÖRNSJÖ 1969, SONESSON 1970) or sociations (NORDHAGEN 1937, KALLIOLA 1939) with reference to their species composition, especially the diagnostic species, into 'alliances' or 'federations.' Often neutral terms like 'vegetation,' 'series,' or common names as 'carr,' 'grassland,' 'heath,' etc. are used instead of a technical designation, thus avoiding definite ranking of the community-type. NORDHAGEN (1937, 1943, cf. also 1928:90 and 465—467) is one of the few Scandinavians who has worked out a complete hierarchic system including not only alliances but also orders and classes.

The reverse principle, viz., a successive division of higher vegetational units into lower ones, has also been used. 'Any plant community and its natural delimitation is established through direct field studies of the unit in question, ... not ... through arranging them' (i.e., the lower units) 'in groups with reference to their greater or smaller similarity,' (Du RIETZ 1942a: 7, see also 1949: 299). This approach has been most highly developed for bog vegetation, a panformation within which, for example, the main formation of the boreal bogs is divided into two formations, viz., Ombrosphagnetea and Sphagno-Drepanocladetea (Du RIETZ 1949, 1954). These units correspond to classes in the sense of the BRAUN-BLANQUET school and are subdivided into subformations (or orders) alliances, suballiances, etc. SJÖGREN (1961, 1964) in a similar manner has distinguished first the federations of bryophyte communities and then within them the unions.

Plant communities (mainly representing sociations and associations) occurring close together within an area, for example on a bog, have often been brought together in higher units called complexes ('Assoziationskomplexen' in Du RIETZ et al. 1918, Du RIETZ 1921, OSVALD 1923; 'Phytocoenosenkomplexen' in Du RIETZ 1930b; 'zonations' in MALMER 1962). The floristic and physiognomic relationships between the communities within a complex may be rather weak in spite of their close ecological relationship. Several of the higher units in the bog vegetation system of Du RIETZ referred to above have more the character of complexes than floristically distinguished units (MALMER 1968). Du RIETZ (1936, 1957c) even argues for regarding a phytocoenose as a vegetational complex of two or more synusiae.

Most authors have regarded all community-types they describe, as well as their indicator species, as of only local significance, viz., representing only the investigated district. Knowledge of related vegetation in other areas is usually not considered when defining the vegetation units (e.g. SJÖRS 1948). The studies on the regional differentiation of the plant communities over larger areas therefore usually are carried out through comparisons with other authors and the units distinguished by them when treating similar vegetation. This is in accordance with the opinion put forward by DU RIETZ (1942a, 1949) that vegetation regions should be regarded as phytocoenose complexes of very high rank, and that the subdivision of their phytocoenoses should be independent, using different units for each region. Many times there may exist close floristic relationships between units widely separated in this way. The approach is in contrast with that of the school of BRAUN-BLANQUET, in which not regional vegetation complexes or patterns, but a formal hierarchy that should be consistent in application to different regions, is given primary emphasis.

Several authors, especially those dealing with bog vegetation, have instead of putting the lower units of vegetation into a hierarchial system followed TUOMIKOSKI (1942) and arranged them according to directions of variation in the vegetation (SJÖRS 1948, PERSSON 1961, MALMER 1962) or vegetational gradients (MALMER 1965, 1968, MÖRNSJÖ 1969, ANDERSSON 1970, SONESSON 1970; see also HAVAS 1961). This gives a multidimensional, coordinate framework in relation to which it is possible to place each vegetational unit. Especially in mosaic complexes, such a system is very informative. It may reflect very well the characteristic features and variation of the vegetation and offer important advantages for studies on the relationship between vegetation and environment. Such a coordinate system can be changed into a conventional hierarchy by using the different gradients, one by one, as bases of dividing the vegetation into units (TUOMIKOSKI 1942).

NILSSON (1902) divided terrestrial vegetation into higher units called 'series,' which have been of great importance for several later Swedish authors. From the beginning they were characterized as ecosystems, but today they are treated as purely vegetational units each including both wooded and non-wooded vegetation types. SJÖRS (1956) recognized four units of this type. The heath series comprises plant communities dominated by dwarf-shrubs and grasses with dry and narrow leaves (e.g. *Deschampsia flexuosa*, *Nardus stricta*) together with mosses and lichens. In the meadow series dwarf-shrubs and lichens are few or lacking, herbs usually common and the grass species broad-leaved. The steppe series (cf. also STERNER 1922 and BÖCHER 1945: 123) comprises the few units of vegetation found in Sweden related to the south-east European steppe and limestone cliff vegetation. All the mire vegetation in the sense of DU RIETZ (1954) is brought together in the mire series.

A terminology related to NILSSON's system, is the common separation of 'poor' and 'rich' vegetation types. These terms are used only to characterize vegetation types according to their species composition and do not state anything about the habitat conditions or the number of species (cf. SJÖRS 1948: 281, MALMER 1960: 111 – 112, 1962: 33, PERSSON 1961: 106 ff). Following this system DU RIETZ (1945a) divided lake vegetation into poor lake vegetation (corresponding to the *Lobelia*-lakes or the oligotrophic lakes) and rich lake vegetation (*Potamogeton*-lakes or the eutrophic lakes) and also divided corticolous lichen vegetation (DU RIETZ 1945b, ALMBORN 1948) into poor bark vegetation (e.g. Physodion) and rich bark vegetation (e.g. Xanthorion). According to this terminology the heath series includes poor vegetation types, the meadow and steppe series rich vegetation types. The mire series includes both poor vegetation types (bogs and poor fen vegetation) and rich vegetation types (rich fen vegetation, cf. DU RIETZ 1949, 1954, MALMER 1965, SONESSON 1970).

One of the most striking features in the Scandinavian vegetation is the strong difference between areas with soils poor in lime (mainly Archean areas) where the number of species found is much less than in areas with soils rich in lime. This is consistently the case from the south Scandinavian lowlands to the mountains in the north. The division of vegetation into a heath series and a meadow series or into poor and rich vegetation types well reflects this important vegetational difference. This poor – rich gradient is, together with the vegetational gradient in response to the water regime of the site, often used as a coordinate system for a first arrangement of phytocoenoses (cf. e.g. GJAEREVOLL 1956) – see Fig. 1.

The lack of clearly defined higher units in Scandinavian vegetation science results partly from incomplete knowledge of the vegetation. One of the drawbacks of the Scandinavian school is that the time-consuming methods, especially in vegetation types rich in species, limit the extent of investigations. For this reason it takes a long time before a general survey is obtained. For several vegetation types and regions in Sweden there are, however, up-to-date articles summarizing results of the investigations of the different vegetation types with references and tables. As examples we may mention studies on forest vegetation (SJÖRS 1965), mire vegetation (MALMER 1965, SJÖRS et al. 1965, PERSSON 1965), mountain vegetation (GJAEREVOLL & BRINGER 1965, PERSSON 1965), steppe vegetation (ALBERTSSON 1950), salt marshes (GILLNER 1965, TYLER

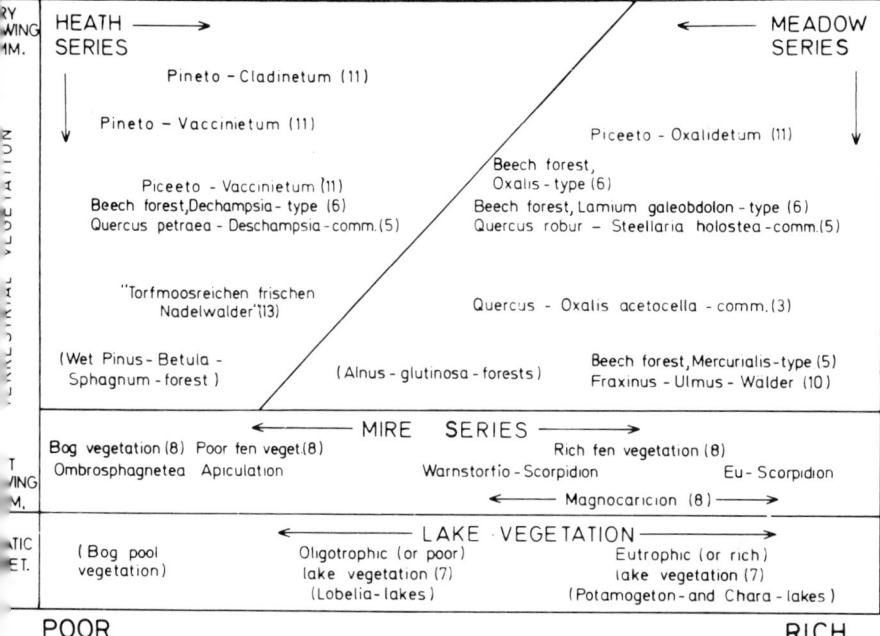

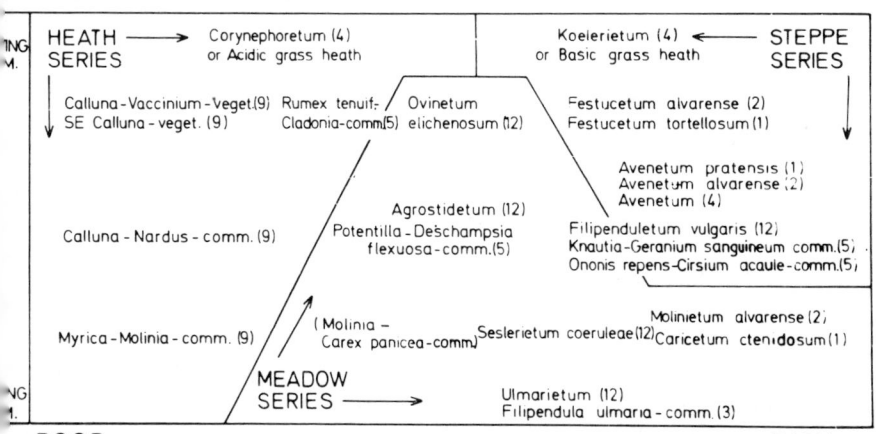

Fig. Tentative, schematic figure illustrating some vegetational units of low and middle from the Nemoral and Boreo-Nemoral regions of South Sweden in relation to vegetation along the moisture gradient and the poor-rich nutrient gradient of vegetation. See r discussion in the text! Explanations: The figures after the names of the vegetation units to the following literature (either original articles with tables or survey articles with r references), viz., 1 and 2 to ALBERTSSON 1946 and 1950; 3 to F. ANDERSSON 1970; . ANDERSSON 1950, 5 to BERGLUND 1963, 6 to LINDGREN 1970, 7 to LOHAMMAR 1965, 8 to ER 1965, 9 to MALMER et al. 1965, 10 to SJÖGREN 1964, 11 to SJÖRS 1965, 12 to STEEN nd 13 to WALDHEIM 1944a. For communities within parentheses no modern literature references are available.

227

1969b), epilithic vegetation (KRUSENTSJERNA 1965), dwarf shrubs (MALMER et al. 1965), and the different vegetation types of coastal landscapes (BERGLUND 1963, HALLBERG & IVARSSON 1965).

18.5 **Other Scandinavian and Baltic Approaches**

18.5.1 THE FINNISH SCHOOL

A. K. CAJANDER of Finland ranks with POST, HULT, and SERNANDER as one of the pioneers of Scandinavian and Baltic phytocoenology. CAJANDER (1903a, 1903b, 1903c) used an early form of the compiled tables of vegetation samples which appear in most later studies, and he established the manners of naming associations (1903a) and sociations (1903c). CAJANDER recognized 'associations' by dominant species or combinations of dominants; and on the basis of characteristics of other layers he distinguished 'facies' within these, corresponding more nearly to the sociations of Uppsala (CAJANDER 1903a, 1922, CAJANDER & ILVESSALO 1921). The concept of 'ecological series' or chains of communities along environmental gradients, characteristic of much later Finnish and Russian work, (articles 5 and 15.5) appeared in his study of the relations of alluvial vegetation to water level (CAJANDER 1903a). The contribution most closely associated with his name, however, is the system of forest site-types based on the relations of undergrowth vegetation to biotope conditions (CAJANDER 1909, 1949). The approach is discussed further in article 15.

18.5.2 THE ESTONIAN SCHOOL

DU RIETZ reached the conclusion that the layers of phytocoenoses are relatively independent units and should be classified as such. Together with GAMS (1918, 1927, 1933) DU RIETZ became an advocate of the synusial approach to communities through units comprising species of a single stratum that are relatively homogeneous in their life-forms and ecological relations. Whereas DU RIETZ advocated that both whole phytocoenoses and stratal societies should be classified, separately, the Estonian school of LIPPMAA (1933, 1935, 1938, 1939, VAGA 1940, article 16) fully shifted emphasis to the stratal units. The resulting synusial classification has not been extensively applied to vascular plant communities, but has been influential in Sweden (18.4.3) and elsewhere, particularly for the classification of thallophyte societies and aquatic plant communities (article 16). More recent Estonian applications of the synusial approach are discussed by TRASS (1964); for further review of contributions to phytocoenology in the Estonian S.S.R. see TRASS (1965a, 1970), MASING & TRASS (1963), and MASING (1969).

18.5.3 DANISH SCHOOLS

The early influence of Denmark was exerted through two of the great plant geographers, WARMING and RAUNKIAER. Along with his contribution to the physiognomic study of vegetation (WARMING 1895, 1896), WARMING (1909) had a major influence in fixing the concept of the 'formation' as a physiognomic unit, defined not by species but by plant growth-forms. RAUNKIAER (1909-10, 1910, 1934) was one of the pioneers of quantitative approaches to vegetation, and shared with the Uppsala school the emphasis on vegetational statistics and the use of small vegetation units. In contrast to WARMING, RAUNKIAER (1909-10, 1918) applied the term 'formation' to narrowly defined units, in general characterized and named by their dominant species; the formation was later (1928) defined as comprising stands agreeing in the occurrence of 'frequency-dominants,' species present in more than 80 percent of the quadrats representing a formation. Formations could be grouped into 'series' of formations by their dominant life-forms (1909-10); and a hierarchy of 'groups,' 'classes,' and 'branches' of formations was also suggested (1918). The 'formations' of RAUNKIAER and Danish phytosociologists following him correspond more nearly to the sociation than to the formations and associations of other schools. Convergence with other schools was sought by IVERSEN (1936), who identified the formation of RAUNKIAER with the sociation of DU RIETZ and grouped sociations into 'Soziation-Verbände' corresponding more nearly with the associations of BRAUN-BLANQUET. SØRENSEN (1948) sought to relate the various units through quantitative comparison of vegetation samples, using an index (article 6, formula 40) that has been widely applied.

In early work of BÖCHER (1933) in Greenland, community-types were defined by frequency dominants and termed associations, and these were grouped into formation-types in the sense of RÜBEL (1930). In later work BÖCHER (1943, BÖCHER et al. 1946) has used the sociation as a unit, together with 'types' and 'series' characterized by dominating species, physiognomy, and ecology. BÖCHER (1933, 1938) has been concerned also with the geographic relations of species and developed this approach in his monograph (1954) on vegetation complexes of Southwest Greenland. In place of the character-species of BRAUN-BLANQUET, four other types of diagnostic species are recognized: (i) Area species, or area-geographical differential species, that occur in comparable communities and situations in different areas, and by their presence and absence distinguish the communities of these different areas. (ii) Climatic species or climatic indicators that indicate, by their broader geographic distribution and climatic relations, something of the affinities of the community in which they occur. (iii) Habitat species or ecological differential species that in a given area are associated with particular edaphic and micro-climatic conditions, and hence may be character-species for the communities of these. (iv) Ecogeographical guiding species, that are of special significance for their indication of both local site or biotope, and broader geographic or climatic, affinities of a community in which they occur. These concepts were used for definition of higher units in a hierarchy from sociations through sociation groups, vegetational types, and vegetational complexes, to vegetational regions.

18.5.4 ICELANDIC VEGETATION

The Arctic tundra is usually thought a single formation or formation-type, but within this treeless vegetation there is diversity of physiognomy with changing dominance of growth-forms – dwarf-shrub, lichen, moss, and graminoid (grass and sedge) – in response to differences in biotope. The criteria of 'formations,' as major units characterized by similarity of physiognomy and broad environmental relations (article 13), can be applied to define within the tundra smaller-scale units that we might perhaps call 'microformations.' Thus in the predominantly tundra vegetation of Iceland OSTENFELD (1889, 1905) distinguished his vegetation types or 'formations' by combinations of biotope characteristics and physiognomy. Within these types lower-level 'facies' and 'associations' were characterized by their major species. The ideas of OSTENFELD and his successors were combined with those of the school of RAUNKIAER in the work of the Danish author MØLHOLM HANSEN (HANSEN 1930). Quantitative samples were taken by the methods of RAUNKIAER and classified into 'formations' as recognized by Icelandic authors. Species were classified into seven geographic groups or areal types, and 'formations' were characterized by life-form and areal-type percentages, or spectra. The distinctive Icelandic classification of arctic vegetation was applied in a series of studies by STEINDÓRSSON (1936, 1945, 1946, 1954) recognizing 'formations' and within these narrowly defined dominance-types, defined by one or more major species, as 'associations' or sociations. The work of HANSEN was, as observed by WHITTAKER (1967) a notable early venture in direct gradient analysis. 'Formations' were arranged in ecological series – by moisture conditions, snow-cover persistence, and elevation – in extensive tables that showed the changes in species populations, species diversities, and representation of RAUNKIAER life-forms and areal types along these gradients. The Danish school thus possessed, in the quantitative methods of RAUNKIAER, use of ecological series by HANSEN (1930, 1932) and similarity measurements by SØRENSEN (1948), the beginnings of what might have become a development of gradient analysis.

18.5.5 NORWEGIAN MOUNTAIN VEGETATION

In the nineteen-twenties the outstanding Norwegian phytocoenologist Rolf NORDHAGEN was one of DU RIETZ' opponents who criticized the rule of constancy, the concept of sharp boundaries, and other views of the latter (NORDHAGEN 1923, 1924, 1928; cf. the responses of DU RIETZ 1925a, 1928). NORDHAGEN (1954: 81) has said about the development of his views that up to 1928 he distinguished vegetation units on the basis of life-forms of the dominating physiognomic species, while afterwards he came to the conclusion 'that such grouping in many ways was unnatural, especially with regard to the ecological or causal side of the problems.' Having been on joint excursions to North Africa, the Alps, and the Carpathian mountains with BRAUN-BLANQUET, SZAFER, and other Middle European phytosociologists, NORDHAGEN got acquainted with basic principles of the Zürich-Montpellier school and associated himself with them. He introduced character- and differential-species as well as the fidelity concept into his classification of vegetation units – associations and alliances. At the same time NORDHAGEN (1954: 82), however, keeps to his standpoint that '... the lower or basic vegetation units must be founded upon *dominance* alone, not only of phanerogams, but of lichens and mosses as well.' In his works (NORDHAGEN 1923, 1937, 1943, 1954) he describes in great detail numerous plant communities, particularly those of the mountain areas. It is noteworthy that Nordhagen attaches much importance to *alliances* as relatively broad groupings regarded as natural units.

We have seen that the views of DU RIETZ and NORDHAGEN – as well as the latter's colleagues FAEGRI (1934, 1937), GJAEREVOLL (1949, 1956), and DAHL (1957) – on the problems of vegetation classification draw nearer and largely coincide from the nineteen-thirties on. This circumstance provided Du RIETZ with sufficient grounds to speak of a Swedish-Norwegian school. At the same time differences in minor issues survive, e.g., the application of synusial method, as well as in classification problems. DAHL's (1957) work on mountain vegetation, 'Rondane,' used the BRAUN-BLANQUET classification, but employed both dominants and differential-species for the characterization of units on the level of the association. DAHL also discussed the occurrence of discontinuity in some, and intergradation in other parts of his alpine vegetation, applied the SØRENSEN (1948) similarity index as an aid to classification, and treated problems of species diversity and homogeneity in communities (see also DAHL & HADAČ 1941, 1949, DAHL 1960).

18.6 Conclusion

What is finally to be said on interpretation of Scandinavian and Baltic phytocoenology and the extent of its contribution? There is, first, the effect that WHITTAKER (1962) terms the ecology of ecological schools – the effect of the kind of vegetation studied and the kinds of problems encountered on a school's choice of ideas and techniques. Dealing with northern vegetation that is on the whole relatively poor in species, with marked species dominance and often well-defined stratal differentiation, the northern authors naturally took directions different from those pursued in southern Europe. The emphases of small samples, careful quadrat technique, species dominance, and stratal structure common as themes to several of the northern schools may be thought responses to these vegetational circumstances. Yet northern vegetation is itself diverse enough; and effects of this diversity are evident (along with those of the ideas of leaders of schools) in the development of quite different approaches to the mountain vegetation of Sweden and Norway, the lowland forests of Finland and Estonia, the tundra of Iceland and Greenland, and the heaths and commons of Denmark. Looking superficially at the history of the Northern and Southern Traditions in European phytosociology, there is a marked contrast: in the North, a scientific effort branching into numerous schools; in the South a number of schools among which that of BRAUN-BLANQUET increasingly prevailed, unifying the efforts of phytosociologists in a great achievement.

Two approaches to an object of study may be distinguished: that of composition, which may be represented in the listing of components, and that of structure, emphasizing the quantitative and spatial relations of those components. It cannot be said that the northern schools neglected composition or that the southern schools were uninterested in structure; yet the difference in emphasis is significant. The school of BRAUN-BLANQUET concerned itself with composition and the use of diagnostic species; from the fact that among many experiments in classification this was, over all, the most successful has resulted the ascendancy of that school. The northern schools emphasized various different aspects of structure; and their approaches to structure proved less adaptable to other vegetation than the approach through composition. The approach of BRAUN-BLANQUET could be accommodated, with change of emphasis, to northern vegetation more reasonably than approaches adapted to the structure of northern vegetation could be used in other areas.

The influence of the northern schools has been exerted in other directions. From the North primarily have come the development of physiognomic ecology, quantitative techniques and quadrat studies, the study of life-forms, synusial approaches to communities, detailed forest typology, and beginnings of gradient analysis. The South has prevailed in its construction a of formal system of classification; but the North has produced a wide range of approaches including classification but extending beyond this to many ways of studying and describing communities that have affected the study of vegetation throughout the world. They have influenced Russian phytocoenology profoundly (ALEKSANDROVA 1969, article 18) and English-language ecology substantially, and have not been without influence on the Southern Tradition. A northern phytocoenologist may thus both grant the achievement of the school of BRAUN-BLANQUET, and value contributions from his own tradition that have served well for the study of northern vegetation and much influenced the study of plant communities elsewhere.

18.7 SUMMARY

One of the main lines of development in the study of northern vegetation led from HULT and POST through SERNANDER to DU RIETZ and the school of Uppsala. The Uppsala school used small sampling areas of defined size for detailed study of vegetation, and classified these samples into relatively narrow units defined (in later practice) by the dominant species of the different strata. While these units have been called 'associations' by some earlier authors, the internationally accepted term for them is now 'sociation.'

The Uppsala school developed, in connection with these units, a theory of the structure of vegetation. Sociations were considered

(i) to be characterized by a nucleus of species that were 'constant' (occurring in 90% or more of the samples of a sociation) and were sharply distinguished from other, accompanying species, (ii) to possess a definite minimum area as a sample size sufficient to include the constant species, and (iii) to be separated by sharp boundaries from other sociations. Research on these concepts forced abandonment of the theory.

Swedish phytosociology is now more a tradition in the approach to northern vegetation than a distinct school. Vegetation study in this tradition is rather more concerned with analysis of small sample plots, with stratal structure, and with quantitative relations and species dominance. The sociation was earlier used as a unit for detailed study, but today more interest is focused on associations comparable with those of BRAUN-BLANQUET. Swedish phytosociology has been on the whole little concerned with formal classification, and a number of different approaches have been used for grouping communities on higher levels.

Other Scandinavian and Baltic approaches include the use of forest site-types in Finland (CAJANDER), stratal or synusial units in Estonia (LIPPMAA), classification of Norwegian mountain vegetation (NORDHAGEN, DAHL), development in Denmark of quantitative approaches (RAUNKIAER) and study of species distributions (BÖCHER, classification by small-scale tundra 'formations' in Iceland (STEINDÓRSSON), and work anticipating gradient analysis (CAJANDER & ILVESSALO, MØLHOLM HANSEN, SØRENSEN, TUOMIKOSKI). The northern schools have produced no general system of classification as widely successful as that of BRAUN-BLANQUET, but have made major contributions through development of classifications appropriate to northern vegetation and approaches to the structure of communities that have been influential throughout the world.

Editor's note

To represent northern phytosociology in this book separate articles by Professors TRASS and MALMER were requested by the editor, and the general editor and publishers. The editor has shortened both articles and combined them into this version. Section 18.4 was contributed by MALMER and the rest of the article primarily by TRASS, while the editor also has contributed to some sections.

The two authors deserve any credit, and the editor will accept any blame, for the result of this collaboration at a distance between Sweden and the Estonian S.S.R., Germany and The Netherlands, and Ithaca, New York.

REFERENCES

ALBERTSSON, N., – 1946 – Österplana hed, ett alvarområde på Kinnekulle. (Germ. summ.). *Acta phytogeogr. suec.* 20: 1—267.
ALBERTSSON, N., – 1950 – Das grosse südliche Alvar der Insel Öland: eine pflanzensoziologische Übersicht. *Svensk bot. Tidskr.* 44: 269—331.
ALEKSANDROVA, V. D., – 1965 – On the problem of distinguishing phytocoenoses in a vegetative continuum (In Russian with Engl. summ.) *Bot. Zh. SSSR* 50 (9):1248—1259.
ALEKSANDROVA, V. D., – 1969 – Classification of vegetation: Principles of classification and classification systems of various phytocoenological schools (In Russian). 'Nauka', Leningrad. 276 pp.
ALMBORN, O., – 1948 – Distribution and ecology of some South Scandinavian lichens. *Bot. Notiser Suppl.* 1 (2): 1—252.
ALMQUIST, E., – 1929 – Upplands vegetation och flora. *Acta phytogeogr. suec.* 1: 1—622.
ANDERSSON, F., – 1970 – Ecological studies in a Scanian woodland and meadow area, southern Sweden. I. Vegetational and environmental structure. *Op. bot. Soc. bot. Lund* 27: 1—190.
ANDERSSON, O., – 1950 – The Scanian sand vegetation—a survey. *Bot. Notiser* 1950: 145—172.
ARCHIBALD, E. E. A., – 1949 – The specific character of plant communities. II. A quantitative approach. *J. Ecol.* 37: 274—288.
ARNBORG, T., – 1940 – Der Vallsjö-Wald, ein nordschwedischer Urwald. *Acta phytogeogr. suec.* 13: 128—154.
ARNBORG, T., – 1943 – Granberget: En växtbiologisk undersökning av ett sydlappländskt granskogsområde med särskild hänsyn till skogstyper och föryngring. *Norrl. Handbibl.* 14: 1—282.
ARRHENIUS, O., – 1921 – Species and area. *J. Ecol.* 9: 95—99.
ASHBY, E., – 1936 – Statistical ecology. *Bot. Rev.* 2: 221—235.
BECKING, R. W. – 1957 – The Zürich-Montpellier school of phytosociology. *Bot. Rev.* 23: 411—488.
BERGLUND, B., – 1963 – Vegetation på ön Senoren. II. Landvegetationen. *Bot. Notiser* 116: 31—80.
BLOMGREN, N. & E. NAUMANN – 1925 – Untersuchungen über die höhere Vegetation des Sees Straken bei Aneboda. *Handl. K. Fysiogr. Sällsk. Lund*, N.F. 36 (6): 1—51, *Acta Univ. Lund.* Avd. 2, 21 (6): 1—51.
BÖCHER, T. W., – 1933 – Studies on the vegetation of the east coast of Greenland between Scoresby Sound and Angmagssalik (Christian IX. s Land). *Meddr Grønland* 104 (4): 1—132.
BÖCHER, T. W., – 1936 – Om en metode til undersøgelse af konstans, skudtaethed og homogenitet. *Bot. Tidsskr.* 43: 278—304.
BÖCHER, T. W. – 1938 – Biological distributional types in the flora of Greenland: a study on the flora and plant-geography of South Greenland and East Green-

235

land between Cape Farewell and Scoresby Sound. (Danish summ.) *Meddr. Grønland* 106 (2): 1—339.

BÖCHER, T. W., – 1943 – Studies on the plant geography of the North-Atlantic heath formation. II. Danish dwarf shrub communities in relation to those of northern Europe. *Biol. Skr.*, K. danske Vidensk. Selsk. 2 (7): 1—130.

BÖCHER, T. W., – 1945 – Beiträge zur Pflanzengeographie und Ökologie dänischer Vegetation. II. Über die Waldsaum- und Graskrautgesellschaften trockener und halbtrockener Böden der Insel Seeland mit besonderer Berücksichtigung der Strandabhänge und Strandebenen. (Engl. summ.) *Biol. Skr.*, K. danske Vidensk. Selsk. 4 (1): 1—163.

BÖCHER, T. W., – 1954 – Oceanic and continental vegetational complexes in Southwest Greenland. *Meddr Grønland* 148 (1): 1—336.

BÖCHER, T. W., T. CHRISTENSEN & M. S. CHRISTIANSEN, – 1946 – Slope and dune vegetation of North Jutland. I. Himmerland. *Biol. Skr.*, K. danske Vidensk. Selsk. 4 (3): 1—78.

BOOBERG, G., – 1930 – Gisselåsmyren. En växtsociologisk och utvecklingshistorisk monografi över en jämtländsk kalkmyr. *Norrl. Handbibl.* 12: 1—329.

BRAUN-BLANQUET, J., – 1925 – Zur Wertung der Gesellschaftstreue in der Pflanzensoziologie. *Vjschr. naturf. Ges. Zürich* 70: 122—149.

BRAUN-BLANQUET, J., – 1928a – Pflanzensoziologie: Grundzüge der Vegetationskunde. Springer, Berlin. 330 pp.

BRAUN-BLANQUET, J., – 1928b – À propos d'associations végétales. *Archs Bot., Bull. mens.*, Caen 2 (4): 67—68.

BRAUN-BLANQUET, J., – 1932 – Die Pflanzensoziologie in Forschung und Lehre. I. Pflanzensoziologische Forschungsprobleme. *Biologe* 1: 175—180, and *Communs Stn. int. Géobot. médit. alp.*, Montpellier 14.

BRAUN-BLANQUET, J., – 1951 – Pflanzensoziologie: Grundzüge der Vegetationskunde. 2nd ed. Springer, Wien. 631 pp. 3rd ed., 1964, 865 pp.

BROCKMANN-JEROSCH, H., – 1907 – Die Pflanzengesellschaften der Schweizeralpen. I. Die Flora des Puschlav (Bezirk Bernina, Kanton Graubünden) und ihre Pflanzengesellschaften. Leipzig, Englemann. 438 pp.

CAIN, S. A., – 1934 – Studies on virgin hardwood forest: II. A comparison of quadrat sizes in a quantitative phytosociological study of Nash's Woods, Posey County, Indiana. *Am. Midl. Nat.* 15: 529—566.

CAIN, S. A., – 1936 – Synusiae as a basis for plant sociological field work. *Am. Midl. Nat.* 17: 665—672.

CAIN, S. A., – 1938 – The species-area curve. *Am. Midl. Nat.* 19: 573—581.

CAIN, S. A., – 1943 – Sample–plot technique applied to alpine vegetation in Wyoming. *Am. J. Bot.* 30: 240—247.

CAJANDER, A. K., – 1903a (1906) – Beiträge zur Kenntnis der Vegetation der Alluvionen des nördlichen Eurasiens. I. Die Alluvionen des unteren Lena-Thales. *Acta Soc. Sci. fenn.* 32 (1): 1—182.

CAJANDER, A. K., – 1903b (1906) – Studien über die Vegetation des Urwaldes am Lena-Fluss. *Acta Soc. Sci. fenn.* 32 (3): 1—40.

CAJANDER, A. K., – 1903c – Beiträge zur Kenntnis der Vegetation der Hochgebirge zwischen Kittilä und Muonio. *Fennia* 20 (9): 1—37.

CAJANDER, A. K., – 1909 – Ueber Waldtypen. *Acta for. fenn.* 1 (1): 1—175.

CAJANDER, A. K., – 1922 – Zur Begriffsbestimmung im Gebiet der Pflanzentopographie. *Acta for. fenn.* 20 (2): 1—8.

CAJANDER, A. K., – 1949 – Forest types and their significance. *Acta for. fenn.* 56 (4): 1—71.

CAJANDER, A. K. & Y. ILVESSALO, – 1921 – Über Waldtypen II. *Acta for. fenn.* 20 (1): 1—77.

DAHL, E., – 1957 – Rondane: mountain vegetation in South Norway and its

relation to the environment. *Skr. norske Vidensk-Akad.*, Mat.-naturv. Kl., 1956 (3): 1—374.
DAHL, E., – 1960 – Some measures of uniformity in vegetation analysis. *Ecology* 41: 805—808.
DAHL, E. & E. HADAČ, – 1941 – Strandgesellschaften der Insel Ostøy im Oslofjord: eine pflanzensoziologische Studie. *Nytt Mag. Naturvid.* 82: 251—312.
DAHL, E. & E. HADAČ, – 1949 – Homogeneity of plant communities. *Studia bot. čechoslov.* 10: 159—176.
DAHLBECK, N. – 1945 – Strandwiesen am südöstlichen Öresund. *Acta phytogeogr. suec.* 18: 1—168.
DU RIETZ, G. E., – 1917 Några synpunkter på den synekologiska vegetationsbeskrifoningens terminologi och metodik. (Germ. summ.) *Svensk bot. Tidskr.* 11: 51—71.
DU RIETZ, G. E., – 1921 – Zur methodologischen Grundlage der modernen Pflanzensoziologie. *Akad. Abhandl., Uppsala.* Holzhausen, Wien. 267 pp.
DU RIETZ, G. E., – 1922 – Die Grenzen der Assoziationen: Eine Replik an John Frödin. *Bot. Notiser* 1922: 90—96.
DU RIETZ, G. E., – 1923 – Der Kern der Art- und Assoziationsprobleme. *Bot. Notiser* 1923: 235—256.
DU RIETZ, G. E., – 1924 – Studien über die Vegetation der Alpen, mit derjenigen Skandinaviens vergleichen. *Veröff. geobot. Inst. Rübel, Zürich* 1: 31—138.
DU RIETZ, G. E., – 1925a – Zur Kenntnis der flechtenreichen Zwergstrauchleiden im kontinentalen Südnorwegen. *Handl. Svenska Växtsociol. Sällsk.* 4: 1—80.
DU RIETZ, G. E., – 1925b – Die regionale Gliederung der skandinavischen Vegetation. *Handl. Svenska Växtsociol. Sällsk.* 8: 1—60.
DU RIETZ, G. E., – 1928 – Kritik an pflanzensoziologischen Kritikern. *Bot. Notiser* 1928: 1—30.
DU RIETZ, G. E., – 1929 – The fundamental units of vegetation. *Proc. 4th Int. bot. (Plant Sci.) Congr.*, Ithaca 1926, 1: 623—627.
DU RIETZ, G. E., – 1930a – Classification and nomenclature of vegetation. *Svensk bot. Tidskr.* 24: 489—503.
DU RIETZ, G. E., – 1930b (1932) – Vegetationsforschung auf soziationsanalytischer Grundlage. *Handb. biol. ArbMeth.* 11,5 (2): 293—480.
DU RIETZ, G. E., – 1932 – Zur Vegetationsökologie der ostschwedischen Küstenfelsen. *Beih. bot. Zbl.* 49 (Erg.-Bd.): 61—112.
DU RIETZ, G. E., – 1935 – Classification and nomenclature of vegetation units 1930—1935. *Proc. Zesde Int. bot. Congr.*, Amsterdam 1935, 2: 104—105.
DU RIETZ, G. E., – 1936 – Classification and nomenclature of vegetation units 1930-1935. *Svensk bot. Tidskr.* 30: 580—589.
DU RIETZ, G. E., – 1940 – Das limnologisch-thalassologische Vegetationsstufensystem. *Verh. int. Verein. theor. angew. Limnol.* 9: 102—110.
DU RIETZ, G. E., – 1942a – Växtgeografins grunder (Mimeographed). (This paper has been rewritten several times but is only partly printed as Du Rietz 1957c. It has been available for most authors dealing with Swedish vegetation and has influenced their work. For this paper the last edition from 1961 has been used.)
DU RIETZ, G. E., – 1942b – Rishedsförband i Torneträskområdets lågfjällsbälte. (Germ. summ.) *Svensk bot. Tidskr.* 36: 124—146.
DU RIETZ, G. E., – 1942c – Linné som fjällväxtgeograf. *Årsskr. Svenska Linné-Sällsk.* 25:
DU RIETZ, G. E., – 1945a – Nitella Nordstedtiana i två uppländska sjöar. *Svensk bot. Tidskr.* 39: 83—94.
DU RIETZ, G. E., – 1945b – Om fattigbark- och rikbarksamhällen. *Svensk bot. Tidskr.* 39: 147—150.

Du Rietz, G. E., – 1949 – Huvudenheter och huvudgränser i svensk myrvegetation. (Engl. summ.) *Svensk bot. Tidskr.* 43: 274—309.
Du Rietz, G. E., – 1954 – Die Mineralbodenwasserzeigergrenze als Grundlage einer natürlichen Zweigliederung der nord- und mitteleuropäischen Moore. *Vegetatio* 5/6: 571—585.
Du Rietz, G. E., – 1957a – Linnaeus as a phytogeographer. *Vegetatio* 7: 161—168.
Du Rietz, G. E., – 1957b – Linné som myrforskare. *Uppsala Univ. Årsskr.* 5: 1—80.
Du Rietz, G. E., – 1957c – Vegetation analysis in relation to homogeneousness and size of sample areas. *Compt. Rend. Rapp. Commun., Int. bot. Congr.*, Paris 1954, sect. 7—8: 24—40. (see also Du Rietz 1942a).
Du Rietz, G. E., T. C. E. Fries & T. Å. Tengwall, – 1918 – Vorschlag zur Nomenklatur der soziologischen Pflanzengeographie. *Svensk bot. Tidskr.* 12: 145—170.
Du Rietz, G. E., T. C. E. Fries, H. Osvald & T. Å. Tengwall, – 1920 – Gesetze der Konstitution natürlicher Pflanzengesellschaften. *Vetensk. prakt. Unders. Lappl.* (Luossavaara-Kiirunavaara Aktiebolag), Flora och Fauna 7: 1—47.
Du Rietz, G. E. & H. Gams, – 1924 – Zur Bewertung der Bestandestreue bei der Behandlung der Pflanzengesellschaften. *Vjschr. naturf. Ges. Zürich* 69: 269—280.
Du Rietz, G. E., A. G. Hannerz, G. Lohammar, R. Santesson, and M. Waern 1939 – Zur Kenntnis der Vegetation des Sees Tåkern. *Acta phytogeogr. suec.* 12: 1—65.
Ellenberg, H., – 1954 – Zur Entwicklung der Vegetationssystematik in Mitteleuropa. *Angew. PflSoziol.*, Wien, Festschr. Aichinger 1: 134—143.
Ellenberg, H., – 1956 – Grundlagen der Vegetationsgliederung I. Aufgaben und Methoden der Vegetationskunde. In, 'Einführung in die Phytologie' ed. H. Walter 4(1): 1—136. Ulmer, Stuttgart.
Faegri, K., – 1934 – Über die Längenvariationen einiger Gletscher des Jostedalsbre und die dadurch bedingten Pflanzensukzessionen. *Bergens Mus. Årbok*, Naturv. rekke 1933 (7): 1—255.
Faegri, K., – 1937 – Some recent publications on phytogeography in Scandinavia. *Bot. Rev.* 3: 425—456.
Fries, R. E., – 1950 – A Short History of Botany in Sweden. Almqvist & Wiksell, Uppsala. 162 pp.
Fries, T. C. E., – 1913 – Botanische Untersuchungen im nördlichsten Schweden: Ein Beitrag zur Kenntnis der alpinen und subalpinen Vegetation in Torne Lappmark. *Vetensk. prakt. Unders. Lappl.* (Luossavaara-Kiirunavaara Aktiebolag), Flora och Fauna 2: 1—361.
Frödin, J., – 1922 – Les limites des associations: une réponse à Einar Du Rietz. *Bot. Notiser* 1922: 149—154.
Gams, H., – 1918 – Prinzipienfragen der Vegetationsforschung: Ein Beitrag zur Begriffsklärung und Methodik der Biocoenologie. *Vjschr. naturf. Ges. Zürich* 63: 293—493.
Gams, H., – 1927 – Von den Follatères zur Dent de Morcles: Vegetationsmonographie aus dem Wallis. *Beitr. geobot. Landesaufn. Schweiz* 15: 1—760.
Gams, H., – 1933 – Die Stellung der Waldtypen im Vegetationssystem. *Forstarchiv* 9: 53—59.
Gillner, V., – 1960 – Vegetations- und Standortsuntersuchungen in den Strandwiesen der schwedischen Westküste. *Acta phytogeogr. suec.* 43: 1—198.
Gillner, V., – 1965 – Salt marsh vegetation in Southern Sweden. *Acta phytogeogr. suec.* 50: 97—104.
Gjaerevoll, O., – 1949c – Snøleievegetasjonen i Oviksfjellene. (Engl. summ.) *Acta phytogeogr. suec.* 25: 1—106.

GJAEREVOLL, O., – 1950 – The snow-bed vegetation in the surroundings of Lake Torneträsk, Swedish Lappland. *Svensk bot. Tidskr.* 44: 387—440.
GJAEREVOLL, O., – 1956 – The plant communities of the Scandinavian alpine snow-beds. *Skr. norske Vidensk-Akad.* 1956 (1): 1—405.
GJAEREVOLL, O. & K. G. BRINGER, – 1965 – Plant cover of the alpine regions. *Acta phytogeogr. suec.* 50: 257—268.
GLEASON, H. A., – 1926 – The individualistic concept of the plant association. *Bull. Torrey bot. Club* 53: 7—26.
GOODALL, D. W., – 1952 – Quantitative aspects of plant distribution. *Biol. Rev.* 27: 194—245.
GOODALL, D. W., – 1953 – Objective methods for the classification of vegetation. I. The use of positive interspecific correlation. *Aust. J. Bot.* 1: 39—63.
GOODALL, D. W., – 1954 – Minimal area: a new approach. *Compt. Rend. Rapp. Commun., Huit. Int. bot. Congr.*, Paris 1954, 7—8: 19—21.
GOODALL, D. W., – 1961 – Objective methods for the classification of vegetation. IV. Pattern and minimal area. *Aust. J. Bot.* 9: 162—196.
GOODALL, D. W., – 1963 – The continuum and the individualistic association. (French summ.) *Vegetatio* 11: 297—316.
GREIG-SMITH, P., – 1964 – Quantitative Plant Ecology. 2nd ed. Butterworths, London. 256 pp.
HALLBERG, H. P., – 1971 – Vegetation auf den Schalenablagerungen in Bohuslän, Schweden. *Acta phytogeogr. suec.* 56: 1—149.
HALLBERG, H. P. & R. IVARSSON, – 1965 – Vegetation of coastal Bohuslän. *Acta phytogeogr. suec.* 50: 111—122.
HANSEN, H. MØLHOLM, – 1930 – Studies on the vegetation of Iceland. In, 'The Botany of Iceland,' ed. J. L. A. KOLDERUP ROSENVINGE & E. WARMING 3 (pt. I, no. 10): 1—186. Frimodt, Copenhagen.
HANSEN, H. MØLHOLM, – 1932 – Nørholm Hede, en formationsstatistisk Vegetationsmonografi. (Engl. summ.) *K. danske Vidensk. Selsk. Skr.*, Nat. Math. Afd., Ser. 9, 3 (3): 99—196.
HAVAS, P. – 1961 – Vegetation und Ökologie der ostfinnischen Hangmoore. *Ann. bot. Soc. Zool.-Bot. Fenn. 'Vanamo' (Suomal. eläin-ja kasvit. Seur. van. kasvit. Julk.)* 31 (2): 1—188.
HULT, R., – 1881 – Försök till analytisk behandling av växtformationerna. *Meddn Soc. Fauna Flora fenn.* 8: 1—155.
HULT, R., – 1885 – Blekinges vegetation: ett bidrag till växtformationernas utvecklingshistoria. *Meddn Soc. Fauna Flora fenn.* 12: 163—252.
IVARSSON, R. – 1962 – Lövvegetationen i Mollösunds socken. (Germ. summ.) *Acta phytogeogr. suec.* 46: 1—197.
IVARSSON, R., – 1971 – Lövvegetationen på Lindö och Kalvö i norra Bohuslän. I. Allmän översikt. *Svensk bot. Tidskr.* 65: 1—38.
IVERSEN, J., – 1936 – Biologische Pflanzentypen als Hilfsmittel in der Vegetationsforschung: ein Beitrag zur ökologische Charakterisierung und Anordnung der Pflanzengesellschaften. *Meddr Skalling-Lab.*, København 4: 1—224.
JULIN, E., – 1948 – Vessers udde, mark och vegetation i en igenväxande löväng vid Bjärka-Säby. (Germ. summ.) *Acta phytogeogr. suec.* 23: 1—186.
KAARET, P. – 1953 – Wasservegetation der Seen Orlången und Trehörningen. *Acta phytogeogr. suec.* 32: 1—64.
KALELA, A., – 1939 – Über Wiesen und wiesenartige Pflanzengesellschaften auf der Fischerhalbinsel in Petsamo Lappland. *Acta for. fenn.* 48 (2): 1—523.
KALLIOLA, R., – 1939 – Pflanzensoziologische Untersuchungen in der Alpinen Stufe Finnisch-Lapplands. *Ann. bot., Soc. Zool.-Bot. Fenn. 'Vanamo' (Suomal. eläin-ja kasvit. Seur. van. kasvit. Julk.)* 13 (2): 1—328.

KATZ, N. J., – 1928 – Zur Kenntnis der Niedermoore im Norden des Moskauer Gouvernments. *Beih. Repert. Spec. nov. Regni veg.* 56: 1—79.

KIELLAND-LUND, J., – 1967 – Zur Systematik der Kiefernwälder Fennoscandiens. *Mitt. flor.-soz. ArbGemein.* N.F., Stolzenau 11/12: 127—141.

KONOVALOV, N. A., – 1935 – On representative area of some oak associations (In Russian), *Trudy leningr. Obshch. Estest.* 64 (2): 40—55.

KONOVALOV, N. A. & V. A. POVARNITZYN, – 1927 – On the methods of statistical phytosociological study of forest associations (In Russian). *Izv. Lesn. Inst.* 35: 1—20.

KRUSENTJERNA, E. VON, – 1945 – Bladmossvegetation och bladmossflora i Uppsalatrakten. (Engl. summ.) *Acta phytogeogr. suec.* 19: 1—250.

KRUSENSTJERNA, E. VON, – 1965 – The growth on rock. *Acta phytogeogr. suec.* 50: 144—148.

KYLIN, H., – 1923 – Växtsociologiska randanmärkningar. *Bot. Notiser* 1923: 161—234.

KYLIN, H., – 1926 – Über Begriffsbildung und Statistik in der Pflanzensoziologie. *Bot. Notiser* 1926: 81—180.

LAGERBERG, T., – 1916 – Markflorans analys på objektiv grund. *Meddn St. SkogsförsksAnst.*, Stockholm 11: 129—200 & XV—XXIV.

LENOBLE, F., – 1926 (1927) – À propos des associations végétales. *Bull. Soc. bot. Fr.* 73: 873—893.

LENOBLE, F. – 1928 – Associations végétales et espèces. *Archs Bot., Bull. mens.*, Caen 2 (1): 1—14.

LEVY, E. B. & E. A. MADDEN, – 1933 – The point method of pasture analysis. *N.Z. Jl. Agric.* 46: 267—279.

LINDGREN, L., – 1970 – Beech forest vegetation in Sweden—a survey. *Bot. Notiser* 123: 401—424.

LINDQUIST, B., – 1931 – Den skandinaviska bokskogens biologi. (Engl. summ.) *SkogsvFör. Tidskr.*, Stockholm 29: 179—532.

LINDQUIST, B., – 1938 – Dalby Söderskog: en skånsk lövskog i forntid och nutid. (Germ. summ.) *Acta phytogeogr. suec.* 10: 1—273.

LINDQUIST, B., – 1954 – Ein Waldtypenschema für die skandinavischen Buchenwälder. *Angew. PflSoziol.*, Wien, Festschr. Aichinger 2: 965—970.

LIPPMAA, T., – 1931 – Pflanzensoziologische Betrachtungen. *Acta Inst. Horti bot. Univ. tartu.* 2 (3/4): 1—32, and *Loodusuur. Seltsi Aruanded* 38 (1/2): 1—32.

LIPPMAA, T., – 1933 – Taimeühingute uurimise metoodika ja Eesti taimeühingute klassifikatsiooni põhijooni (Eston. with Germ. summ.: Grundzüge der pflanzensoziologischen Methodik nebst einer Klassifikation der Pflanzenassoziationen Estlands) *Acta Inst. Horti bot. tartu* 3 (4): 1—169, and *Loodusuur. Seltsi Aruanded* 40 (1/2): 1—169.

LIPPMAA, T., – 1935 – Une analyse des forêts de l'île estonienne d'Abruka (Abro) sur la base des associations unistrates. *Acta Inst. Horti bot. tartu.* 4 (1/2, art. 5): 1—97, and *Acta Commen˙. Univ. tartu.* A 28 (1): 1—97.

LIPPMAA, T., – 1938 – Areal und Altersbestimmung einer Union (Galeobdolon-Asperula-Asarum-U.) sowie das Problem der Charakterarten und Konstanten. *Acta Inst. Horti bot. tartu.* 6 (2/3): 1—152.

LIPPMAA, T., – 1939 – The unistratal concept of plant communities (the unions). *Am. Midl. Nat.* 21: 111—145.

LOHAMMAR, G., – 1965 – The vegetation of Swedish lakes. *Acta phytogeogr. suec.* 50: 23—47.

LÜDI, W., – 1928 – Der Assoziationsbegriff in der Pflanzensoziologie, erläutert am Beispiel der Pflanzengesellschaften des Tansbodengebietes im Lauterbrunnental. *Biblthca bot.* 96: 1—93.

LUTHER, H., – 1949 – Vorschlag zu einer ökologischen Grundeinteilung der Hydrophyten. *Acta bot. fenn.* 44: 1—15.

MALMER, N., – 1960 – Some ecological studies on lakes and brooks in the South Swedish uplands. *Bot. Notiser* 113: 87—116.

MALMER, N., – 1962 – Studies on mire vegetation in the archean area of southwestern Götaland (South Sweden). I. Vegetation and habitat conditions on the Åkhult mire. *Op. bot. Soc. bot. Lund* 7 (1): 1—322.

MALMER, N., – 1965 – The southern mires. *Acta phytogeogr. suec.* 50: 149—158.

MALMER, N.,– 1968 – Über die Gliederung der Oxycocco-Sphagnetea und Scheuchzerio-Caricetea fuscae: Einige Vorschläge mit besonderer Berücksichtigung der Verhältnisse in S-Schweden. (Engl. summ.) In, 'Pflanzensoziologische Systematik,' ed. R. Tüxen, *Ber. Symp. int. Vereinig. Vegetationskunde,* Stolzenau/ Weser 1964, 8: 293—305.

MALMER, N., B. E. BERGLUND, J. ERICSON, L. PÅHLSSON, & G. RASMUSSON, – 1965 – The south-western dwarf shrub heaths. *Acta phytogeogr. suec.* 50: 123— 130.

MALMSTRÖM, C., – 1923 – Degerö stormyr. En botanisk, hydrologisk och utvecklingshistorisk undersökning över ett nordsvenskt myrkomplex. *Meddn St. SkogsförsAnst.,* Stockholm 20: 1—206.

MALMSTRÖM, C., – 1937 – Tönnersjöhedens försökspark i Halladnt. Et bidrag till kännedomen om sydvästra Sveriges skogar, ljunghedar och torvmarker. *Meddn St. SkogsförsAnst.,* Stockholm 30: 323—528.

MASING, V., – 1969 – Structural analysis of plant cover and classification problems. In, 'Plant Taxonomy, Geography and Ecology in the Estonian S.S.R.' pp. 49—59. 'Valgus,' Tallinn.

MASING, V. & H. TRASS, – 1963 – Elaboration of some theoretical problems in the works of Estonian geobotanists (In Russian with Engl. summ.). *Bot. Zh. SSSR* 48 (4): 473—485.

MATVEYEVA, E. A., – 1967 – The Meadows of Soviet Pribaltic (Comparative Analysis). 'Nauka,' Leningrad. 336 pp.

MC INTOSH, R. P., – 1967 – The continuum concept of vegetation. *Bot. Rev.* 33: 130—187.

MELIN, E., – 1917 – Studier över de norrländska myrmarkernas vegetation med särskild hänsyn till deras skogsvegetation efter torrläggning. *Norrl. Handbibl.* 7: 1—426. Almqvist & Wicksells, Uppsala.

MIRKIN, B. M., – 1965 – On the ecological classifications of meadow vegetation in foodplains (In Russian with Engl. summ.). *Bot. Zh. SSSR* 50 (3): 324—334.

MIRKIN, B. M., – 1968 – The criteria of dominants and determinants in the classification of phytocoenoses (In Russian with Engl. summ.). *Bot. Zh. SSSR* 53 (6): 767—778.

MÖRNSJÖ, T., – 1969 – Studies on vegetation and development of a peatland in Scania, south Sweden. *Op. bot. Soc. bot. Lund* 24: 1—187.

NEGRI, G., – 1914 – Le unità ecologiche fondamentali in fitogeografia. *Atti (R.) Accad. Sci., Torino* 49: 1089—1105, 1174—1198.

NICHOLS, G. E., – 1929 – Plant associations and their classification. *Proc. 4th Int. bot. (Plant Sci.) Congr.,* Ithaca 1926, 1: 629—641.

NILSSON, A., – 1902 – Svenska växtsamhällen. *Tidskr. Skogshushållning* 30: 127— 147.

NITZENKO, A. A., – 1961 – On the phytotopological classification of the plant cover (In Russian). *Trudy Inst. Biol., Ural. fil., Sverdlovsk* 27: 29—37.

NITZENKO, A. A., – 1969 – The problem of continuity and intermittence of vegetation cover (In Russian with Engl. summ.). *Zh. obshch. Biol.* 30 (4): 387—397.

NORDHAGEN, R., – 1923 – Vegetationsstudien auf der Insel Utsire im westlichen Norwegen. *Bergens Mus. Årb.*, Naturv. rekke 1920—21 (1): 1—149.
NORDHAGEN, R., – 1924 – Om homogenitet, konstans och minimiareal. *Nytt Mag. Naturvid.* 61: 1—51.
NORDHAGEN, R., – 1928 – Die Vegetation und Flora des Sylenegebietes I. Die Vegetation. *Skr. norske Vidensk-Akad.*, Mat.-naturv. Kl. 1927 (1): 1—612.
NORDHAGEN, R., – 1937 – Versuch einer neuer Einteilung der subalpinen-alpinen Vegetation Norwegens. *Bergens Mus. Årb.*, Naturv. rekke 1936 (7): 1—88.
NORDHAGEN, R., – 1943 – Sikilsdalen og Norges Fjellbeiter: en plantesosiologisk monografi. *Bergens Mus. Skr.* 22: 1—607.
NORDHAGEN, R., – 1954 – Vegetation units in the mountain areas of Scandinavia. *Veröff. geobot. Inst. Rübel, Zürich* 29: 81—95.
NOSKOVA, T. A., – 1928 – Minimum area in the forest associations (In Russian). *Dnevnik Vses. s'esda bot.* : 253—254.
OSTENFELD, C. H., – 1889 – Skildringer af Vegetationen i Island. I-II. *Bot. Tidsskr.* 22: 227—253.
OSTENFELD, C. H. – 1905 – Skildringer af Vegetationen i Island. III-IV. *Bot. Tidsskr.* 27: 111—122.
OSVALD, H., – 1923 – Die Vegetation des Hochmoores Komosse. *Handl. Svenska Växtsociol. Sällsk.* 1: 1—436.
OSVALD, H., – 1925 – Zur Vegetation der ozeanischen Hochmoore in Norwegen. *Handl. Svenska Växtsociol. Sällsk.* 7: 1—106.
PAVILLARD, J., – 1927 – Les tendences actuelles de la phytosociologie. *Archs Bot., Bull. mens.*, Caen 1 (6): 89—112.
PAVILLARD, J., – 1928 – Espèces et associations. *Archs Bot., Bull. mens.*, Caen 2 (4): 68—72.
PAVILLARD, J., – 1935 – The present status of the plant association. *Bot. Rev.* 1: 210—232.
PEARSALL, W. H., – 1924 – The statistical analysis of vegetation: a criticism of the concepts and methods of the Upsala school. *J. Ecol.* 12: 135—139.
PERSSON, Å., – 1961 – Mire and spring vegetation in an area north of Lake Torneträsk, Torne Lappmark, Sweden. I. Description of the vegetation. *Op. bot. Soc. Bot. Lund* 6 (1): 1—187.
PERSSON, Å., – 1965 – Mountain mires. *Acta phytogeogr. suec.* 50: 249—256.
PETTERSSON, B., – 1958 – Dynamik och konstans i Gotlands flora och vegetation. *Acta phytogeogr. suec.* 40: 1—288.
POORE, M. E. D., – 1956 – The use of phytosociological methods in ecological investigations. IV. General discussion of phytosociological problems. *J. Ecol.* 44: 28—50.
POST, H. VON, – 1851 – Om vextgeografiska skildringer. *Bot. Notiser* 1851: 110—127, 161—187.
RAMENSKY, L. G., – 1915 – The problem of quantitative study of herb cover (In Russian). *Materialy po organiz. i kulture kormov. ploshchadi* 12: 105—140.
RAMENSKY, L. G., – 1918 – Study of meadows in Voronesh gouvernements (In Russian). *Mat. po jest-ist. issl. Voron. gub.* : 63—93.
RAMENSKY, L. G., – 1924 – Main regularities of the plant cover and their study (In Russian). *Vêstnik opytnogo dêla*, Voronezh, pp. 37—73.
RAMENSKY, L. G., – 1937 – Calculation and Description of the Vegetation (on the Basis of Projection Method) (In Russian). Moscow. 100 pp.
RAMENSKY, L. G., – 1938 – Introduction to the Complex Pedologic-Geobotanic Investigation of Landscapes (In Russian). Selkhozgiz, Moscow. 620 pp.
RAMENSKY, L. G., – 1952 – On some principal positions in contemporary geobotany (In Russian). *Bot. Zh. SSSR* 37 (2): 181—201.

RAUNKIAER, C., – 1909-10 – Formationsundersøgelse og Formationsstatistik. *Bot. Tidsskr.* 30: 20—132.
RAUNKIAER, C., – 1910 – Statistik der Lebensformen als Grundlage für die biologischen Pflanzengeographie. *Beih. bot. Zbl.*, Abt. 2, 27: 171—206d.
RAUNKIAER, C., – 1912 – Measuring apparatus for statistical investigations of plant-formations. *Bot. Tidsskr.* 33: 45—48.
RAUNKIAER, C., – 1918 – Recherches statistiques sur les formations végétales. *Biol. Meddr*, K. danske Vidensk. Selsk. 1 (3): 1—80.
RAUNKIAER, C., – 1928 – Dominansareal, Artstaethed og Formationsdominanter. *Biol. Meddr*, K. danske Vidensk. Selsk. 7 (1): 1—47.
RAUNKIAER, C., – 1934 – The Life Forms of Plants and Statistical Plant Geography. Clarendon, Oxford. 632 pp.
RÜBEL, E., – 1925 – Betrachtung über einige pflanzensoziologische Auffassungsdifferenzen: Verständigungsbeitrag Schweden-Schweiz. *Veröff. geobot. Inst. Rübel, Zürich.* Beibl. 2: 1—12.
RÜBEL, E., – 1927 – Einige skandinavische Vegetationsprobleme. *Veröff. geobot. Inst. Rübel, Zürich* 4: 19—41.
RÜBEL, E., – 1930 – Pflanzengesellschaften der Erde. Huber, Bern-Berlin. 464 pp.
SAMUELSSON, G., – 1917a – Studien über die Vegetation der Hochgebirgsgegenden von Dalarne. *Nova Acta. R. Soc. Scient. upsal.*, Ser. 4, 4 (8): 1—252.
SAMUELSSON, G., – 1917b – Studien über die Vegetation bei Finse im inneren Hardanger. *Nytt Mag. Naturvid.* 55: 1—108.
SCHALYT, M. S., – 1935 – Rule of constancy and minimum area in steppes of the U.S.S.R. (In Russian). *Sov. Bot.* 1: 8—36.
SCHENNIKOV, A. P., – 1937 – Theoretical geobotany during last 20 years (In Russian). *Sov. Bot.* 5: 58—94.
SERNANDER, R., – 1894 – Studier öfver den gotländska vegetationens utvecklingshistoria. *Diss., Uppsala.* 112 pp.
SERNANDER, R., – 1898 – Studier öfver vegetationen i mellersta Skandinaviens fjälltrakter. 1. Om tundraformationer i svenska fjälltrakter. *Öfvers. Förhandl., K. (Svenska) Vetensk-Akad.* 6: 325—367.
SHIMWELL, D. W., – 1971 – Description and Classification of Vegetation. Sigdwick & Jackson, London. 322 pp.
SJÖGREN, E., – 1961 – Epiphytische Moosvegetation in Laubwäldern der Insel Öland (Schweden). *Acta phytogeogr. suec.* 44: 1—149.
SJÖGREN, E., – 1964 – Epilithische und epigäische Moosvegetation in Laubwäldern der Insel Öland (Schweden). *Acta phytogeogr. suec.* 48: 1—184.
SJÖRS, H., – 1948 – Myrvegetation i Bergslagen. (Engl. summ.) *Acta phytogeogr. suec.* 21: 1—299.
SJÖRS, H., – 1954 – Slåtterängar i Grangärde finnmark. (Engl. summ.) *Acta phytogeogr. suec.* 34: 1—135.
SJÖRS, H., – 1956 – Nordisk växtgeografi. Scandinav. Univ. Books, Oslo, København, Stockholm, Helsingfors. 229 pp. 2nd ed. 1967, 240 pp.
SJÖRS, H., – 1965 – Forest regions. *Acta phytogeogr. suec.* 50: 48—63.
SJÖRS, H., F. BJÖRKBÄCK & NORDQVIST, – 1965 – Regional ecology of mire sites and vegetation. *Acta phytogeogr. suec.* 50: 180—188.
SMITH, H., – 1920 – Vegetationen och dess utvecklings-historia det centralsvenska högfjällsområdet. *Norrl. Handbibl.* 9: 1—238. Almqvist & Wiksells, Uppsala.
SONESSON, M. – 1970 – Studies on mire vegetation in the Torneträsk area, Northern Sweden. III. Communities of the poor mires. *Op. bot. Soc. bot. Lund* 26: 1—120.
SØRENSEN, T. A., – 1948 – A method of establishing groups of equal amplitude in plant sociology based on similarity of species content and its application to

analyses of the vegetation on Danish commons. *Biol. Skr.*, K. danske Vidensk. Selsk. 5 (4): 1—34.

STEEN, E., – 1954 – Vegetation och mark i en uppländsk beteshage med särskild hänsyn till betesgångens inverkan. *Meddn St. JordbrFörs.*, Stockholm 49: 1—146

STEINDÓRSSON, S., – 1936 – Om Vegetationen paa Melrakkasljetta i det nordøstlige Island. *Bot. Tidsskr.* 43: 436—483.

STEINDÓRSSON, S., – 1945 – Studies on the vegetation of the central highland of Iceland. In, 'The Botany of Iceland,' ed. J. L. A. KOLDERUP ROSENVINGE & E. WARMING 3 (pt. 4, no. 14): 345—547. Munksgaard, Copenhagen.

STEINDÓRSSON, S., – 1946 – Contributions to the plant-geography and flora of Iceland. IV. The vegetation of Ísafjarðardjúp, northwest Iceland. *Acta nat. islandica* 1 (3): 1—32.

STEINDÓRSSON, S., – 1954 – The coastline vegetation at Gásar in Eyjafjörður in the North of Iceland. *Nytt Mag. Bot.* 3: 203—212.

STERNER, R., – 1922 – The continental element in the flora of South Sweden. *Geogr. Annlr* 1922: 221—444.

SUKATSCHEV, V. N., – 1964 – Fundamental concepts in forest biogeocoenology (In Russian). In, 'Osnovy lesn. biogeotsenologii' (Fundamentals of Forest Biogeocoenology), pp. 5—49. 'Nauka,' Moscow.

TENGWALL, T. Å., – 1920 – Die Vegetation des Sarakgebietes. I. In, 'Naturwiss. Untersuch. des Sarakgebirges in Schwed.-Lappland, geleitet von Dr. Axel Hamberg' 3 (4): 269—436. Fritzes, Stockholm.

THUNMARK, S., – 1931 – Der See Fiolen und seine Vegetation. *Acta phytogeogr. suec.* 2: 1—198.

TRASS, H., – 1963a – Probleme des Vegetationsklassifikationen der waldlosen Nieder- und Übergangsmoore Estlands (In Russian with Germ.summ.). *Tartu Riikl. Ülik. Toim.* 145, Bot.-al. Tööd 7: 60—73.

TRASS, H., – 1963b – On the typology of the dominants of plant communities (In Russian with Engl. summ.). *Byull. mosk. Obshch. Ispyt. Prir.*, otd. biol. 68 (5): 29—36.

TRASS, H., – 1964 – Application and validity of the synusial method in phytocoenology (In Russian with Engl. summ.). In, 'Izuchenie rastit. ostrova Saaremaa,' pp. 82—111. Tartu.

TRASS, H., – 1965a – Trends of development in the Estonian ecology and some theoretical problems (In Finnish). *Luonnon Tutk.* 69 (3): 110—120.

TRASS, H., – 1965b – On significance of dominants of plant communities in classification of vegetation (In Russian). *Problemy sovrem. Bot.* 1: 247—250.

TRASS, H., – 1966 – On vegetation discontinuity and continuity (In Russian with Engl. summ.). *Trudy mosk. Obshch. ispyt. Prir.*, Otd. biol. 27: 167—182.

TRASS, H., – 1970 – Coenoelements in plant communities (In Russian with Engl. summ.). *Trudy mosk. Obshch. ispyt. Prir.*, Otd. biol. 38: 184—193.

TUOMIKOSKI, R., – 1942 – Untersuchungen über die Vegetation der Bruchmoore in Ostfinnland. I. Zur Methodik der pflanzensoziologischen Systematik. (Finnish summ.). *Ann. bot. Soc. Zool.-Bot. Fenn. 'Vanamo' (Suomal. eläin-ja kasvit. Seur. van. kasvit. Julk.)* 17 (1): 1—203.

TÜXEN, R., – 1970 – Einige Bestandes- und Typenmerkmale in der Struktur der Pflanzengesellschaften. (Engl. summ.). In, 'Gesellschaftsmorphologie (Strukturforschung),' ed. R. Tüxen, *Ber. Symp. int. Vereinig. Vegetationskunde*, Rinteln 1966, 10: 76—107.

TYLER, G., – 1969a – Studies in the ecology of Baltic sea-shore meadows. II. Flora and vegetation. *Op. bot. Soc. bot. Lund* 25: 1—101.

TYLER, G., – 1969b – Regional aspects of Baltic shore-meadow vegetation. *Vegetatio* 19: 60—86.

URANOV, A. A., – 1925 – Materialien zu einer phytosoziologischen Beschreibung der Hegesteppe in Gouvernement Pensa im Lichte des Gesetzes der Konstant (In Russian with Germ. summ.). Otd. ohr. prir. glavnauki NKP 7, Tr. no izutsch. zapov. : 1—40.
URANOV, A. A., – 1966 – Number of species and area (In Russian with Engl. summ.). *Trudy mosk. Obshch. ispyt. Prir.*, Otd. biol. 27: 183—204.
VAGA, A., – 1940 – Fütotsönoloogia põhiküsimusi (Eston. with Engl. summ.: On some fundamental problems in phytocoenology). *Acta Inst. Horti bot. tartu.* 7 (1/2): 1—152, and *Acta Comment Univ. tartu.*, Ser. A, 35 (6): 1—152.
VASSILEVITSCH, V. I., – 1966 – A study of vegetational continua (In Russian with Engl. summ.). *Trudy mosk. Obshch. ispyt. Prir.*, Otd. biol. 27: 59—69.
VESTAL, A. G., – 1949 – Minimum areas for different vegetations: their determination from species-area curves. *Illinois biol. Monogr.* 20 (3): 1—129.
WALDHEIM, S., – 1944a – Die Torfmoosvegetation der Provinz Närke. *Handl. K. fysiogr. Sällsk. Lund*, N.F. 55 (6): 1—91, *Acta Univ. Lund*, N.F., Avd. 2, 40 (6): 1—91.
WALDHEIM, S., – 1944b – Mossvegetationen i Dalby-Söderskogs nationalpark: ett bidrag till kännedomen om Skånes bryofytvegetation. *K. svenska Vetensk. Akad. Avh. Naturskydd.* 4: 1—142.
WALDHEIM, S., – 1947 – Kleinmoosgesellschaften und Bodenverhältnisse in Schonen. *Bot. Notiser Suppl.* 1 (1): 1—203.
WANGERIN, W., – 1925 – Beiträge zur pflanzensoziologischen Begriffsbildung und Terminologie. I. Die Assoziation. *Beih. Repert. Spec. nov. Regni veg.* 36: 3—59
WARÉN, H., – 1926 – Untersuchungen über sphagnumreiche Pflanzengesellschaften der Moore Finlands unter Berücksichtigung der soziologischen Bedeutung der einzelnen Arten. *Acta Soc. Fauna Flora fenn.* 55 (8): 1—133.
WARMING, E., – 1895 – Plantesamfund: Grundtraek af den økologiske Plantegeografi. Philipsens, København. 335 pp.
WARMING, E., – 1896 – Lehrbuch der ökologischen Pflanzengeographie: Eine Einführung in die Kenntnis der Pflanzenvereine. Borntraeger, Berlin. 412 pp.
WARMING, E., – 1909 – Oecology of Plants: An Introduction to the Study of Plant-Communities. Oxford Univ. Press, Oxford. 422 pp.
WASSÉN, G., – 1966 – Gardiken: Vegetation und Flora eines lappländischen Seeufers. *K. svenska Vetensk. Akad. Avh. Naturskydd.* 22: 1—142.
WHITTAKER, R. H., – 1956 – Vegetation of the Great Smoky Mountains. *Ecol. Monogr.* 26: 1—80.
WHITTAKER, R. H., – 1962 – Classification of communities. *Bot. Rev.* 28 (1): 1—239.
WHITTAKER, R. H., – 1967 – Gradient analysis of vegetation. *Biol. Rev.* 42: 207—264.

19 NUMERICAL CLASSIFICATION

DAVID W. GOODALL

Contents

19.1	Introduction	249
19.2	General Concepts	249
19.2.1	Classification Principles	249
19.2.2	The Geometrical Model of Classification	251
19.2.3	How Many Classes?	253
19.2.4	Hypotheses and Classification	254
19.2.5	Heterogeneity and Discontinuity	255
19.3	Variables and Weighting	255
19.4	Sample Size Effects	258
19.5	Clustering Procedures	258
19.5.1	Divisive Methods	263
19.5.1.1	Association Analysis	263
19.5.1.2	Elimination of Outliers	265
19.5.1.3	Method of EDWARDS and CAVALLI-SFORZA	266
19.5.1.4	MCNAUGHTON-SMITH's Method	267
19.5.1.5	Information Analysis	267
19.5.2	Agglomerative Methods	269
19.5.2.1	Simple Similarity Methods	269
19.5.2.2	ORLOCI's Agglomerative Technique	270
19.5.2.3	HALL's Heterogeneity Method	271
19.5.2.4	A Probabilistic Approach	272
19.5.2.5	Information Analysis	272
19.6	Cluster Shape and Dendrogram Shape	273
19.7	Classification of Groups of Samples	274
19.8	Classification and Ordination	276
19.9	Evaluation and Comparison	277
19.10	Conclusions	279
19.11	Summary	282

19. NUMERICAL METHODS OF CLASSIFICATION

19.1 Introduction

Though most classification of vegetation has been based on more or less subjective approaches (see, for instance, the review by WHITTAKER, 1962), there has been an increasing tendency in the past two or three decades to use numerical methods. This has largely been due to the greater objectivity that could be claimed for them. Let it be understood that this objectivity does not imply that an element of choice does not enter into the selection of methods (LAMBERT & DALE 1964); it means that, given the same stands and the same data from them, a procedure can be defined which can be applied unequivocally by anyone who understands it, with the assurance that the same result will be obtained. In other words, subjectivity is removed from an important part of the whole operation.

The development of numerical methods for classifying vegetation has been greatly encouraged by the development and ready availability of digital computers, which have taken much of the drudgery out of the extensive calculations often involved. The rapid development of numerical taxonomy over the past fifteen years (see SOKAL & SNEATH 1963) has contributed to the development of parallel methods in plant ecology. Numerical methods for the classification of vegetation are discussed in a number of recent reviews and monographs, including WILLIAMS & DALE (1965), GOUNOT (1969), PIELOU (1969), VASILEVICH (1969), and GOODALL (1970).

19.2 General Concepts

19.2.1 CLASSIFICATION PRINCIPLES

Classes can be defined by extension or intension — by a listing of the items included, or a definition of their range of variation. Except in the higher echelons (to which I shall return later), extensive classification has no place in the classification of vegetation — or indeed in biological classifications of any sort. The number of members, or potential members, of any class is virtually unlimited, and in consequence it is only through their attributes (in the broadest sense) that the class membership can be defined. In general, the classification aims at defining a set of classes in which members of the same class shall be 'as alike as possible,' members of different classes 'as unlike as possible,' to quote WILLIAMS & DALE (1965). And in vegetation, as in most biological subject matter, similarity is a matter of degree. No class of interest is likely to be homogeneous — it will only be less heterogeneous than the set of objects as a whole.

By dividing an assemblage of objects into classes, one has the possibility of making a wide variety of statements about them with great economy. By saying that an object in the room falls into the class *dog*, one saves oneself the trouble of specifying that it has four legs and a tail, claws, teeth, fur, warm blood and all the other features which distinguish members of that class from the classes *table, carpet, lamp, man, canary* and so forth. A special-purpose classification ('external classification' — WILLIAMS & DALE, 1965) is directed to prediction of a limited and specified range of attributes; others are of interest only in so far as they facilitate the identification of the class to which an unknown object belongs. A general-purpose classification ('internal classification' — WILLIAMS & DALE, 1965), at which one is usually aiming in biology, is intended to predict as wide a range of attributes as possible.

In classifying vegetation, the attributes one wishes to predict certainly include the floristic composition (qualitative and quantitative) and many features of the abiotic environment — climate and soil. The vegetational characteristics should be predictable at any season from the class membership. Predictions concerning the zoological and microbiological components of the ecosystem may also be needed. Thus a general-purpose classification is usually the aim.

It should be remembered that a general-purpose classification, like the proverbial jack-of-all-trades, is master of none — it is suboptimal for the prediction of any particular variable. An optimal general-purpose classification is optimal only in relation to a particular weighted function of the various variables to be predicted; if the weights are changed, the optimal classification may well be different.

In the primary classification of vegetation samples or stands, we are not interested in the samples in their own right. We are interested in them only *qua* samples — as representatives of a hypothetical larger population of similar stands or parts of the landscape that might equally well have been sampled — and our conclusions about their classification are of interest only if the conclusions can be applied to the population as a whole. This means that an extensive classification is not wanted except in so far as it can be translated into intensive terms, and hence become applicable to the parent population.

Once a set of primary classes has been established, the situation becomes different. A further (secondary) classification of these classes is concerned with this complete definable and delimited population of classes, and in consequence sampling questions do not arise. Thus, at this level an extensive classification is quite appropriate, and it is not essential that it should be expressed in terms of attribute values.

It is sometimes said that stands or samples of vegetation are 'over-defined' — that there is much 'redundancy' in their description; the observations made on them are more numerous than the minimum theoretically needed to allot them among a limited number of classes. This implies, however, a failure to distinguish the problem of allocation from that of classification. It is true than the number of variables and their alternative values by which a vegetation sample could be defined is far greater than are ever likely to be measured, and far greater than are needed to distinguish it from any others in the set. If, however, the set of samples is regarded as representative of a population, the same principles apply as in sampling practice generally — every additional sample contributes to and improves the estimation of population parameters, and there is no true redundancy. A particular classification procedure may fail to make use of some part of the data; these data may indeed then be regarded as redundant for that procedure—but this implies that the procedure is sub-optimal, not that the stands are over-defined.

19.2.2 THE GEOMETRICAL MODEL OF CLASSIFICATION

If one considers the set of stands to be classified as a set of points in a multidimensional space, the axes of which represent the different variables specifying the stands, it is clear that any subdivision of this space generates a classification. The stands within each cell thus delimited constitute a class (and, indeed, any classification of the stands (assumed intensive) can be so represented). The value of such a classification will then depend on the precision with which the values of the variables can be predicted for an arbitrary stand of which nothing is known but its class membership. Where classification is based on presence or absence of one or more species, or on the dominance of a particular species (in which case the dividing hypersurfaces are perpendicular to one of the axes), the underlying assumption is that numerous other variables are correlated with these species, so that class allocation by this one (or these few) species will mean that many other variables tend to be defined more or less precisely at the same time. In the same way, in a successful ordination (see articles 10, 11), the main axes are highly correlated with many of the variables, so that a division into classes by hyperplanes at right angles to the main ordination axes would permit high predictability for many variables.

In the preceding paragraph no account has been taken of the distribution in the hyperspace of the points representing stands. They may, with suitable scaling of the variables, be distributed in a hypersphere. This is quite improbable in a practical situation, for it would imply that the variables are uncorrelated. Much more likely is a hyperellipsoid with its axes inclined to those of the coordinate system, or some more complicated hypersolid. Wherever the points are clustered around a single centre, with their density decreasing progressively away from it, there is no 'natural' subdivision, and a choice of hypersurfaces dividing the stands into classes must be arbitrary. It may be, however, that the points representing the stands are distributed in several clusters, each separately hyperspheroidal or hyperellipsoidal, and separated by zones where the density of points is relatively low. In such a case, division by hypersurfaces passing through these zones of low density would give a classification which would have a claim to being 'natural' in some sense, and which would, on average, be expected to give greater predictability than any other arbitrary division into the same number of classes. (This appeals intuitively, but no demonstration of its truth has ever been seen; and indeed an unequivocal demonstration of the existence, in empirical data, of the partial discontinuities[1] postulated is far from straightforward, as will be shown below.)

If the stands to be classified constitute a representative sample of an underlying population, and the population has discontinuities in its distribution, then the sample may be expected to reflect these discontinuities. A classification of the sample set based on discontinuities is then likely to lead to optimal conclusions regarding the classification of the population itself. A selective sampling procedure, on the other hand, may generate discontinuities which do not correspond with discontinuities in the parent population, and the same classification technique may in this case give results which are far from optimal when applied to the population.

Even where the parent population shows discontinuities which are reflected in the sample set, one should interpret them with cir-

---

1) WILLIAMS (see LAMBERT & DALE 1964) has proposed a definition of discontinuity which would require each class of stands to have no species in common with any other, and LAMBERT & DALE would use 'heterogeneity' for other cases. This appears an unduly restrictive definition of discontinuity, however, and very few instances listed in the literature of plant sociology would qualify.

cumspection. The discontinuities in the vegetation may be the result of varying representation of the biotope space in the area sampled, so that there are some intermediate habitat types which are underrepresented, while those on either side are well represented. The vegetation characteristic of the latter types of habitat will then appear as clusters separated by a discontinuity. For purpose of prediction within the area sampled, the resulting classification will be unexceptionable; but one must not proceed to deductions about the interpretation of vegetational discontinuity, ignoring the habitat discontinuity on which it was based.

19.2.3 How Many Classes?

For the primary classification of a set of entities, there are three types of decision to be made, though these decisions may be linked. They are:
(i) How many classes should be recognized?
(ii) How are they to be delimited?
(iii) To which class is each entity to be allotted?
If the approach is that outlined in the preceding paragraphs, in which the classes are defined intensively, the answer to (iii) automatically follows from (ii); if, as often happens, the class is *extensively* defined within the particular set of samples studied, allocation of new samples may require a new decision procedure.

If the data show no inherent discontinuities, the number of classes to be recognized (question (i)) is an arbitrary decision, though the arbitrariness may be masked by throwing the blame on to the 'stopping rule' of a divisive procedure (see below). In principle, it is clear that, as one increases the number of classes, the predictions that can be based on class membership become more precise, but the allocation of a stand to its appropriate class becomes more difficult. As the size of classes decreases, the variance or information between classes increases at the expense of that within classes, so that in the extreme, when each class consists of a single stand, all the variance is between classes. Where no natural discontinuities exist the choice of how far to subdivide is one of practical convenience rather than of principle. On the other hand, where class delimitation is based on discontinuities in the distribution of points, the number of classes need not be an arbitrary decision, but may be decided by the data. Even here, however, the discontinuities may themselves vary continuously in their distinctness, so that, even though the first one or two are perfectly clear, one comes to a point where decision as to whether or not a discontinuity exists is arbitrary. The arbitrariness may, indeed, be pushed one stage further off by defining a statistical test to determine whether a discontinuity will be recognized or not, and in this case the arbitrary element will be the choice of significance level.

19.2.4 Hypotheses and Classification

Where, as in the primary classification of vegetation samples, we are interested in drawing conclusions as to the structure of a parent population, it is often appropriate to set up a null hypothesis and to test it before proceeding to classify the data. The appropriate null hypothesis (GOODALL 1966b, c) is clearly that the samples were drawn from a single population, with variation but without discontinuity. If this null hypothesis is upheld by the test, classification must be regarded as an arbitrary procedure, serving practical ends only, and probably following different rules from a 'natural' classification based on discontinuities in the underlying population.

As will be seen below, many of the classification procedures proposed in the literature have no statistical element and do not lend themselves to the testing of null hypotheses. In such cases, it has been said that the procedures are to be regarded as ways, not of testing hypotheses, but of generating them (e. g. LAMBERT & DALE 1964). However, too little attention has been given to the type of hypothesis which might be generated, or how it might be tested. A very suitable type of hypothesis would be

i) If a new sample is collected, and is allotted to this or that class by such and such criteria, then the values of such and such other attributes in it will have expected values specified in one vector, with error variance specified in another vector; or

ii) The population from which these samples are drawn is composed of such and such classes, the attributes in each of which are defined by specified multivariate distributions.

The first of these hypotheses is purely operational, and implies nothing about the structure of the parent population, whereas the second explicitly describes the parent population. Either can be tested by taking new samples from the defined population. It should be emphasized that if numerical classification of vegetation or other objects is regarded as a procedure for generation of hypotheses, then the hypotheses generated should be made explicit in a testable form, and their generation should be followed by the performance of a test.

Where one is considering the classification process in terms of hypotheses, and hopes to develop objective tests, it is wise to use separate sets of data for building the hypotheses and for testing them. If this is not done, there is a considerable risk that the test will be invalid. For certain restricted types of hypothesis, statistical procedures exist whereby some part of the formation of a hypothesis (estimation of parameters) can make use of the same data as are used to test it; this applies, for instance, to regression analysis but is rarely likely to be true for procedures in vegetation classification, and the safest plan is to use separate sets of data for the two processes.

19.2.5 HETEROGENEITY AND DISCONTINUITY

As stated above, if the data contain discontinuities they can serve as the basis for a 'natural' classification; if on the other hand they are continuous, a classification is bound to be arbitrary. The approach suited to these two situations is likely to be different. Accordingly, a first step in primary classification should be to determine whether or not the data are homogeneous. As suggested by BRISSE & GRANDJOUAN (1971), this indicates the desirability of ordination as a prelude to classification. Ordination will not only enable the continuity or discontinuity of the data to be studied, but will also give information on the number of clusters ('natural' classes) to be recognized, and on their shape.

The recognition of discontinuities in multivariate data is far from simple. With continuous variables, an acceptable approach might be through a null hypothesis of multivariate normal distributions. The estimation of parameters in the latter case would require a provisional allocation of samples among the two classes, and this allocation should be that which, in some sense, maximizes the variance between classes and minimizes that within them.

For binary attributes (such as species presence) it might be best to reduce the dimensionality by an ordination, which at the same time will replace the original binary variables by a smaller number of quasi-continuous variables. These new variables may then be treated in the same way as described in the preceding paragraph. Even for continuous variables, ordination in fewer dimensions may prove a convenient preliminary to tests for discontinuity.

19.3 **Variables and Weighting**

As in vegetation classification generally, a wide range of variables can be used in numerical classification. Most commonly they are floristic variables, the commonest of all being the presence and absence of each species — a binary variable. Any variable ('importance value') expressing the quantity in which each species occurs in the vegetation can also be used, relative or absolute — cover, basal area, density, biomass, etc. — or any combined measure such as the synthetic importance value of the Wisconsin school (e. g. CURTIS & McINTOSH 1951). For simplicity in computation, continuous variables such as cover may be converted into binary variables, by allotting the alternative values according to whether or not the quantities exceed a specified threshold (e.g. NOY-MEIR et al. 1970). Rankings have rarely been used; they deserve more attention, for subjective rankings of species can be observed quickly and perhaps more reliably than quantitative assessments of the proportional or absolute contribution of a species to the vegetation.

It has often been suggested (LAMBERT & DALE 1964, 1967, ORLOCI 1968, NOY-MEIR et al. 1970, GREIG-SMITH 1971) that species presence is an appropriate variable to use where the set of stands to be classified is highly heterogeneous, and that in such cases quantitative measures add little useful information. In more restricted and less heterogeneous sets of data, on the other hand, variables expressing the quantitative contribution of the different species to the vegetation may be preferable.

Species variables are not the only way in which vegetation can be described. The species can, for instance, be divided into ecological groups, responding similarly to factors of the environment (article 2). Or they can be grouped in terms of growth-form, and the proportion of each growth-form in terms of cover, biomass, etc. used as variables in the classification process; or the division of species could be in terms of geographical origin, or disseminule type; or physiognomic characters, closely linked with the proportions of different life-forms, could be used (WEBB et al. 1970).

Finally, the environmental factors associated with the vegetation could also be used as variables in the classification of vegetation by numerical methods, as they have often been used in nonnumerical methods at the formation level (article 13).

The results of classification procedures may depend greatly on the weighting accorded to the different variables. It seems intuitively likely that this dependence may be unimportant if the classification is based on discontinuities in the data. In other cases, however, the effects may be considerable.

Weighting may be implied by the variable used and the method of analysis, without any explicit consideration. If, for instance, the variables used are species presence and absence only, many methods of analysis tacitly allot equal weight to each species. The same is true if the variables are standardized measures of quantity. If they are unstandardized, on the other hand, the tacit weighting may be proportional to the average quantity in which the species occurs.

An extreme case of differential weighting is to use data for some of the species only, and ignore the rest. Not infrequently, for computational convenience, species occurring in few samples have been ignored (e.g. GOODALL 1954, WEBB et al. 1967b) — though this may make the recognition of distinct vegetation types represented by few samples more difficult (GOODALL 1969). In species-rich vegetation, where each vegetation type is well represented in the sample set, however, sampling of the species present by life form, for instance, or at random, is unlikely greatly to affect the resulting classification (WEBB et al. 1967b, 1970, NOY-MEIR et al. 1970).

In numerical taxonomy, emphasis has often been placed on the 'Adansonian principle' of equal weight for all attributes. This goal is illusory when the attributes to be recorded are selected subjectively — which is usual in taxonomy proper. In vegetation, at least the different species present constitute an objectively definable list of variables; and if we adopt a common measure for them — biomass, cover, etc. — their respective weighting is decided in a biologically meaningful way.

WILLIAMS & DALE (1965) have claimed that the scale of the axes for continuous variables may affect the results of classification. It is clear that a proportional change in scale cannot generate discontinuities where there were none, though this could easily happen with some types of non-linear transformation. If, on the other hand, the change in scale affects the weighting used for maximization, the optimum division into arbitrary classes would be affected. And the results of the various classification procedures without significance tests described below may also depend on scale.

It has been claimed with some justice (e.g. LAMBERT & DALE 1964) that absence of a species from a sample should not be considered simply as the zero limit of a quantitative variable, but is qualitatively different from any positive value. Absence may arise in a sample in two ways. Either one may be sampling a population from which the species is absent; or the species may be present in the population but happen to be absent from the sample. The problem is akin to that of the 'contamination' of a Poisson distribution by additional zeros. The presence of a species is positive evidence that environmental conditions fall within the range of tolerance of that species; its absence is not positive evidence to the contrary.[1])

It has been suggested that the situation be handled by using two

---

[1]) Similarly, HALL (1970) suggests that species present in high quantity in the sample are those more relevant to its classification.

variables for each species — one a binary variable expressing presence or absence only, the other a real variable taking positive values only, and indeterminate when the first variable is zero. This is done, for instance, in the matrix partition method of WILLIAMS & DALE (1962). The probabilistic methods of GOODALL (1966a, d, 1968, 1971), which are distribution-free and do not depend on 'distance' measures, can also deal satisfactorily with this difficulty.

19.4 Sample Size Effects

Since vegetation is structured, the results of classification may depend on the size of the samples classified. If samples are not less than the *minimal area* for the vegetation (defined as that of a square with the side equal to the distance at which variance between samples ceases to be a function of their spatial separation — see GOODALL 1961), then the results of classification should be independent of their size (NOY-MEIR et al. 1970). If they are smaller, then the classification will in part reflect the mosaic structure of the vegetation, in addition to differences among vegetation types. The assumption that a minimal area exists — that the vegetation is homogeneous at a certain scale — cannot however be taken for granted (GOODALL 1961), and where it is untrue one must accept the dependence of classification on sample size at all levels.

19.5 Clustering Procedures

Methods used to identify the clusters which should be distinguished (together often with the interrelations among them) are commonly called 'clustering procedures'.

One of the earliest of these procedures was that of CZEKANOWSKI (1909), originally applied to the classification of ethnological material, but used for vegetation classification by Polish ecologists in particular (e. g. MOTYKA 1947, MATUSZKIEWICZ & MATUSZKIEWICZ 1956).

The first step is to calculate a matrix of similarities between pairs of stands. Any index of similarity (see article 6) may be used, but MOTYKA's choice was that of KULCZYŃSKI (1928). The order of the stands in the matrix is then rearranged by inspection so that as many as possible of the high similarities are close to the main diagonal (article 7). When this is done, it will often be found that the stands fall into groups, similarities within which are high, while those between different groups are mainly low. These groups then constitute the classes of the classification.

Re-ordering of a matrix is also an important step in the Zürich-Montpellier method of vegetation analysis (see article 20), but here the matrix is one of species × stands (relevés). Re-ordering is normally carried out by inspection, but numerical methods for the purpose have been devised (MOORE et al. 1967, LIETH & MOORE 1971), and lead to the arrangement of stands in a sequence where breaks (separating classes of stands) can easily be distinguished.

Both these approaches have the drawback of assuming that the classes can be arranged in a one-dimensional sequence. Otherwise it will not be possible to re-order the matrix so that similarities decrease regularly away from the main diagonal. If the classes are highly discrete, this may not matter — all similarities between classes will be much lower than those within classes, and the classes will be readily recognizable as separate groups even though their order is uncertain.

An alternative numerical basis for subjective classification of a set of vegetation samples can be provided by an ordination. Any of the standard methods of ordination (see article 11) can be used in this way, the purpose being to reduce the dimensionality of the system, and hence make its characteristics more readily open to inspection.

There are rather few examples in the literature where an ordination has been divided arbitrarily to give classes. BROWN & CURTIS (1955) divided a forest gradient into five ranges of their weighted-average index, and LOOMAN (1963) divided a one-dimensional ordination of grassland samples into seven classes.

The preparation of an ordination in two or three dimensions and its graphical representation often bring clustering and discontinuities to light, and thus form the basis for a 'natural' classification. An example is the distinction between swale and dune communities in GOODALL's (1954) study of the Victorian Mallee.

In contrast to the mixed numerical and subjective procedures outlined above, purely numerical procedures are usually organized in step-wise fashion. They lead to an arrangement of the samples in a hierarchical pattern which can be expressed as a tree diagram or *dendrogram*, reflecting both the successive steps and the presumptive relationships among the classes distinguished (Figs. 1, 2).

It is clear that a hierarchical system or dendrogram is not logically equivalent to a distribution of samples into classes. It establishes a pattern of closeness of relationship among the samples, without specifying dividing lines which distinguish classes. To distinguish a set of classes on the basis of a dendrogram requires the establishment of a 'stopping rule', which will define a point beyond which distal branches of the dendrogram will be regarded as mem-

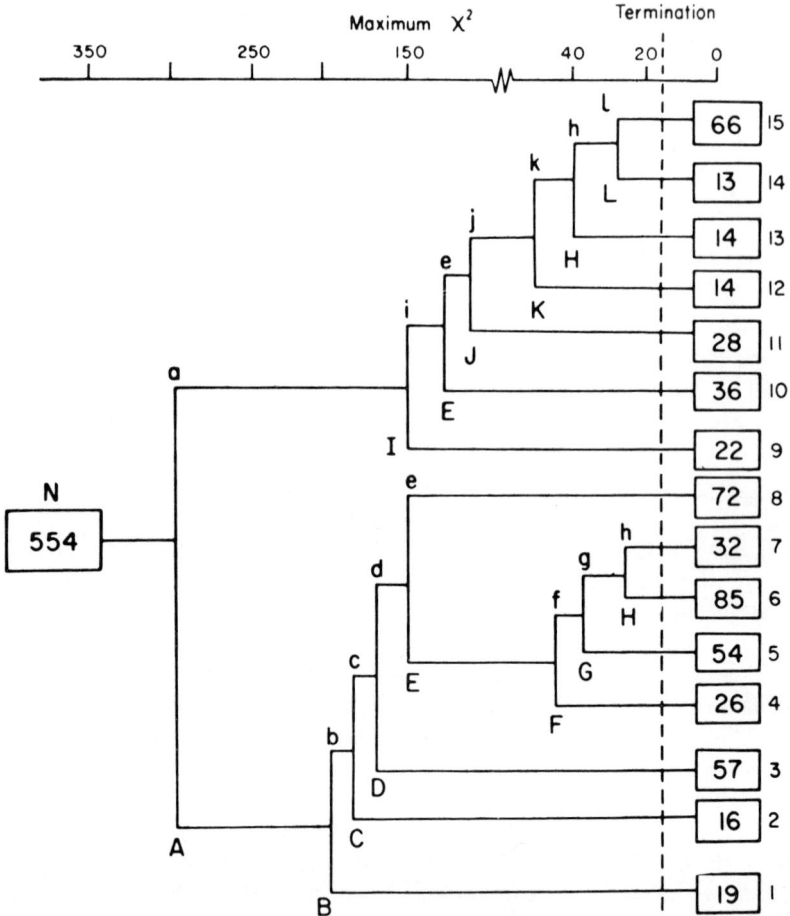

Fig. 1. A dendrogram, representing classification of 554 quadrats comprising 98 species from wet-land vegetation on the Isle of Arran, Scotland, by normal association analysis (CRAWFORD & WISHART 1967, Fig. 1). The numbers in the boxes are numbers of quadrats in the terminating classes; the numbers outside the boxes are reference numbers for these classes. The letters at the nodes represent the species on which a given division of the quadrats was based, with presence of a species indicated by a capital letter and its absence by a small letter: A, *Molinia caerulea*; B, *Anthoxanthum odoratum*; C, *Cirsium palustre*; D, *Rhacomitrium lanuginosum*; E, *Sphagnum* spp.; F, *Juncus acutiflorus*; G, *Myrica gale*; H, *Calluna vulgaris*; I, *Iris pseudacorus*; J, *Glaux maritima*; K, *Sagina procumbens*; L, *Alnus glutinosa*.

bers of the same class. For classifying a set of samples representing one or more underlying populations, the stopping rule should be probabilistic (WEBB et al. 1967a). It may, for instance, depend on a measure of the homogeneity of the set of samples considered as

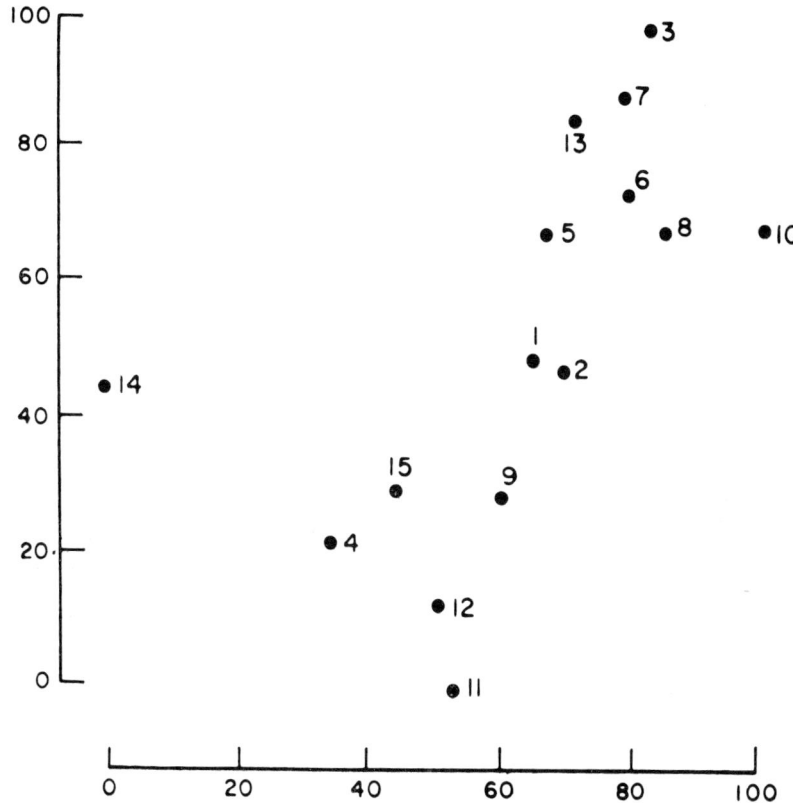

Fig. 2. A Wisconsin comparative ordination of the numbered classes of quadrats derived by association analysis as shown in Fig. 1. (CRAWFORD & WISHART 1967, Fig. 3). The x axis is based on classes 14 and 10, the y axis on classes 11 and 3. In this case the association analysis, as a monothetic technique, does not effectively reproduce the relationships indicated in the ordination, as a polythetic technique (cf. GREIG-SMITH et al. 1967, FLENLEY 1969, WARD 1971).

belonging to a single class. Often, however, the stopping rule is much more arbitrary, and may depend on the value of some statistic calculated from the samples without regard to probabilistic considerations. CZEKANOWSKI (1909), for instance, used the minimum similarity within a set as a stopping rule. WILLIAMS & LAMBERT (1960) used a fixed value for max. χ^2 in association analysis (see below) in the same way, whereas in the earlier form of this procedure GOODALL (1953) had used χ^2 as a probabilistic stopping rule, and NOY-MEIR et al. (1970) seem to have adopted a rather similar practice.

A variety of techniques is available in deriving a hierarchical structure or dendrogram from a primary matrix of attribute values for each of the samples to be classified. The first step is usually to convert this primary matrix into another which is simpler in the sense that either the distinctions between samples, or those between attributes, have been eliminated. That is, one may first convert the primary matrix of samples-by-attributes into one of attributes-by-attributes, and classify the stands on the basis of the interrelations shown among the attributes. These are often known as R-techniques (the 'transposed-structuring' techniques of LAMBERT & DALE (1964), or the 'derived-structuring' methods of WILLIAMS & DALE (1965)). Other methods derive a secondary sample-by-sample matrix in which each cell represents some measure of similarity or 'distance' between the two samples in question (see article 6); these methods are known as Q-techniques ('self-structuring' techniques of LAMBERT & DALE, or 'direct-structuring' methods, according to WILLIAMS & DALE).

In considering numerical procedures for classification a distinction is often made between *monothetic* and *polythetic* methods. In the former, attributes are considered singly, and each dividing hyperplane, in our geometrical model, is consequently perpendicular to one of the axes. In a polythetic procedure, on the other hand, a number of different attributes are taken into account simultaneously, and the dividing hypersurface is consequently likely to be oblique to the axes, and may be curved. In numerical classification, a well-known method using the monothetic approach is that known as 'association analysis' (section 19.5.1.1 below); a descriptive analogue is classification of vegetation by the dominant species (article 14), or strict application of the presence or absence of a single characteristic species. Since the absence of a particular species may be a matter of chance, there is a substantial random element in the allocation of a particular sample by monothetic criteria (NOY-MEIR et al. 1970). If the species is annual, there may also be important seasonal differences in classifications based on this principle (HOPKINS 1968). Polythetic classifications are much less affected by such influences.

Another distinction often made in the literature of numerical taxonomy is that between *divisive* and *agglomerative* approaches. A divisive approach is one which begins with the complete set of entities to be classified, and divides them progressively into smaller and smaller groups. An agglomerative approach proceeds from a single entity, considers which other entity is most 'similar' to it in some defined sense, then which third entity is most 'similar' to these two, and thus proceeds to build up a cluster of entities around that which was first selected. At some point it may be decided that none of the remaining entities is similar 'enough' (a 'stopping rule') to be associated with the cluster being formed, and a new cluster is initiated. It will be noted that both types of procedures develop *extensively* — the classes are defined in the first instance by their membership, though afterwards attempts may be made to define the classes intensively, and thus provide a principle of allocation for any future samples.

In considering the various classification procedures that have been proposed and used for vegetation, we shall take first those that are mainly divisive, and then those which are mainly agglomerative. The distinction is not a firm one — a procedure may include steps of both types, and it is likely that combined procedures of this sort will be favoured as numerical classification becomes more sophisticated.

WILLIAMS et al. (1966), in considering criteria for choice of a classification procedure, suggest that the use of a hierarchical system implies that the user is specifically interested in 'the actual path of fusion'. In the present discussion, however, we are distinguishing the primary recognition of clusters from consideration of their interrelations, so that the hierarchical system is merely regarded as a convenient means to the end of cluster recognition, and the path of fusion is not a matter of interest nor a suitable criterion for choice of method.

A drawback of the hierarchical approach (LAMBERT & WILLIAMS 1966) is that the decisions made are irrevocable. If the rules o the clustering algorithm, at a particular point in the process, lead to a certain division or a certain fusing of groups, this can never be corrected by subsequent actions within the strictly hierarchic procedure. This has led some workers to amend it by allowing for the possibility of fusion among clusters that have been separated and have become associated with different branches of the dendrogram (GOODALL 1953, GREIG-SMITH et al. 1967, CRAWFORD & WISHART 1968).

19.5.1 DIVISIVE METHODS

These include both monothetic and polythetic procedures. Some have — or can have — a statistical basis. At any stage one may set up a null hypothesis that an existing group of samples could have arisen by sampling from a single population with attributes distributed in some specified fashion — perhaps independently of one another. Then this grouping is divided further only if the null hypothesis is rejected at the chosen significance level.

19.5.1.1 *Association Analysis*

This is a monothetic divisive approach which had its origin in the concept that, within a homogeneous community, species are uncorrelated as between different samples. Consequently, criteria for division were sought which would give classes within which interspecific correlations were as few as possible.

As originally described by GOODALL (1953), the interspecific associations were tested for each pair of species by 2×2 tables (see section 6.2), using either the χ^2 test or the exact hypergeometric test. Of those species which had significant associations with others, that which occurred in the largest number of samples was selected, and the samples were divided into two groups according to whether this species was present or absent. The same test was then applied to the set of samples in which the first criterion species was present, and if it still proved heterogeneous it was divided on the presence of a second criterion species; the process was continued until the final group had a matrix of associations such as might have arisen by chance. The remaining samples, lacking one or more of the criterion species for the first group, were then reconsidered in the same way, and the process was continued until the samples had been divided into a number of groups, each of which had no associations significant by the test used. Finally, the groups were re-combined by pairs; if no significant associations re-emerged, the combined group was retained.

This technique was modified by WILLIAMS & LAMBERT (1959, 1960) in four ways. In the first place, the null hypothesis of a homogeneous group of stands was abandoned, and with it the testing of significance of a 2×2 table. Instead, a fixed level of χ^2 was adopted as a stopping rule, further subdivision being abandoned if the maximum value of χ^2 was less than this. Second, at each stage, the species selected as the criterion for division was that which had the maximum sum of χ^2 values with all other species. As WILLIAMS & LAMBERT pointed out, this maximized the effect of the subdivision in reducing other associations within the resulting groups. Thirdly, the procedure was made strictly hierarchical in that the group in which a criterion species was absent, as well as that in which it was present, was divided further without recombination. Finally, WILLIAMS & LAMBERT abandoned tests for possible recombination of homogeneous groups at the end—which fitted in with their abandonment of a significance criterion, and made the procedure purely divisive. Fig. 1 shows a dendrogram from association analysis.

NOY-MEIR et al. (1970) point out that association analysis is likely to be much more satisfactory as a classification procedure in species-rich vegetation than where only a few species are present. Like other monothetic procedures, the risk of misclassification is considerable if the criterion species happens to be absent by chance.

A variant of this method was proposed by CRAWFORD &

WISHART (1967) under the title 'group analysis.' Successive divisions are based on the presence or absence of particular species, but these species are selected according to the frequency with which they occur in the presence of numerous others. If S_i is the proportion of the total species which occurs in the ith sample, $\sum_j$ the sum of S_i for all samples in which species j occurs, P_j is the proportion of such samples, and $\sum.$ the sum of S_i over all samples, then the statistic that forms the criterion for choice of which species should serve as basis for dividing the set of samples is

$$\mu'^2 = 4\ (\textstyle\sum_j - P_j\sum.)^2$$

Division ceases when $\sum.$ exceeds half the number of samples in the set. CRAWFORD & WISHART tested this method on a random set of data; no clear-cut classes were found, but the data divided into very small groups — which, they claimed, served to indicate its value.

It will be noted that this method does not provide for test of a null hypothesis, and the 'stopping rule' is an arbitrary one. The rationale of the method in giving special weight to species occurring mainly in species-rich samples seems open to question. Some species-poor communities in pioneer habitats are at least as distinctive as those in more favoured situations.

In a subsequent paper (CRAWFORD & WISHART 1968), an adjustment of the original classification was introduced by examining the values of S_i for each sample within the group to which it was allotted. Where S_i was less than an arbitrary value β, the corresponding value for this sample was recalculated on the assumption that it was reclassified with each of the other groups; it was then attached to the group for which S_i was maximal. Another rule limited the minimum size of a group; this resulted in a smaller number of groups which were more compact than a similar number of groups would have been without the reclassification process.

While reclassification to avoid wrong allocation by chance seems to have merit, particularly in a monothetic procedure, there seems no particular reason for limiting the minimum size of a group, and thus not recognizing the rarer (but perhaps very distinctive) communities included in the sampling.

19.5.1.2 *Elimination of Outliers*

Among divisive procedures the simplest is the removal from the collection of a single sample which for some reason is regarded as differing most markedly from the norm represented by the rest. If the null hypothesis is entertained that the collection of samples can be regarded as taken from a single population in which the presence and variations in quantity of the different species are uncorrelated, a method has been described (GOODALL 1966a) for determining the probability that any particular sample will have a specific composition differing as widely from the norm as it does; this probability is termed the *deviant index*. If a sample is found to differ 'significantly' from the rest by this test, it is reasonable to regard it as representing a different type of vegetation, and hence to exclude it from the class formed, presumptively, by the rest. If this test is applied repeatedly, the bulk of the samples can be reduced to a homogeneous core in which no one sample need be considered as an outlier by this test (cf. article 5). Possible relationships among the outliers successively eliminated may then be tested.

It is of course possible, in a small group, that a subset of samples differ from the rest, but that none individually has a significant deviant index because the homogeneous part of the group is 'contaminated' by other members of the subset. A method has been described (GOODALL 1966d) for testing whether any two samples differ together from the norm of the group as a whole; and the same test may be generalized to larger subsets.

A simpler version of the deviant index (GOODALL 1969) depending on presence and absence only was applied to the removal of outliers from a set of vegetation samples in which a large number of species were recorded only in one or two samples.

19.5.1.3 *Method of Edwards and Cavalli-Sforza*

EDWARDS & CAVALLI-SFORZA (1965) described a divisive polythetic method for determining optimal dichotomous subdivisions of a set of samples. All possible dichotomies are considered, and the intragroup sum of squared distances determined:

$$\sum_{i=1}^{M-1} \sum_{j=i+1}^{M} \sum_{k=1}^{S} (x_{ik} - x_{jk})^2 \qquad (2)$$

where x_{ij} is the value of the j th attribute in the i th sample, and there are S attributes with M samples in the group in question. That dichotomy for which this sum is minimal is chosen, and the process is repeated on the group thus divided.

This is a laborious process, because of the number of possible dichotomies to be considered, and the authors do not suggest it be used where the total number of samples exceeds 15. It can be applied to either quantitative or binary data (and is considerably simplified in the latter case), but is sensitive to the scale used for the different attributes.

No probabilistic treatment was proposed to determine whether clusters should be separated, and it is clear that the authors envisaged it as a method of classifying the elements of a defined population rather than a set of samples which might not in fact need classifying at all.

The procedure leads to an analysis of variance. But, since the classification is not defined *a priori* but developed as a result of the operation of the procedure, the usual tests for comparison of variance do not apply. Presumably it would be possible (perhaps by Monte Carlo methods) to develop criteria for the partition of variance, optimal in the sense described, based on the null hypothesis of a single population; but so far this has not been done.

19.5.1.4 *MacNaughton-Smith's Method*

MACNAUGHTON-SMITH et al. (1964) proposed a rather similar polythetic divisive method, which seems to have been little applied. A measure of dissimilarity between samples or groups of samples is defined — in fact, the weighted Euclidean distance proposed by the same authors (WILLIAMS et al. 1964). That sample which is most dissimilar from the rest is identified, and any other samples which are more similar to this sample than to the rest of the set are added to it, one by one, to form a distinct class. When no more samples are more similar to this segregate group than to the rest, the process is repeated within each of the two classes thus distinguished.

The drawback of this procedure, as in that of EDWARDS & CAVALLI-SFORZA, is that no significance tests are incorporated. This is an improvement which could, however, be introduced with little difficulty, by making use of the probabilistic indices described below (section 19.5.2.4) instead of weighted Euclidean distance.

19.5.1.5 *Information Analysis*

If one considers a set of M samples in which S species are recorded, the information content or entropy can be expressed (WILLIAMS et al. 1966) as:

$$I = SM \log M - \sum_{j=1}^{S} [a_j \log a_j + (M - a_j) \log (M - a_j)] \qquad (3)$$

where a_j is the number of samples in which the j th species is recorded. Within any sub-set of m' samples, the information content will be less, and if the set of M samples is divided into two, the sum of the information content within the two sub-sets will be less than that of the set as a whole. The reduction in information content through subdivision can thus be regarded as a measure of the 'efficiency' of the subdivision. Successive divisions like this will result in a hierarchical system or dendrogram.

Successive divisions, at each stage maximizing the reduction in information content, could thus constitute an algorithm for divisive polythetic classification. This, however, comes against the same difficulty as the procedure of EDWARDS & CAVALLI-SFORZA above — that the number of partitions to be tested is very large. LANCE & WILLIAMS (1968) accordingly proposed a monothetic procedure using the concept of information. A division of samples according to the presence or absence of each species is considered, the reduction of information content calculated in each case, and that division selected which gives the greatest reduction.

If the information content of a set of M samples is I, and division of the set into two portions at random gives the information content of the two sub-sets as I', so that the decrease in information content through their separation (or the increase through their fusion) is:

$$\Delta I = I - I' \qquad (4)$$

then the quantity $2\Delta I$ is distributed as χ^2 with n degrees of freedom. LAMBERT & WILLIAMS (1966) suggest that this could be used as a probabilistic stopping rule. Since, however, the hierarchical divisions are chosen so as to maximize ΔI at every step, a significance test of a null hypothesis of homogeneity of the group before subdivision would need to be based on the distribution of the *maximum* of this quantity within a set of the size in question.

Information-theory methods have usually been applied to binary data — the presence and absence of species. As PIELOU (1969) points out, they could equally be applied to quantitative data if the different variables (species quantity, for instance) can all be expressed in terms of a common integral measure — such as dry weight to the nearest gram, cover to the nearest unit per cent, number of individuals (ORLOCI 1969), or frequency in sub-samples of a fixed size (ORLOCI 1970, DALE 1971). On the other hand, the earlier suggestion by ORLOCI (1968) that arbitrarily chosen ranges of a continuous variable could be treated as alternative values of a qualitative variable involves loss of the relevant ordering of these values, and in consequence seems unlikely to be satisfactory.

It should be borne in mind that information-theory methods

contain built-in assumptions regarding weighting, depending on the particular variables used. In ORLOCI's (1970) method, for instance, species occurring only at low frequency will have rather little effect on clustering.

19.5.2 AGGLOMERATIVE METHODS

In contrast with the methods described above, these proceed from the individual samples, and combine them in progressively increasing groups of decreasing internal similarity until the whole set of samples is combined. Again, a dendrogram results.

In applying the agglomerative procedures, the clusters may be developed according to various rules. One of the commonest is that of the 'nearest neighbour' — that is, decisions as to whether a new sample should be added to a cluster, or two clusters should be joined, are based on the similarity of the two most similar samples involved in the fusion, irrespective of the size or extent of the cluster. An alternative often used is a centroid measure — that is, the individual samples in the cluster are ignored and are replaced by their centroid, a hypothetical sample for which each variable has the mean value of all samples in the cluster. Yet another approach is to apply whatever pairwise measure of similarity of distance is selected to every possible pair involved, and to minimize the mean of these measures in the new cluster. Finally, there is the rather rarely used 'furthest-neighbour' criterion, in which the most dissimilar members of the two clusters to be amalgamated are considered, thus minimizing the maximum dissimilarity or distance within the resulting larger cluster.

19.5.2.1 *Simple Similarity Methods*

The agglomerative procedure used by SNEATH (1957) for bacterial classification deserves mention for its historical importance in numerical taxonomy, though it has only rarely been used to classify vegetation (GILBERT & WELLS 1966). It is polythetic, being based on a similarity coefficient between the entities to be classified (the JACCARD (1901) 'coefficient of community', (39) in article 6, was used, but the procedure is not tied to any particular measure of similarity). That pair of entities with the highest similarity are combined, and combination proceeds using successively smaller similarities to associate each as yet unallotted entity, either with another single entity or with a group already recognized. It is a nearest-neighbour observed, or a greater, proportion of the variance should be accounted for by the first division; that the observed, or a greater, proportion of the *residual* variance should be accounted for by the second division; and so forth. These questions are not on all fours with those normally asked in an analysis of variance, and no formal answer appears yet to have been produced. As an approximate test, however, one may determine the ratio

$$(n-g)Q_i/Q_r \qquad (7)$$

where a total of n samples have already been partitioned into g groups with an intra-group dispersion of Q_r, and Q_i is the increment in Q which results from the proposed amalgamation. If this ratio does not exceed the variance ratio ($n_1=1$, $n_2=n-g$) corresponding with a probability of

$$1-(1-a)^{2/\{g(g-1)\}} \qquad (8)$$

where a is the protection level selected, then the amalgamation is acceptable.

19.5.2.3 *Hall's Heterogeneity Method*

A similar method proposed by HALL (1967a, 1967b) is based on a heterogeneity function expressing the differences in attribute values (species quantity, etc.) within a group of samples, the contribution of each species to the heterogeneity function being weighted by its average presence in that group, as a proportion of its average presence in all samples. The clustering procedure is agglomerative, association of samples or groups at each stage being such as will minimize the increase in heterogeneity within the group. Class distinctions are recognized when the increase in heterogeneity at a given link is abruptly greater than usual. It is a drawback of this, like many other clustering methods, that no satisfactory test of a null hypothesis is possible, the distribution of the heterogeneity function being unknown. It is, of course, conceivable that a nonparametric test for the increase in heterogeneity at a given linkage could be developed. So far this has not been done.

In applying this method, HALL comments (1967b) that association of a single stand with an existing cluster is much more likely to occur than a combination of two existing clusters of about equal size, even if the single stand is much more unlike either cluster than they are like one another. Thus there is a tendency to 'chaining' — the association of stands one by one with existing clusters — rather than the progressive expansion of several clusters of comparable size as the process of agglomeration proceeds.

19.5.2.4 *A Probabilistic Approach*

A method for classification based on probabilistic considerations has been described by GOODALL (1966a, 1966d, 1968, 1971), though applications to the classification of vegetation have not yet been published. It is based on the null hypothesis that the samples are derived from a single population in which the attributes are uncorrelated. Methods are derived for calculating the probability that the set of attributes in any individual will deviate from the norm as much as they do ('deviant index'), that the sets in any two individuals will deviate similarly from the norm as much as they do ('similarity index'), and that the set of attributes of an individual will resemble those of a specified group as much as they do ('affinity index'). Each of these indices can handle qualitative, ordered or quantitative variables and combine the evidence from them.

With the aid of these tools, it is possible to examine a set of stands and determine whether they can (within the significance limit selected) be regarded as a sample from a homogeneous population. If not, the most deviant samples may be removed one by one until the residue is homogeneous. Alternatively, if two samples have a signifiant similarity index, they can be treated as the nucleus of a cluster, and the affinity of other samples with this cluster nucleus is tested. Those which show a significant affinity are added to the cluster, and the process continues, again until the residue is homogeneous, with separation of additional clusters if indicated. As mentioned, it is an assumption of this method that the variables are uncorrelated; if this is untrue for the variables as observed, they may be replaced by a new set of variables derived from them by transformation and rotation of axes.

19.5.2.5 *Information Analysis*

A divisive method using the concept of information has been described above (section 19.5.1.5). An agglomerative treatment (WILLIAMS et al. 1966, LAMBERT & WILLIAMS 1966, WEBB et al. 1967a, ORLOCI 1971) may be based on the same principle. One may ask what combination of samples into clusters (or of clusters already recognized into larger clusters) will result in a minimum increase in information within the resulting larger groups — that is, in effect, which combination of samples or groups will unite those which are most similar. This procedure can also be continued, of course, at each stage combining two single samples, one sample with a group, or two groups, choosing whichever fusion gives the minimum increase in intra-group information of all the possible alternatives. Thus again a hierarchy of groups is built up, comparable with that obtained by the divisive approach.

ORLOCI(1970) suggests two alternative algorithms for clustering in this way, using functions respectively taking into account the difference in distribution of each species considered separately

$$\sum_i \sum_j f_{ij} \log_e \frac{c f_{ij}}{f_{i\cdot}} \qquad (9)$$

or the differences if the species are considered in the context of the overall records for each stand

$$\sum_i \sum_j f_{ij} \log_e \frac{f_{\cdot\cdot} f_{ij}}{f_{i\cdot} f_{\cdot j}} \qquad (10)$$

where f_{ij} is the number of sub-samples in which the i th species was recorded in the j th stand (out of a total of c), and the dot subscript indicates summation. If the first expression is used, differences between stands in the sum of frequencies for all species would contribute to the clustering decisions, in the latter case they would depend only on their *relative* frequencies. In the example given by ORLOCI, however, the dendrograms resulting from use of the two functions were very similar.

LAMBERT & WILLIAMS(1966) point out that their information-analysis technique has a tendency to associate with one another samples which differ widely from the larger groups present, even though their characteristics may be very diverse, and though they may share more features with one or other of the larger groups than they have in common. It also tends to combine in one group a number of samples which agree only in containing a small number of species. These features suggest that the results of information analysis should not be accepted uncritically.

19.6 Cluster Shape and Dendrogram Shape

In the geometric model we have been using, the cluster of points representing a class may take a wide variety of shapes. In the simplest case, the density of points will decrease outwards from a centre in a nested series of hyperspherical envelopes. In other cases, they will be hyperellipsoids. This distinction will depend on the scaling and orientation of the axes; by appropriate rotation and scaling any hyperellipsoid can be transformed into a hypersphere. Other figures are more distinctive — hyperellipsoids with one or more axes curved, for instance; or star-shaped figures, with a dense central body from which a number of extensions radiate.

The shape of a cluster representing a vegetational class may have considerable interest to the plant sociologist, and is well worth study. This can only be done effectively if numerous samples are available; it calls for an ordination to be applied separately to the samples allocated to each class (GREIG-SMITH et al. 1967), followed by an inspection of the results of ordination, or formal tests if any particular hypotheses about cluster shape appear to be of interest.

Cluster shape should not be confused with the shape of a dendrogram. Comments have frequently been made on the occurrence of 'chaining' in some dendrograms — in agglomerative procedures, a cluster may tend to grow by accretion of individual samples one by one, rather than by fusing with other clusters. This may indeed arise if the cluster is a very elongated one; but this cluster shape is by no means a necessary condition for dendrogram chaining, and chaining seems to be more a feature of the technique than of the data (LANCE & WILLIAMS 1967). Objections to it seem to be practical (in terms of increased computational effort) rather than theoretical.

19.7 Classification of Groups of Samples

In principle, the first step in classification is the recognition of classes among which the samples are allocated, without regard to any hierarchical or other relationship among the classes thus delimited. In practice, as we have seen, the two operations are often combined. In association analysis or information analysis, for instance, no distinction is made between the procedure for distinguishing ultimate groups (if indeed there is any stopping rule at which ultimate groups are delimited, as often there is not) and that for combining the ultimate groups into broader ones. This lack of distinction between cluster recognition and cluster classification is to be regretted, for the same procedures are not necessarily optimal for the two processes. The considerations involved may be different, and in any case some rescaling of the variables may be appropriate (WILLIAMS & DALE 1965).

If the primary classes or groups of samples represent sub-populations separated from one another by discontinuities, then the question of the relationship among these classes is a substantive scientific one. If they are in any case artificial segregates of a single population, then the question of their interrelationship may have little interest.

One common way in which the relationship between classes is expressed is in terms of a higher-order classification — a hierarchy of classes. A set of classes, A_i, form a hierarchy if

(1) for every A_i, A_j, there is an A_k for which

$$(A_i \subset A_k) \wedge (A_j \subset A_k) \qquad (11)$$

and

(2) $\quad (A_i \subset A_k) \wedge (A_i \subset A_j) \rightarrow (A_j \subset A_k) \vee (A_k \subset A_j) \qquad (12)$

Thus, in a purely hierarchical arrangement, the relationship between two classes is expressed in terms of class membership, and not quantitatively.

A distinction has sometimes been drawn between hierarchical and multidimensional or reticulate relationships among vegetation classes (e.g. POORE 1955). These views of the relationships between classes are not mutually exclusive. A system in multidimensional space could be expressed as a hierarchy, but in so doing one loses information regarding the distance and directional relationships among entities classified.

A hierarchical structure for a set of classes can be established by any of the step-wise procedures used to identify primary classes and delimit their membership. Since, however, we are concerned with a defined population rather than a set of samples, probabilistic considerations are hardly relevant.

In a hierarchical system the question may arise of the comparability of levels of subdivision in the different branches of the dendrogram. This can be — and often is — decided by arbitrary levels of the similarity or distance index on which the hierarchy is based; if the two groups into which a higher group is being divided have a similarity less than a certain arbitrary value, then they are at a higher hierarchical level ('order,' for instance); if they are more similar the groups are at a lower level ('alliance,' say). For such decisions, however, the arbitrariness of the choice of similarity value for each hierarchical level, its essential incomparability in different contexts, and the further arbitrary choice of a similarity index itself, should be emphasized.

In the spatial model, a higher-order class is simply a group of clusters in spatial proximity. This suggests that, rather than develop a hierarchical arrangement for the classes, it might be more informative and more fruitful to express their spatial relations in quantitative terms. This is done, for instance, in the various plexus models discussed elsewhere (article 7), though an adequate representation in this form would require almost as many dimensions as there are classes.

Spatial relations among classes can often be represented more simply if one first applies ordination to the primary classes. Ordination procedures are discussed elsewhere in this volume (articles 10, 11), and any making use of quantitative variables could be applied to the values specifying the centroids of each class. Often only the first two or three axes of the ordination are statistically significant or meaningful, and in this case the plexus construction can be based on them and the residual variation can be treated as an error term.

Yet another method of representing the relations among classes, which however has been little used, is by cross-classification. This could take the form of two or more quite independent hierarchical classifications of the same set of classes — which might, for instance be based respectively on habitat features and on successional status.

19.8 Classification and Ordination

In the early days of applying numerical methods to vegetational classification, ordination methods were also being developed concurrently, and there was much confusion as to their relative roles. It is now, however, generally recognized that the two approaches can be appropriately applied to the same data (Figs. 1, 2), and that to some extent the choice is a matter of taste, and should be guided by the intended use of the results. For vegetational mapping, for instance, classification is greatly to be preferred. Where the vegetational characteristics of a site are required in terms of a simplified vector of real variables — for further statistical analysis, say — ordination is clearly the treatment of choice.

It is difficult to grasp effectively the results of an ordination where the number of axes is large. It has accordingly been suggested that classification is appropriate in the first instance where very heterogenous data are included in the same analysis, but that ordination should then be applied to each of the classes thus distinguished. On the other hand, arguments have been adduced in favour of an initial ordination as a guide to classification (e.g. BRISSE & GRAND-JOUAN 1971). On balance, it seems best to use a rough ordination in few dimensions to indicate discontinuities, cluster shape and number, and perhaps to show where dividing lines between classes should be drawn. Ordination should then be applied afresh to these more manageable groups separately; as GREIG-SMITH (1971) points out, their main axes may not be parallel, so that to treat them in a single analysis may be unfruitful.

19.9 Evaluation and Comparison

Numerical procedures for the classification of vegetation have often been judged by comparison with accepted subjectively-developed classifications (e.g. HALL 1970). WILLIAMS et al. (1966) similarly state that the results of numerical classification should 'suggest groupings which are ecologically meaningful when tested by appeal to other relevant information from outside.' These points of view indicate appropriately that numerical classification, of vegetation as of other objects, is still at an early stage of development, and has not yet reached the point where it can be freed from the leading-strings of common sense and ecological experience.

A formal comparison of the merits of two alternative classifications should be based on the objectives to which they are directed. If they are intended as general-purpose classifications, then the criterion should be the prediction of a randomly-chosen attribute in a randomly-chosen stand; and the basis of their evaluation should be the success with which they do so. Since, however, this success may vary greatly over the domain covered (between classes; between variables; and between different ranges of the same variable), a weighting function to be applied to the differences between true and predicted values would be required for a rigorous treatment.

Closely akin to this criterion is the suggestion often made that the classification should be able to predict, for the samples used in developing the classification, values of variables which have not been involved in it. If the classification was based on only the tree species present, for instance, its evaluation might be based on the precision with which it predicts the composition of the field layer. There is a risk of bias here in that the variables used for the test are likely to be correlated with those used in developing the classification. And, as in the previous case, a weighting function would be needed.

Where comparison of two classifications is not for the purpose of evaluation and choice, but simply to express the extent to which they agree or disagree, an information-theory approach seems promising; the joint information in the two classifications could be compared with that peculiar to the two classifications (cf. ESTABROOK 1967). An alternative approach by WILLIAMS et al. (1969) was to determine the common elements in a set of twelve classifications by enumerating the times each pair of stands (among ten) occurred in the same class out of four. One of these pairs occurred in the same class in all twelve classifications, indicating that this pair at least were consistently associated throughout — for the chance of such an event by random assortment of stands among classes would be quite small. This approach could clearly be developed and extended to cover other comparisons between classifications; but tests based on information theory seem, in principle, likely to be more sensitive.

In the long run, evaluation of classification *procedures* should be based on evaluation of the classifications they generate: by their fruits ye shall know them. One more direct suggested criterion, however, has been based on stability. If the procedure is applied independently to two sets of samples, of which the larger includes the smaller, the class membership of samples in the smaller set should coincide in the two classifications. While one may agree that this property is desirable, there seems no reason to regard it as a sufficient criterion.

Comparative studies of classification procedures have generally been concerned with those producing a hierarchical structure, and have been limited to visual and descriptive comparisons of the shape of the dendrogram, and the arrangement of the samples classified at the ends of the branches. Since few of these procedures provide for significance tests of a suitable null hypothesis, tests of significance of differences between them are also lacking.

Even where clustering procedures are non-probabilistic and so are inappropriate as a basis for conclusions regarding a parent population from which a sample was drawn, they are likely in fact to be used for this purpose. Consequently, it might be appropriate to set up an artificial population with known properties — consisting perhaps of several distinct classes, each with specified means and distributions for each variable; take repeated random samples from this artificial population; apply several classificatory procedures to these samples; and study how well the results agreed with the structure of the known population from which the samples were drawn. By applying this approach to a number of artificial populations differing in structure, it might be possible to arrive at conclusions as to the general value of these procedures when applied to samples from unknown populations.

If classification procedures are regarded primarily as methods of generating hypotheses, then the appropriate test of the value of alternative procedures should be based on tests of the validity of the hypotheses they respectively generate. These tests should clearly be based on the same material, and applied to structurally similar hypotheses generated by the different procedures. So far, this has not been done. Different procedures have been applied to the same data, but the classifications resulting have not been put in the form of alternative hypotheses suitable for comparative testing, and *a fortiori* no such comparative tests have performed.

19.10 Conclusions

Classification of vegetation differs from classification of organisms in that, in most groups of organisms, the process of evolution has led to the formation of natural classes (species). The organisms in each species share a common gene pool and hence have a limited range of phenotypic variation; since gene flow between species is limited or absent, there tend to be sharp discontinuities in phenotype. In vegetation, on the other hand, there is no comparable genetic basis for differences in attribute sets between different types of stand, the barriers to movement of individuals between adjacent stands are much less marked than barriers to gene flow between species, and in consequence discontinuities between stand types do not arise as clearly from the logic of vegetation organization as they do from the logic of speciation.

The purpose of numerical classification of vegetation is the same as that of classification in general — to identify sets of stands for which many of the same statements will be true, and thus to reduce redundancy and economize effort in describing the variety of the stands classified. This purpose may be facilitated if there are discontinuities in the distribution of the attributes which characterize the stands; if they exist, these discontinuities are likely to be discovered and used by any efficient classification, numerical or otherwise.

Numerical classification can claim advantages over more traditional methods if
(i) it is quicker;
(ii) it uses less skilled labour;
(iii) it is more objective;
(iv) it more consistently identifies and uses natural discontinuities; or
(v) it uses optimization procedures.
These possible advantages will be considered *seriatim*.

Though at first glance it would seem likely that use of a digital computer could speed up the classification process, this is not necessarily true. The experienced field ecologist is often, half-consciously, developing and refining his conceptions of classification during the actual process of collecting stand data, whereas the computer can only start work after these data have been collected and converted to the form required for input. On the other hand, some non-numerical methods of classification involve steps which could be mechanized with great advantage. In particular, the work involved in rearranging stand × species tables in studies by the Zürich-Montpellier school (article 20) could be performed better and more quickly by computer.

The use of less skilled labour for numerical classification is a real possible advantage. In the simpler types of vegetation, records of stand composition can be made by relatively unskilled labour. If then the process of analyzing these records and sorting them into classes can be given to a suitably programmed computer, skilled labour may be needed only in programme preparation. The resulting classifications might, however, have value only as evaluated and interpreted in the light of field experience and understanding.

It is also true that numerical classification is more objective than most traditional methods. The use of numerical methods requires that the classification process be defined exactly and in great detail and, once this algorithm has been written, it contains no ambiguity and allows no scope for individual judgement or choice. Anyone applying the same algorithm or computer program to the same data must arrive at the same conclusions. Thus all subjectivity is removed from this part of the classification process — though, of course, the choice of the particular algorithm used was, at an earlier stage, the subject of individual judgement.

An explicit search for discontinuities calls for a probabilistic approach, and consequently most of the techniques for numerical classification, being deterministic, do not meet this need in any direct fashion; in fact the intuitive approach, which will often look for discontinuities sub-consciously, may perform better in this respect. Even some of the probabilistic approaches described above, based on a model of uncorrelated attributes, will not distinguish an elongated cluster from two separate clusters. Consequently, though a consistent and objective search for discontinuities could be an advantage of numerical methods, it is not in fact realized in most of the existing techniques.

Optimization of a classification, in the sense that the classes predict attribute values of their members as precisely as a division into a given number of classes could do, may be approached by some intuitive classifications; but a consistent attainment of the optimum can only result from application of a numerical method. Many of the clustering methods described include optimization elements; but the quantity maximized or minimized is often only rather indirectly related to the intra-class variance. And, in the many hierarchical procedures, optimization of a quantity at each step of the hierarchy may not ensure its optimization in the final groupings. These difficulties seem to be avoided by the procedure of ORLOCI (section 19.5.2.2), which however is unsatisfactory when it comes to a 'stopping rule.'

The theoretical advantages of numerical over traditional classification seem thus to be more in potential than in performance. One may expect further development of probabilistic methods which will be based on natural discontinuities in the data, and will perform an optimal allocation of samples adjacent to the discontinuities. Meanwhile, advocacy of the use of numerical methods must be based more on economy in time and skill than on their theoretical advantages.

It has been mentioned that it may be useful to precede a numerical classification by an ordination, which will make clearer the main directions of variation, and the broad structure of the data collected. It has, moreover, been suggested that classes which have been distinguished should subsequently be ordinated as a means of establishing and presenting their mutual relationships. In this event, it may be asked, what is the function or value of a classification process sandwiched between these two ordination operations? The answer lies simply in the convenience, for human thought, of the class concept. It is much easier to think or write about a group of stands considered simply as members of a class, whose variation within the class can conveniently be ignored, than as individuals placed within a continuum. In a continuum concept, all statements about the composition or behavior of a stand would have to be expressed as a function of position in the continuum, whereas in terms of a class concept they can be absolute for each class. The biggest advantage, however, probably comes in relation to mapping. To map the spatial distribution of a number of mutually exclusive classes is easy. To map position in a continuum in relation to the two-dimensional space of the earth's surface would require a series of overlay contour diagrams, one for each axis of the continuum, and would be far more difficult to interpret.

Though preliminary ordination is desirable, it is likely that many investigators will prefer to proceed directly to classification, using one of the existing clustering methods. Choice among them should depend, among other things, on computing costs (LAMBERT & DALE 1964); but the risks of chance misclassification with monothetic methods (see above) argue for one of the polythetic methods, despite their usually greater computing load. The polythetic methods have the advantage of greater flexibility; and, in fact, they include monothetic division as a possible, but improbable, outcome. Where quantitative variables are used, the best choice seems to be either the agglomerative method of ORLOCI (section 19.5.2.2) or the divisive method of MAC NAUGHTON-SMITH et al. (section 19.5.1.4); for binary variables (presence and absence of species) one may use the agglomerative method based on information theory (section 19.5.2.5); and for mixed variables the probabilistic method described in section 19.5.2.4 may be appropriate. There is great need for extensive trials of the power of these different techniques to detect known discontinuities in artificial populations, and the differences in power revealed could then be balanced against differences in cost.

As for the choice of variables, there seems reason to support the view that, where the number of species and the range of variation to be covered are large, species presence may supply all the information needed for classification, and quantitative variables may even cloud the issue. If only a fraction of the species is being used in the analysis, care should be taken in their selection. It cannot be assumed that the less frequent species can safely be ignored (GOODALL 1969); a better rule might be to omit from the analysis those species which show little association with others.

Where the group of stands to be classified cover only a rather limited range of variation, or where the vegetation is poor in species, quantitative variables will generally be needed for the additional information they contain. Thus, it may often be appropriate to perform some inital subdivisions on the basis of species presence and absence, and then to examine these classes in greater detail, using quantitative measures of species participation.

It should be emphasized again, though, that direct application of clustering methods should be regarded as a rather makeshift procedure. As further experience in this field accumulates it is to be expected that techniques will develop involving ordination in the first instance, which will permit discontinuities and clusters to be identified in a single operation, rather than by a hierarchical procedure, and based on tests of successively more complex null hypotheses regarding the cluster structure of the underlying population of stands from which a sample has been taken.

19.11 SUMMARY

Considering stands of vegetation to be represented by points in a multidimensional system of which the axes are the variables by which the stands may be described, classification amounts to the division of this space into discrete cells by hypersurfaces. If the distribution of points is interrupted by discontinuities or regions of low density, the dividing hypersurfaces should follow these discontinuities. Even if there are no such discontinuities, division of vegetation space into arbitrary classes may still serve practical purposes.

It is recommended that the distribution of points in vegetation space — simplified by an ordination — should be studied as a preliminary to numerical classification. If the distribution is continuous, an arbitrary number of classes may be chosen, and dividing hypersurfaces selected which will minimize the total intra-class variance. If discontinuities are revealed, they should be identified, and the classes delimited in reference to them.

The various 'clustering' methods in vogue should be regarded as approximations to the ideal procedure described above. They are usually hierarchic, proceeding either by successive divisions of the complete group of stands, or by succesive amalgamation of smaller, then larger, groups. Monothetic divisive procedures, in which each successive division is based on the value of a single variable, are usually simplest and most expeditious, but are less reliable than the polythetic methods (depending on several variables simultaneously) which, for practical reasons, are usually agglomerative. All these clustering methods may be improved by abandoning the strictly hierarchical principle, and at the end of the process re-examining the resulting classes for possible recombination or for transfers of stands from one class to another. Since the stands classified are only a sample of the population of stands to which the classification is intended to apply, probabilistic tests are highly desirable as part of the classification procedure.

Any relevant variables may be used for classification, but the floristic variables have most usually been employed — either some measure of the quantity of each species present, or a binary variable expressing its presence or absence only. For classifications covering a wide range of stands the latter is recommended; but when the focus narrows quantitative variables are more useful.

Once primary classes of stands have been recognized the interrelations among the classes may be investigated, either by ordination or by a further classification procedure, which should now be non-probabilistic.

Some further, recent developments in numerical classification cited elsewhere in the book are: nodal component analysis and other continuous multivariate techniques (11.6), numerical techniques in phytosociology (20.8.6 and 20.12), problems of discontinuous data sets (10.4.2.4, 11.3.4 and 11.6.1), and studies combining classification and ordination (10.4.1.3, 11.4.4, 20.9 and 20.12).

REFERENCES

BRISSE, H. & G. GRANDJOUAN, – 1971 – Adaptation d'une méthode de classification multivariable par similitudes à l'écologie végétale en milieu naturel. I. Exposé de la méthode. *Oecol. Pl.* 6: 163—187.
BROWN, R. T. & J. T. CURTIS, – 1952 – An ordination of the upland forest communities of southern Wisconsin. *Ecol. Monogr.* 22: 217—234.
CRAWFORD, R. M. M. & D. WISHART, – 1967 – A rapid multivariate method for the detection and classification of groups of ecologically related species. *J. Ecol.* 55: 505—524.
CRAWFORD, R. M. M. & D. WISHART, – 1968 – A rapid classification and ordination method and its application to vegetation mapping. *J. Ecol.* 56: 385—404.
CURTIS, J. T. & R. P. MCINTOSH, – 1951 – An upland forest continuum in the prairie-forest border region of Wisconsin. *Ecology* 32: 476—496.
CZEKANOWSKI, J., – 1909 – Differentialdiagnose der Neandertalgruppe. *Korresp Bl. dt. Ges. Anthrop.* 40: 44—47.
DALE, M. B., – 1971 – Information analysis of quantitative data. In, 'Statistical Ecology,' ed. G. P. PATIL, E. C. PIELOU & W. E. WATERS. Pa. State Univ., University Park 3: 133—148.
EDWARDS, A. W. F. & L. L. CAVALLI-SFORZA, – 1965 – A method for cluster analysis. *Biometrics* 21: 362—375.
ESTABROOK, G. F., – 1967 – An information theory model for character analysis. *Taxon* 16: 86—97.
FLENLEY, J. R., – 1969 – The vegetation of the Wabag region, New Guinea Highlands: a numerical study. *J. Ecol.* 57: 465—490.
GILBERT, N. & T. C. E. WELLS, – 1966 – Analysis of quadrat data. *J. Ecol.* 54: 675—685.
GOODALL, D. W., – 1953 – Objective methods for the classification of vegetation. I. The use of positive interspecific correlation. *Aust. J. Bot.* 1: 39—63.
GOODALL, D. W., – 1954 – Objective methods for the classification of vegetation. III. An essay in the use of factor analysis. *Aust. J. Bot.* 2: 304—324.
GOODALL, D. W., – 1961 – Objective methods for the classification of vegetation. IV. Pattern and minimal area. *Aust. J. Bot.* 9: 162—196.
GOODALL, D. W., – 1966a – Deviant index - a new tool for numerical taxonomy. *Nature, Lond.* 210: 216.
GOODALL, D. W., – 1966b – Classification, probability and utility. *Nature, Lond.* 211: 53—54.
GOODALL, D. W., – 1966c – Hypothesis testing in classification. *Nature, Lond.* 211: 329—330.
GOODALL, D. W., – 1966d – A new similarity index based on probability. *Biometrics* 22: 882—907.
GOODALL, D. W., – 1968 – Affinity between an individual and a cluster in numerical taxonomy. *Biométr.-Praxim.* 9: 52—55.
GOODALL, D. W., – 1969 – A procedure for the recognition of uncommon species combinations in sets of vegetation samples. *Vegetatio* 18: 19—35.
GOODALL, D. W., – 1970 – Statistical plant ecology. *Ann. Rev. Ecol. Syst.* 1: 99—124.
GOODALL, D. W., – 1971 – Cluster analysis using similarity and dissimilarity. *Biométr.-Praxim.* 11: 34—41.

GOUNOT, M., – 1969 – Méthodes d'étude quantitative de la végétation. Masson, Paris. 314 pp.
GREIG-SMITH, P., – 1971 – Application of numerical methods to tropical forests. In, 'Statistical Ecology,' ed. G. P. PATIL, E. C. PIELOU & W. E. WATERS. Pa. State Univ., University Park 3: 195—204.
GREIG-SMITH, P., M. P. AUSTIN, & T. C. WHITMORE, – 1967 – The application of quantitative methods to vegetation survey. I. Association-analysis and principal component ordination of rain forest. *J. Ecol.* 55: 483—503.
HALL, A. V., – 1967a – Methods for demonstrating resemblance in taxonomy and ecology. *Nature, Lond.* 214: 830—831.
HALL, A. V., – 1967b – Studies in recently developed group-forming procedures in taxonomy and ecology. *J. S. Afr. Bot.* 33: 185—196.
HALL, A. V., – 1970 – A computer-based method for showing continua and communities in ecology. *J. Ecol.* 58: 591—602.
HOPKINS, B., – 1968 – Vegetation of the Olokemji Forest Reserve, Nigeria V. The vegetation on the savanna site with special reference to seasonal changes. *J. Ecol.* 56: 97—115.
JACCARD, P., – 1901 – Distribution de la flore alpine dans le Bassin des Dranses et dans quelques régions voisines. *Bull. Soc. vaud. Sci. nat.* 37: 241—272.
KULCZYŃSKI, S., – 1928 – Die Pflanzenassoziationen der Pieninen. *Bull. int. Acad. pol. Sci. Lett.*, Cl. Sci. Math. Nat., Ser. B., 1927 (Suppl. 2): 57—203.
LAMBERT, J. M. & M. B. DALE, – 1964 – The use of statistics in phytosociology. *Adv. ecol. Res.* 2: 59—99.
LAMBERT, J. M. & W. T. WILLIAMS, – 1966 – Multivariate methods in plant ecology. IV. Comparison of information-analysis and association-analysis. *J. Ecol.* 54: 635—664.
LANCE, G. N. & W. T. WILLIAMS, – 1967 – A general theory of classificatory sorting strategies. I. Hierarchical systems. *Comput. J.* 9: 373—380.
LANCE, G. N. & W. T. WILLIAMS, – 1968 – Note of new information-statistic classificatory program. *Comput. J.* 11: 195.
LIETH, H. & G. W. MOORE, – 1971 – Computerized clustering of species in phytosociological tables and its utilization for field work. In, 'Statistical Ecology,' ed. G. P. PATIL, E. C. PIELOU & W. E. WATERS. Pa. State Univ., University Park 1: 403—422.
LOOMAN, J., – 1963 – Preliminary classification of grasslands in Saskatchewan. *Ecology* 44: 15—29.
MACNAUGHTON-SMITH, P., W. T. WILLIAMS, M. B. DALE, & L. G. MOCKETT, – 1964 – Dissimilarity analysis: a new technique of hierarchical sub-division. *Nature, Lond.* 202: 1034—1035.
MATUSZKIEWICZ, W. & A. MATUSZKIEWICZ, – 1956 – Materiały do fitosocjologicznej systematyki ciepłolubnych dąbrów w Polsce (Germ. summ.: Zur Systematik der Quercetalia pubescentis-Gesellschaften in Polen). *Acta Soc. bot. Pol.* 25: 27—72.
MOORE, G. W., W. S. BENNINGHOFF, & P. S. DWYER, – 1967 – A computer method for the arrangement of phytosociological tables. *Proc. Ass. Comput. Mach.* 297—299.
MOTYKA, J., – 1947 – O zadaniach i metodach badań geobotanicznych (French summ.: Sur les buts et les méthodes des recherches géobotaniques). *Annls. Univ Mariae Curie-Skłodowska*, Lublin, Sect. C. Suppl. 1: 1—168.
NOY-MEIR, E., N. H. TADMOR, & G. ORSHAN,– 1970 – Association analysis of desert vegetation. *Israel J. Bot.* 19: 561—591.
ORLOCI, L., – 1967 – An agglomerative method for classification of plant communities. *J. Ecol.* 55: 193—206.

ORLOCI, L., – 1968 – Definitions of structure in multivariate phytosociological samples. *Vegetatio* 15: 281—291.
ORLOCI, L., – 1969 – Information analysis of structure in biological collections. *Nature, Lond.* 223: 283—284.
ORLOCI, L., – 1970 – Analysis of vegetation samples based on the use of information. *J. theoret. Biol.* 29: 173—189.
ORLOCI, L., – 1971 – Information theory techniques for classifying plant communities. In, 'Statistical Ecology,' ed. G. P. PATIL, E. C. PIELOU & W. E. WATERS. Pa. State Univ., University Park 3: 259—279.
PIELOU, E. C., – 1969 – An Introduction to Mathematical Ecology. Wiley-Interscience, New York. 286 pp.
POORE, M. E. D. ,– 1955 – The use of phytosociological methods in ecological investigations. III. Practical applications. *J. Ecol.* 43: 606—651.
SNEATH, P. H. A., – 1957 – The application of computers to taxonomy. *J. gen. Microbiol.* 17: 201—226.
SOKAL, R. R. & P. H. A. SNEATH, – 1963 – Principles of Numerical Taxonomy. Freeman, San Francisco. 359 pp.
VASILEVICH, V. I., – 1969 – Statistical Methods in Geobotany (In Russian). Nauka, Leningrad. 232 pp.
WARD, S. D., – 1971 – The phytosociology of *Calluna-Arctostaphylos* heaths in Scotland and Scandinavia. I. Dinnet Moor, Aberdeenshire. *J. Ecol.* 58: 847—863.
WEBB, L. J., J. G. TRACEY, W. T. WILLIAMS, & G. N. LANCE, – 1967a – Studies in the numerical analysis of complex rain-forest communities. I. A comparison of methods applicable to site/species data. *J. Ecol.* 55: 171—191.
WEBB, L. J., J. G. TRACEY, W. T. WILLIAMS, & G. N. LANCE – 1967b – Studies in the numerical analysis of complex rain-forest communities. II. The problem of species-sampling. *J. Ecol.* 55: 525—538.
WEBB, L. J., J. G. TRACEY, W. T. WILLIAMS, & G. N. LANCE, – 1970 – Studies in the numerical analysis of complex rain-forest communities V. A comparison of the properties of floristic and physiognomic-structural data. *J. Ecol.* 58: 203—232.
WEST, N. A., – 1966 – Matrix cluster analysis of montane forest vegetation of the Oregon Cascades. *Ecology* 47: 975—980.
WHITTAKER, R. H., – 1962 – Classification of natural communities. *Bot. Rev.* 28: 1—239.
WILLIAMS, W. T. & M. B. DALE, – 1962 – Partitioned correlation matrices for heterogeneous data. *Nature, Lond.* 196: 602—603.
WILLIAMS, W. T. & M. B. DALE, – 1965 – Fundamental problems in numerical taxonomy. *Adv. bot. Res.* 2: 35—68.
WILLIAMS, W. T. & J. M. LAMBERT, – 1959 – Multivariate methods in plant ecology. I. Association-analysis in plant communities. *J. Ecol.* 47: 83—101.
WILLIAMS, W. T. & J. M. LAMBERT, – 1960 – Multivariate methods in plant ecology. II. The use of an electronic digital computer for association-analysis. *J. Ecol.* 48: 689—710.
WILLIAMS, W. T., M. B. DALE, & P. MACNAUGHTON-SMITH, – 1964 – An objective method of weighting in similarity analysis. *Nature, Lond.* 201: 426.
WILLIAMS, W. T., J. M. LAMBERT, & G. N. LANCE, – 1966 – Multivariate methods in plant ecology. V. Similarity analysis and information-analysis. *J. Ecol.* 54: 427—445.
WILLIAMS, W. T., G. N. LANCE, L. WEBB, J. G. TRACEY, & M. B. DALE, – 1969 – Studies in the numerical analysis of complex rain-forest communities. III. The analysis of successional data. *J. Ecol.* 57: 515—535.

20 THE BRAUN-BLANQUET APPROACH

Victor Westhoff and Eddy van der Maarel

Contents

20.1	Introduction	289
20.2	History	290
20.3	General Concepts	293
20.3.1	Concrete vs. Abstract Units	293
20.3.2	Floristic-Sociological Classification Units	295
20.3.3	Diagnostic Species	296
20.3.4	Sociological Progression	298
20.3.5	Natural Classification	299
20.4	Analytical Research Phase	301
20.4.1	Reconnaissance and Choice of Plots	301
20.4.2	Boundaries	303
20.4.3	Minimal Area and Plot Size	305
20.4.4	Description of Structure and Strata	307
20.4.5	Floristic Analysis	308
20.4.5.1	Cover and Abundance	308
20.4.5.2	Sociability	310
20.4.5.3	Vitality and Fertility	311
20.4.5.4	Periodicity	312
20.4.6	Relevé Protocols	312
20.5	Synthetical Research Phase	313
20.5.1	Steps in Synthesis	313
20.5.2	Primary Table	315
20.5.3	Presence, Constancy, Homotoneity	316
20.5.4	Species Weights	319
20.5.5	Table Rearrangements	320
20.5.6	Mechanical Means for Rearrangement	323
20.5.7	Determination of Fidelity	324
20.5.8	Geographical Effects on Fidelity	326
20.5.9	Phytocoenon Characterization	328
20.6	Syntaxonomical Research Phase	329

20.6.1	Syntaxon Table Characterization	329
20.6.2	Characterization of the Association	329
20.6.3	Geographic Problems with Associations	330
20.6.4	Higher Units (Alliance, Order, Class)	332
20.6.5	Units above the Class	334
20.6.6	Horizontal and Vertical Classification	336
20.6.7	Lower Units (Subassociation, Variant, Facies)	336
20.6.8	Nomenclature	339
20.6.8.1	Construction of Names	340
20.6.8.2	Validity, Changes, and Authors' Names	342
20.6.9	Scheme of Syntaxa	343
20.7	Extension of the Approach	344
20.7.1	Other Geographic Areas	344
20.7.2	Biotic Communities	348
20.7.3	Microcommunities and Synusiae	348
20.7.4	Community Complexes	349
20.7.5	Applied Phytosociology	351
20.7.6	Structural Considerations	352
20.7.7	Indicator Groups	353
20.7.7.1	Indicators for Classification	354
20.7.7.2	Indicators for Gradient Relations	355
20.8	Numerical Techniques	355
20.8.1	Storage of Relevés	355
20.8.2	Species Correlation	357
20.8.3	Fidelity Tests	358
20.8.4	Relevé Similarity	359
20.8.5	Measurement of Homotoneity	360
20.8.6	Numerical Classification	360
20.8.7	Table Rearrangement	361
20.8.8	Numerical Syntaxonomy	363
20.9	Ordination	365
20.9.1	Informal Ordination	365
20.9.2	Formal Ordination	366
20.10	Conclusion	371
20.11	Summary	372
20.12	Additions to the Second Edition	374

20 THE BRAUN-BLANQUET APPROACH

20.1 Introduction

We shall, in this last chapter, treat the floristic-sociological or BRAUN-BLANQUET approach to classification and interpretation of communities. Before elaborating the details of the approach, we may state its essence in three ideas:

(i) Plant communities are conceived as types of vegetation, recognized by their *floristic composition*. The full species compositions of communities better express their relationships to one another and environment than any other characteristic.

(ii) Amongst the species that make up the floristic composition of a community, some are more sensitive expressions of a given relationship than others. For practical classification (and indication of environment) the approach seeks to use those species whose ecological relationships make them most effective indicators; these are *diagnostic species* (character-species, differential-species, and constant companions).

(iii) Diagnostic species are used to organize communities into a *hierarchical classification* of which the association is the basic unit. The vast information with which phytosociologists deal must, of necessity, be thus organized; and the hierarchy is not merely necessary but invaluable for the understanding and communication of community relationships that it makes possible.

The reader will see how far the elaboration of this three-part theme has led the members of the 'school of BRAUN-BLANQUET.' The results are (like community relationships) complex. We fear lest the reader should lose sight of the heart of the approach — the floristic perspective — for the abundance of technical details that follow. The latter are only the skeleton of the approach. Furthermore, if we do not press the physical comparison too far, its body is the still-growing corpus of basic and applied research and the resulting knowledge of and insight into communities. This review must deal mainly with the skeleton of the approach; we cannot in the space we have discuss adequately the kinds of understanding of communities that have come from it. We ask the reader to recognize wherein this account must stop short, and to seek if he is interested a further feeling for what the approach offers by observing or reading studies applying it. On the other hand, we shall pay some attention (especially in sections 20.8 and 20.9) to recent developments in phytosociology that are not part of the BRAUN-BLANQUET approach as such but may be of value since they introduce concepts and methods that relate the floristic-sociological approach to other approaches.

Textbooks and reviews on the floristic-sociological approach have been published in most European countries, Argentine, India, Japan and the U.S.A. (see the bibliography by MAAREL, TÜXEN & WESTHOFF, 1970, Exc.[1]). Recent accounts include GOUNOT (1969), WESTHOFF & HELD (1969), and KNAPP (1971). Useful accounts in English are BECKING (1957), WHITTAKER (1962), KÜCHLER (1967) and SHIMWELL (1972).

20.2 **History**

The origin and history of the approach have been reviewed elsewhere and can be stated only briefly here (see GAMS 1918, BRAUN-BLANQUET 1921, and DU RIETZ 1921, as well as BECKING 1957, WHITTAKER 1962, and SHIMWELL 1972). The systematic description of plant communities and the idea of community types can be traced from great students of plant geography, HUMBOLDT (1805), SCHOUW (1823), HEER (1835) and GRISEBACH (1838). From their work two main approaches developed, the physiognomic and the floristic. The physiognomic approach developed as its principal unit the formation, a unit characterized by physiognomy or vegetation structure (POST 1862, GRISEBACH 1872, WARMING 1895; see article 13). Through the work of students dealing with smaller-scale units (esp. LECOQ 1844, 1855, THURMANN 1849, LORENZ 1858, KERNER 1863 and DRUDE 1890, in southern and central Europe, POST 1862, HULT 1881 and CAJANDER 1903 in northern Europe) there developed the essential idea of the floristic-sociological approach: plant communities as units of classification based primarily on species composition. From LORENZ (1863), MÖBIUS (1877) and DAHL (1908) a parallel approach to biotic communities can be traced. (See BALOGH 1958 and WHITTAKER 1962 for review and references.)

Much of the further development leading to the BRAUN-BLANQUET approach was centered in Zürich (STEBLER & SCHRÖTER 1893, SCHRÖTER 1894, SCHRÖTER & KIRCHNER 1902, BROCKMANN-

---

[1] This and many other bibliographies to be mentioned in this contribution have been published in *Excerpta Botanica Sect. B Sociologica* (R. TÜXEN Ed). They will not be listed in the references, but indicated with 'Exc'.

JEROSCH 1907) and Montpellier (FLAHAULT 1893, 1898, 1901, PAVILLARD 1901, 1912). A main outcome of this period was a full hierarchy for vegetation classification—vegetation-type, formation-group, formation, subformation, stand-type down to local variants (Nebentypen) and geographic variants (Fazies). The stand-type was called the 'association' and considered the basic unit. FLAHAULT & SCHRÖTER (1910) agreed on the following definition, presented to the Third International Botanical Congress in Brussels, 'An association is a plant community of definite floristic composition, presenting a uniform physiognomy, and growing in uniform habitat conditions. The association is the fundamental unit of synecology.' (translation following PAVILLARD 1935b).

From this background BRAUN-(BLANQUET) carried out a monographic study of alpine vegetation (1913). In this, and the essay by BRAUN-BLANQUET & FURRER (1913), attention was focused on 'Charakterpflanzen' or character species — species that possess 'fidelity' (relative restriction) to a given association. GRADMANN (1909) had advocated the approach through floristic composition and the use of character species ('Leitpflanzen'). The key ideas in BRAUN-BLANQUET's treatment were: (i) The study of communities should be based on a fundamental unit, comparable to the species. (ii) This unit should be the association, and associations should be defined by their possession of character-species. (iii) Each association consists (like a species) of 'individuals,' and the association (like the species) can be described from samples of its individuals. (iv) Each sample ('Aufnahme,' relevé) should be chosen so as to represent adequately such an individual, and it should include analysis of the complete species assemblage. (v) Associations should be grouped into higher units not by physiognomy, but by floristic composition.

Additions to the approach came in further publications. BRAUN-BLANQUET (1915) added the 'Assoziationsgruppe' (later called the Verband, alliance) as a unit above the association, also defined by character-species. BRAUN-BLANQUET (1918) added the subassociation, as a deviation from the typical association expressed in a constant floristic difference, and the facies as a subordinate unit possessing merely quantitative differences. BRAUN-BLANQUET (1921) outlined essentially the full system, including the analytical scales (20.4.5) and the 'sociological progression' (20.3.4) for arranging communities by their levels of organization. BRAUN-BLANQUET (1925) pressed the claims of fidelity or 'Gesellschaftstreue' as the key to vegetation systematics, against the disagreement of northern schools. In the same essay he introduced the 'characteristic species combination' (charakteristische Artenkombination after SCHMID, 1923) as the ultimate community diagnosis, comprising both character-species and constant companion species. KOCH (1925) and BRAUN-BLANQUET & JENNY (1926) added the concept of differential-species for the characterization of subordinate units, and the order as a unit above the alliance. BRAUN-BLANQUET & PAVILLARD (1922, 1925, 1928) in their *Vocabulaire* codified the analytic and synthetic procedures of the approach. Finally the science of plant sociology and its application were spelled out more fully in the first edition of the textbook of BRAUN-BLANQUET (1928, 1932).[1])

From these beginnings the influence of the approach spread through western and central Europe. We cannot review that spread, but will mention such leaders as PAVILLARD, ALLORGE and MOLINIER in France, TÜXEN and OBERDORFER in Germany, SZAFER and PAWŁOWSKI in Poland, FURRER in Switzerland, Soó in Hungary, KLIKA in Czechoslovakia, HORVAT, HORVATIĆ and WRABER in Yugoslavia, BORZA in Rumania, GIACOMINI and TOMASSELLI in Italy, BOLOS in Spain, LEBRUN in Belgium, and DE LEEUW in the Netherlands. References for these authors are found in BRAUN-BLANQUET (1964, 1968). BRAUN-BLANQUET's own work began at Zürich and continued from 1927 onwards at Montpellier, where he still leads the S.I.G.M.A., Station Internationale de Géobotanique Méditerranéenne et Alpine. A second centre was established 1932 at Stolzenau, West Germany under Reinhold TÜXEN, a pre-eminent leader of the approach and former director of the Zentralstelle (later, Bundesanstalt) für Vegetationskartierung, who has done much to give the approach its value in application. From TÜXEN's centre the approach spread to various non-European countries as well, and notably to Japan.

During its spread the approach has been known by several names: French-Swiss (or Swiss-French) school, Zürich-Montpellier school, Middle European-Mediterranean school, and Sigmatism (TANSLEY 1922, DU RIETZ 1936, BRAUN-BLANQUET 1959, 1968, EGLER 1954, BECKING 1957, WHITTAKER 1962, TÜXEN 1969a). Most of these terms lack specificity or are unclear. Moreover we feel that the time has passed for the word 'school' with its implications of a fixed system. We prefer in this article the most direct name, viz. BRAUN-BLANQUET approach, whilst we may characterize the essential ideas in the term, the 'floristic-sociological approach.'

---

[1]) One notes through this development of the full range of concepts the continuing contribution of BRAUN-BLANQUET himself, that has given the approach its present character (GAMS 1972 to the contrary).

20.3 General Concepts

20.3.1 CONCRETE VS. ABSTRACT UNITS

Much dispute has centered on the nature of the plant community. This will be briefly reviewed as far as the BRAUN-BLANQUET approach is concerned. (See further PAVILLARD 1935, WESTHOFF 1951, 1970, WHITTAKER 1962, BRAUN-BLANQUET 1964.)

Concepts of the plant community include: (i) the organismal concept (CLEMENTS 1936, TANSLEY 1920): the community as a 'superorganism.' (ii) the concept of social structure (PACZOSKI 1930) and many early Russian authors such as SUKATSCHEW (1929). (iii) the individualistic concept (GLEASON 1926): the community as a changeable mixture of 'individualistically' distributed plant species. (iv) the concept of population structure (WHITTAKER 1953, 1962, 1970): the community as a system of interacting species and vegetation as a complex population pattern.

The BRAUN-BLANQUET approach takes a practical, intermediate position that recognizes the heterogeneity of species distributions but emphasizes nonetheless the interactions between plants in the community, which has a certain individuality because of relative discontinuities between communities in the field. Definitions of plant community or phytocoenose range from the more superficial, 'any collection of plants growing together which has as a whole a certain unity' (TANSLEY 1935), to the more profound, 'a plant community (+animal community = biotic community) is a working community; the species composition of which is in a sociologic-dynamical equilibrium in competition for space, minerals, water and energy, in which each component affects all others and which is characterized by harmony between environment and production and phenomena expressing life in form, colour and temporal course' (TÜXEN 1957).

As shown by TÜXEN's definition, the community concept may be broadened from the plant community to the biotic community of plants and animals: producers, consumers and decomposers. In a further broadening of perspective the biotic community (= biocoenose) plus its environment or biotope is treated as a functional unit, the ecosystem. The concept of ecosystem has become most familiar from its expression in English-language ecology by TANSLEY (1935), but the concept is largely identical with FRIEDERICH's (1927, 1958) 'holocoene' and the 'biogeocoenose' of SUKACHEV (e.g. 1954, 1960) and other Russian authors. Though by far the greatest development of the BRAUN-BLANQUET approach has been in application to plant communities, biotic communities are clearly amenable to parallel study (section 20.7.2). TÜXEN (1957, 1965b) has emphasized this possibility and stressed the influence of FRIEDERICHS and THIENEMAN (e.g. 1956) on his own views.

The term plant community and its German equivalent 'Pflanzengesellschaft' have been used in both concrete and abstract senses, which has caused dispute and confusion (see TANSLEY 1920, 1935, DU RIETZ 1921, ALECHIN 1926, PAVILLARD 1935, WESTHOFF 1951, 1965, WHITTAKER 1956, 1962). The floristic-sociological approach has always stressed the distinction between the concrete and the abstract community. The classification units were, naturally, abstract units. Concrete plant communities were often, especially in forestry, referred to as stand (Bestand), whilst concrete representatives of associations were even called association-individuals (PAVILLARD 1912, BRAUN-BLANQUET & PAVILLARD 1922). This term has been much disputed and criticized (most thoroughly by WHITTAKER 1962) and we consider it now as of only historic interest.

WESTHOFF (1951, 1965) has acknowledged the distinction between concrete and abstract communities in separate definitions, which were linked to a general definition of vegetation. For the concrete plant community the term phytocoenose (GAMS 1918) was proposed, for the abstract community the term phytocoenon (MAAREL 1965, specifying the general term 'coenon' proposed by BARKMAN et al. 1958, WESTHOFF et al. 1959).

The term (phyto)coenon may be proposed as a suitable international term which may replace the terms community-type (WHITTAKER 1956, 1962) and nodum. The latter term has been suggested by POORE (1956, 1962), but has been used in a more specific meaning by WILLIAMS & LAMBERT (1961) in their 'nodal analysis' (cf IVIMEY-COOK & PROCTOR 1966).

For biotic communities parallel terms: biocoenose and biocoenon may be used (MAAREL 1965). For partial communities MÖRZER BRUYNS' (1950) term merocoenose can be used with merocoenon as the abstract equivalent. Layer communities could be called, stratocoenose-stratocoenon. For specific subcommunities consisting of plants of the same stratum, life-form and seasonal relations the couple, society-synusia may be reserved. (See further article 16.)

In conclusion we present WESTHOFF's (1951, 1970) definitions in a slightly adapted form. Vegetation is defined as a system of largely spontaneously growing plant populations, growing in coherence with their sites and forming an ecosystem with these sites and all other forms of life occurring in these sites. (Thus are excluded all assemblages of mobile plants and collections of plants growing in arrangements set up by man such as flowerbeds and arboreta.)

A phytocoenose is defined as a part of a vegetation consisting of interacting populations growing in a uniform environment and showing a floristic composition and structure that is relatively uniform and distinct from the surrounding vegetation.

A phytocoenon is defined as a type of phytocoenose derived from the characterization of a group of phytocoenoses corresponding with each other in all characters that are considered typologically relevant.

20.3.2 FLORISTIC-SOCIOLOGICAL CLASSIFICATION UNITS

Phytocoena include such various kinds of vegetation units as formations defined by physiognomy, dominance-types defined by major species, forest site-types defined by undergrowth composition, and noda derived by quantitative comparisons or numerical procedures. The approach of BRAUN-BLANQUET has its own, formal hierarchy of units, of which none of those just mentioned is a part. The fundamental unit of the hierarchy is the *association*, a unit that corresponds in function to the species as the fundamental unit of idiotaxonomy, or the classification of individual organisms. The word 'association' has had a long, complex, and argument-afflicted history as different schools sought to determine its meaning to their preference (WHITTAKER 1962, SHIMWELL 1972, see also articles 14, 17 and 18). In an earlier definition by BRAUN-BLANQUET (1921) 'the association is a plant community characterized by definite floristic and sociological (organizational) features which shows, by the presence of character-species (exclusive, selective, and preferential) a certain independence.' MEIJER DREES (1951) defined the association as ' a plant community identified by its characteristic taxon combination, including one or more (local) character-taxa or differentiating taxa.' A similar conception was agreed on during a Symposium on Plant Sociological Systematics at Stolzenau in 1964 (in TÜXEN 1968b, especially OBERDORFER 1968) and a later colloquium at Rinteln (see DIERSCHKE 1971). We shall return to characterization of the association (20.6.2) but emphasize that, in keeping with the floristic-sociological perspective, the association is defined by its *characteristic species combination* including character- and differential-species as well as companions (Begleiter) with high presence values (over 60%). The Sixth Botanical Congress at Amsterdam, 1935, accepted definition of the association by characteristic or differential species in the sense of BRAUN-BLANQUET as one of three resolutions given in article 18.3.6.

As basic units associations, like species, are to be grouped into a hierarchy of higher units. Associations are classed into alliances, alliances into orders, orders into classes, and classes into divisions (20.6.4-5); associations are divided into lower units of the hierarchy (20.6.7). All these units are coena as defined above; but the coena of the formal system of BRAUN-BLANQUET may be termed *syntaxa* (BARKMAN et al. 1958, WESTHOFF et al. 1959). Thus the parallelism in the classification of organisms and phytocoenoses is realized: to the species as a fundamental unit corresponds the association, to the taxon for units on any level corresponds the syntaxon, to (idio)taxonomy for the practice of classification corresponds syntaxonomy, to (idio)systematics as the broader study of relationships among organisms corresponds synsystematics.

20.3.3 DIAGNOSTIC SPECIES

Syntaxa are defined by diagnostic species (character-species, differential species, and constant companions). *Character-species* are species that are relatively restricted to the stands (or samples) of a given phytocoenon, and therefore characterize it and indicate its environment (Fig. 1). In ideal, a group of character-species is used for the characterization of a syntaxon of the BRAUN-BLANQUET classification on any level from the association to the class. To serve as character-species for an association, a species should have a relatively narrow distribution even if it is not simply restricted to the association. (Degrees of restriction to a given syntaxon, which are termed degrees of *fidelity* or Treue, are discussed in 20.5.6.) Note that the concept does not say that the species need be important in phytocoenoses; very minor species may have diagnostic value.

The German term is 'Charakterart'; this has been variously translated into English as characteristic species (BRAUN-BLANQUET 1932), which seems unsatisfactory, faithful species (POORE 1955a, BARKMAN 1958a, BEEFTINK 1962, MOORE 1962, WESTHOFF 1959), and character-species (BECKING 1957, WHITTAKER 1962). The latter, most direct translation, is our preference. HEIMANS (1939) introduced 'kensoort' in Dutch to avoid the germanism 'karaktersoort'; and the Dutch kensoort returned to German as *Kennart* (TÜXEN 1950: 99). Most phytosociologists writing in German now use that term, which is accepted in the third edition of BRAUN-BLANQUET's (1964) text. Syntaxa are most often defined by the fundamental units of idiotaxonomy — species —, but this is not always the case. Sometimes plant subspecies, varieties, or ecotypes may contribute to the definition of lower-level syntaxa or geographically vicariant

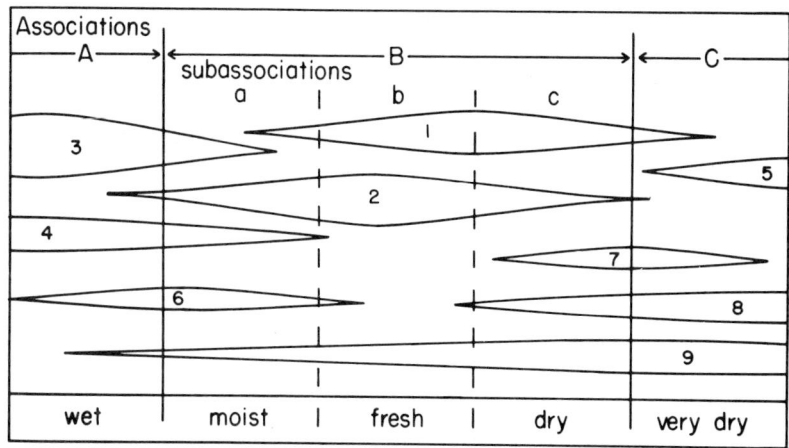

Fig. 1. Diagnostic species along a moisture gradient. Species 1 and 2 are character-species for association *B*, and have their populations centered in (or largely confined to) that association. Species 3 and 4 are character-species for association *A*, and species 5 is a character-species for association *C*. Species 4 and 6 are differential-species for subassociation *a* of association *B*, and species 7 and 8 are differential-species for subassociation *c* of association *B*. In each case presence of the differential-species distinguishes the moist, or dry, subassociation from the "typical" (fresh) subassociation *b*. Species 9 is a more widely distributed species that might help characterize association *B* as a constant companion for that association. Species 9 might also be a character-species for a higher syntaxon, such as an alliance uniting associations *B* and *C* with other associations.

associations. Sometimes a genus or subgenus may be used to define a higher syntaxon, as the alliance Spartinion is defined by the fidelity of any species of the genus *Spartina* (BEEFTINK 1965, 1968, see also PIGNATTI 1968a: 72, WESTHOFF apud PIGNATTI 1968a: 74). It is therefore appropriate to broaden character-species to character-taxon (BARKMAN et al. 1958) as we shall do here following Dutch use of 'kentaxon' (WESTHOFF & HELD 1969). The German equivalent is of course 'Kenntaxon'.

Character-species characterize syntaxa by their normal occurrence in phytocoenoses of that syntaxon, contrasted with their absence less frequent occurrence or smaller total estimate in phytocoenoses of all other syntaxa. It is possible also to distinguish closely related syntaxa by the presence and absence of certain species without concern for the broader distributions of those species. Two subassociations of an association may be characterized by the fact that samples of one normally include species 4 and 6, whereas samples of the other normally lack these species (Fig. 1). These species function as *differential-species* that distinguish the two subassociations. Differential-species, as the second class of diagnostic species, define syntaxa on the basis of the distributional boundaries of species (without regard to fidelity for the syntaxa in question), and are used primarily to define lower syntaxa.

Since being introduced by KOCH (1925) and BRAUN-BLANQUET (1925) as 'Differentialart', the concept has entered English as differential species (BRAUN-BLANQUET 1932, WHITTAKER 1962) or differentiating species (BECKING 1957), whilst in German the term *Trennart* has come into general use (TÜXEN 1950, 1952, BRAUN-BLANQUET 1959, 1964a) in parallel with Kennart. The use of differential-species has been further developed by SCHWICKERATH (1931, 1942, 1954a) and TÜXEN (1937, 1950, 1952). Differential-species and Trennart are synonyms, and these are largely synonymous also with differentiating species; but we shall here use 'differentiating species' for the more informal use of diagnostic species of unassigned syntaxonomic position in the synthetic phase of research (20.5), and speak of differential-species and differential-species groups for the formal characterization of syntaxa (20.6). A further recent development is the concept of 'character-combination' (after the Dutch kencombinatie) introduced by WESTHOFF & HELD (1969) following BEEFTINK (1965, 'differentiating species combination'). The essential concept is the exclusiveness to a syntaxon of the combination of species (or taxa), whilst none of them need be a character-taxon.

A third group of species with diagnostic value are the constant companions (20.5.3) which occur in most relevés of a syntaxon but are not designated as character- or differential-species. Constant companions are added to the character-taxa to form the 'characteristic species (taxon) combination' for associations and higher taxa. Although BRAUN-BLANQUET (1925) introduced this concept rather early and repeatedly (e.g. 1959, see MOORE 1962) stressed its relevance there has been doubt amongst critics (e.g. POORE 1955-1956). Of course, as BRAUN-BLANQUET (e.g. 1932: 68) remarked for associations, these syntaxa are established best which possess a high proportion of both characteristic and constant species.

20.3.4 SOCIOLOGICAL PROGRESSION

BRAUN-BLANQUET (1921) sketched an 'arrangement of the plant communities according to their sociological progression.' The idea of this arrangement was again a parallel with idiobiology, specifically with the arrangement of taxonomic groups by evolutionary level. Levels of structural and organizational development were chosen as criteria for arranging phytocoenoses from moving aero- and hydroplankton communities on the lowest level to the tropical rain forest at the highest level. Other communities are arranged between these according to increasing stratification, complexity, presence of dependent communities, species richness, diversity of growth-forms, and assumed intensity of species interactions. Thus BRAUN-BLANQUET (1921) summed up many of the structural-successional trends much later commented on by ODUM (1969)! The scheme was later modified in detail, and increasing stability was added as a criterion. BRAUN-BLANQUET (1964) arranges higher syntaxa (classes and orders) in the progression.

WAGNER (1968) called the sociological progression a 'purely artificial division principle' (cf. ELLENBERG 1963). Of course, no phylogenetic meaning should be imputed to its sequence. Yet the sociological progression is useful, as a principle at right angle to the levels of the hierarchy, in arranging syntaxa in surveys. Its use is illustrated in vegetation monographs by TÜXEN (1937), OBERDORFER (1957), ELLENBERG (1963), OBERDORFER et al. (1967), WESTHOFF & HELD (1969), and others. It may be of interest that it has appeared also as an axis of broad-scale ordination of syntaxa (MAAREL 1972a, see 20.9.2).

20.3.5 NATURAL CLASSIFICATION

The problem of 'natural' classification has been discussed in another article (12.1.2). It has been accepted by BRAUN-BLANQUET and others that associations are abstract units. A degree of continuity or gradualness in transitions between phytocoena is recognized, but considered not to preclude classification. 'We are convinced that the plant cover of the earth in all its dimensions can be divided into phytosociological groupings of higher and lower rank; their delineation may be either sharp or less sharp and gliding.' (BRAUN-BLANQUET and TÜXEN in comment on GOODALL 1963, translated.)

American research in gradient analysis has led to interpretations that may seem in conflict with those of BRAUN-BLANQUET. The American work of WHITTAKER (1951, 1954, 1956, 1962, 1967, articles 2 and 3) and CURTIS (CURTIS & McINTOSH 1951, CURTIS 1959, McINTOSH 1967, article 7) developed in parallel to Russian work of RAMENSKY (1930, article 17) toward the concept expressed by WHITTAKER (1970) as 'the population structure of vegetation.' In this concept the population structure of the individual phytocoenose may be analysed through dominance-diversity and species-importance relations, and niche differentiation among plant species expressed in stratification and functional differences among them. The phytocoenose is then conceived as a 'system of interacting, niche-differentiated and partially competitive species.' The vegetation of an area, on the other hand, is to be approached through gradient analysis, studying the manner in which species populations are distributed along environmental gradients and combined into phytocoenoses. Species are differently distributed from one another, so that undisturbed phytocoenoses intergrade continuously in most areas. 'In relation to patterns of environmental gradients, communities form complex and largely continuous population patterns.' (WHITTAKER 1970, article 3.9.)

It is further recognized that, 'Because of environmental interruptions and some relative discontinuities inherent in vegetation itself, the pattern may also be considered a complex mixture of continuity and relative discontinuity' (WHITTAKER 1956: 32). Vegetational continuity by no means precludes useful classification (WHITTAKER 1956, 1962, 1970). Gradient analysis implies a vision of vegetation, regarded as a coherent pattern of intergrading phytocoenoses, different from that of phytosociologists who tend to see it in terms of a typology of phytocoena (see, however, 20.7.4 on complexes). Yet we think that the difference between the concept of the gradient approach and BRAUN-BLANQUET's is clearly one of degree — of emphasis of continuity vs. discontinuity where both are present, of species individuality vs. species groupings where both are realistic, and of gradient analysis vs. classification where both are possible. We judge the difference between less extreme students of gradient analysis, and of the BRAUN-BLANQUET approach, to be one of emphasis and perspective, not one of fact or understanding.

Given the individuality of species distributions and some degree of continuity between phytocoenoses, classification is not strictly natural in the sense defined in article 12.1.2. BRAUN-BLANQUET (1951a: 561, 1959: 147) has considered argument on the naturalness of classification pointless. TÜXEN (1955), in response to a criticism of ELLENBERG (1954b), interpreted phytosociological classification through the concept of *types* as ideal concepts, recognized in an empirical way from 'correlation concentrates,' i.e. groups of correlated characters. That which is evident and characteristic of a type is always its nucleus, not its periphery; types are not pigeonholes but foci in a field of variation. GLAHN (1968, see also RAUSCHERT 1969) elaborated the concept of vegetation type, distinguishing as its three aspects: (i) The vegetation type as identity: repetitions of certain observations are approached via intuitive integration, resulting in vegetation-type concepts based on recurring combinations of species. (ii) The vegetation type as maximal correlative concentration: The joint floristic-sociological and ecological approach leads to types as correspondences of recurring species combinations with recurring combinations of environmental factors. (iii) The vegetation type as systematic category: A hierarchic system is feasible as a result of an integrative, inductive process grouping the initial types into higher and higher ranks on the basis of common species combinations. The similarity of this statement to the theory of classification developed at greater length by WHITTAKER (1962) may be noted. Relative naturalness of classification in this perspective is to be judged by success in embodying the maximum number of significant relationships among phytocoena and species in the structure of the hierarchy. We assert that compared with other classifications the BRAUN-BLANQUET approach, with its floristic emphasis and maximum use of species distributional relationships for classification, does not fall short.

20.4 Analytical Research Phase

20.4.1 RECONNAISSANCE AND CHOICE OF PLOTS

An analysis of the vegetation is as a rule preceded by a preliminary survey of the area, when this is little known to the investigator. This reconnaissance (see CAIN & CASTRO 1959) includes a study of the general vegetation pattern, and the establishment of the apparent relations of the various vegetation types with geology, topography and soil conditions. The next step, 'primary survey,' including a superficial description of the major communities, is mostly passed over in the BRAUN-BLANQUET approach.

A vegetation analysis starts with the choice of stands (phytocoenoses) on the basis of the reconnaissance. Within the stand one sample plot is laid down, often covering a large part of the stand. The analysis of this plot is called the *relevé*. The BRAUN-BLANQUET approach has often been criticized for the subjectivity of its sampling procedure. However, subjectivity of stand choice must be accepted in the procedures of many empirical sciences. A selection of relevés is desired that will effectively represent the variation in the vegetation under study, the samples being so chosen that they will not represent different phytocoena disproportionately and will not include mixed, incomplete, or unstable stands. For this purpose of equitable representation of different kinds of communities with most useful relevés, a subjective, 'stratified' sample selection is far superior to sample choice by random points on a map.

Relevés may be undertaken without classification as a purpose. They may then serve for the study of vegetation dynamics on a given area (periodicity, fluctuation, or succession) or another ecological purpose, e.g. the study of the ecology of a certain taxon or ecotype by describing its pattern in relation to the pattern of the vegetation. In most cases, however, relevés are intended to be used for some form of classification or ordination. In that situation, the only preconception in this choice is the demand for uniformity of the stand.

'Uniformity' may better express what is sought than 'homogeneity.' This chapter is not the place for a treatment of the mathematical aspects of homogeneity in vegetation; we may refer to POORE (1955, 1956), ELLENBERG (1956), DAHL (1957, 1960), BECKING (1957), LAMBERT & DALE (1964), BRAUN-BLANQUET (1964), MAAREL (1966b) and GOUNOT (1969) for general considerations, to GOODALL (1952, see also his bibliography, 1962 Exc.), DAHL & HADAČ (1949), DAHL (1957), GREIG-SMITH (1964) for aspects of homogeneity in the distribution of plant individuals, to GOODALL (1961) and GREIG-SMITH (1964), for pattern analysis, to RAUNKIAER (e.g. 1934), GUINOCHET (1955), DAHL (1957), CAIN & CASTRO (1959) for the relation between homogeneity and frequency distribution and to CURTIS (1959), GODRON (1966) and GOUNOT (1969) for infrastand similarities and information measures as approaches to homogeneity.

The first condition is that no obvious structural boundaries or variation in stratification are visible within the stand. The second criterion is uniform floristic composition. It is usual to look for joint patterns of dominant and/or abundant species and then to delimit a stand where qualitative changes in patterns occur, i.e. where one or more species drop out and others come in. In many cases an experienced field worker is able to judge this rapidly. In many cases however the changes in pattern are quantitative rather than qualitative. The species composition differs little from one site to an adjacent one, but the relative proportions of abundance and cover do vary. It is usual to make separate relevés in any case where the abiotic habitat factors show a clear discontinuity or at least a marked transition.

Many species cause pattern heterogeneity only by their growth form or their aggregation or shoot clustering. In such a case one may try at first to study this biotically heterogeneous pattern as one single stand, i.e. with one relevé. However, the situation becomes different as soon as the crowding of one major species leads to the establishment of one or more other species which appear to be locally bound to it. A simple example is the establishment of a dwarf shrub patch in a herbaceous vegetation, e.g. a clone of *Salix repens* in open dune grassland. When one or more species are found particularly bound to that dwarf shrub patch we prefer to consider the latter as a separate phytocoenose which has thus to be analysed with a separate relevé. However, there are exceptions to this rule. E.g. OBERDORFER (1970, Canaries) and WERGER (1973, South Africa) report stable savanna mosaics consisting of a dwarf-shrub and grassland and a low tree and shrub phytocoenose. Such mosaics are considered one vegetation type, since one does not find locally more extensive patches of either phytocoenose without adjacent patches of the other one.

In intermediate or doubtful cases, such as a swamp community with tall tussocks of sedges (*Carex hudsonii*) alternating with wetter hollows, it will be advisable, at least when the type of vegetation pattern is unknown, to follow both procedures, viz. analysing the pattern as a whole as well as relevés of tussocks and hollows separately. It will then turn out later which relevé is more useful for the classification purpose.

20.4.2 BOUNDARIES

As was said before, the recognition of distinct stands may presuppose the occurrence of discontinuities in the field. Although according to the opinion of BRAUN-BLANQUET workers such discontinuities are mostly to be observed, it is obvious that boundaries between stands are less sharp in some cases than in others. In such cases the boundary may be detected by means of a belt transect analysis or a 'line taxation.' Such transects consist of a series of small quadrats laid down at right angles to the extension of the boundary zone. By comparison of quantitative data on species occurrences in the quadrats the boundary zone can be discerned from the uniform phytocoenoses. MAAREL (e.g., MAAREL & LEERTOUWER 1967) and FRESCO (1972) refined boundary analysis by constructing 'differential profiles,' in which can be shown degrees of change between adjacent quadrats along a gradient. For similar techniques see article 5.4.4.

In general we may distinguish between two types of boundary zones which are, according to the relation theory of LEEUWEN (1965–1970, WESTHOFF 1971a, b, WESTHOFF & LEEUWEN 1966, SHIMWELL 1972), the limes convergens and the limes divergens. The *limes convergens* zone, or convergent limit, is characterized by sharp

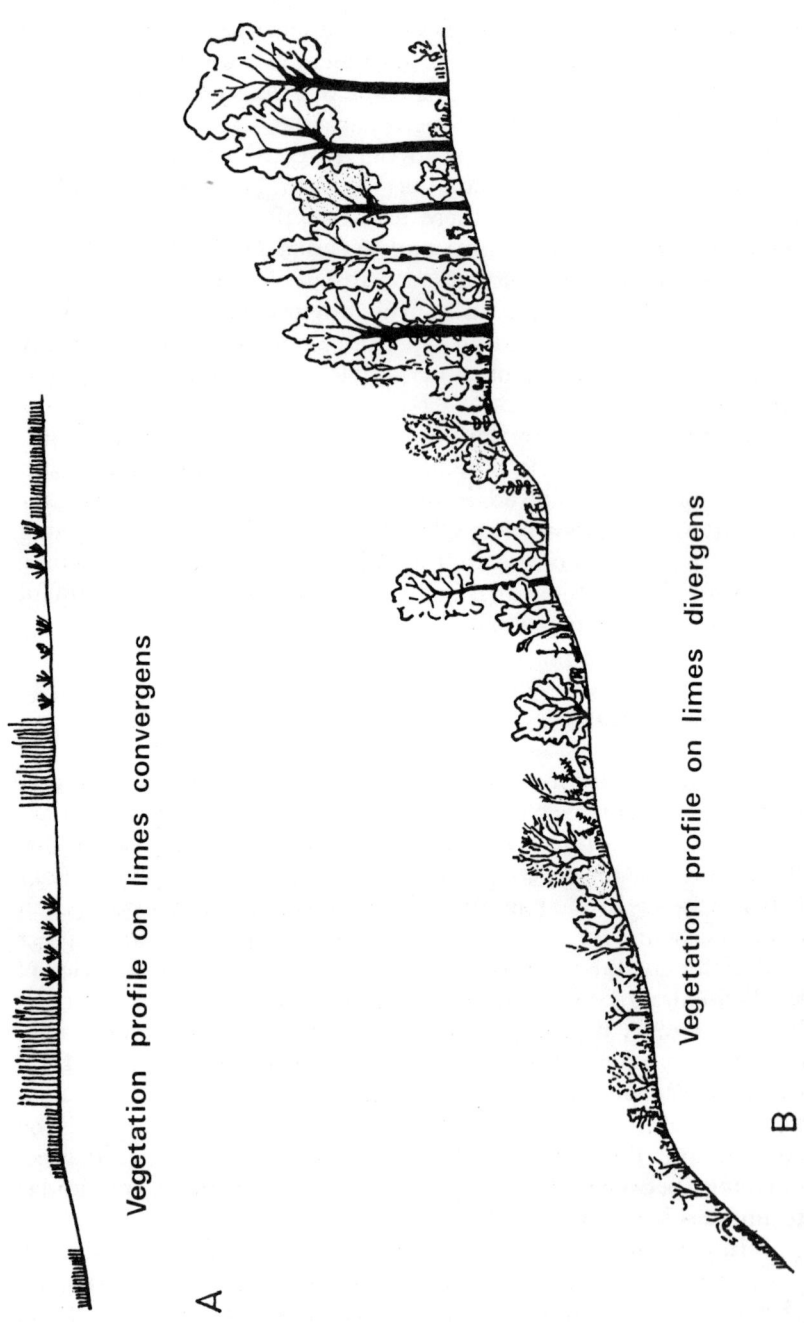

Fig. 2. Vegetation profiles corresponding to the limes convergens and limes divergens zones (Leeuwen 1965).

vegetational boundaries on either side of which uniform phytocoenoses occur, with few species represented, some by many individuals, in coarse-grained patterns. The corresponding environment is unstable, i.e. with sharp fluctuations in vertical direction (e.g. water-table or phreatic level) or in horizontal direction (e.g. deposition of flood marks). A typical synsystematical unit of this type is the Agropyro-Rumicion crispi.

The *limes divergens* zone, or divergent limit, is characterized by numerous small-scale boundaries in phytocoenoses merging continuously into each other, with many species generally represented by few individuals in fine-grained patterns. The corresponding environment is stable and determined by a gradient. A typical series of divergent synsystematical units is Mesobromion-Trifolion medii-Carpinion betuli. Fig. 2 shows a vegetation profile belonging to these types of boundary. According to MAAREL (e.g. 1966a), LEEUWEN (1966) and WESTHOFF (1971a) the ecotone concept of LIVINGSTONE and CLEMENTS (e.g. WEAVER & CLEMENTS 1938, CURTIS 1959) could be applied to the limes convergens zone, whilst the ecocline concept as derived from HUXLEY's cline concept (WESTHOFF 1947, MEIJER DREES 1951) may be considered an equivalent of the limes divergens zone.

The delimitation of stands within ecotones generally gives no difficulties, except for pattern problems already discussed. Gradients of vegetation-and-biotopes (ecoclines) can be approached by belt transects following the direction of variation, which transects have to be divided into sample plots (not necessarily quadrats) small enough to be uniform. Recently JAKUCS (1972) gave examples of this approach in 'grassland-shrub-woodland' transitions within the range of the Quercetea pubescenti-petraeae in Hungary.

20.4.3 MINIMAL AREA AND PLOT SIZE

The size of the sample plot is largely dependent on the structure of the vegetation under study, but may be affected also by the size of the stand. In many cases the stand is sufficiently small to be analysed largely in one relevé. In many other cases, however, this is not possible, and then the plot should be large enough that all species of regular occurrence in the stand should be present in the sample plot. In extremely uniform phytocoenoses which are very poor in species, e.g. salt marshes, where differences in abundance and cover are of major concern, these differences should be considered in establishing the size of the sample plot.

These conditions refer to the minimal area concept (see also articles 5.2.1, and 18.3.3). Without treating this concept in detail we may give some comments relevant to the phytocoenose description. First the minimal area as an analytical concept (for the phytocoenose) has to be distinguished from the synthetic minimal area (for the phytocoenon). The analytic minimal area should be defined for a stand under study as a representative area, e.g. as an adequate sample of species of regular occurrence in the stand; the size decision may, but will not necessarily, be based on the total number of species in the stand or the usual minimal area curve (BRAUN-BLANQUET 1964, CAIN & CASTRO 1959, DU RIETZ 1921, 1930, GOODALL 1952, 1961, GREIG-SMITH 1964, HOPKINS 1957, MAAREL 1966b, TÜXEN 1970c and 1970 Exc., VESTAL 1949). A synthetic minimal area, in contrast, is subject to many current definitions (e.g. BRAUN-BLANQUET 1964, CAIN & CASTRO 1959, and WESTHOFF 1951 — who explicitly defined it as 'the minimal surface which has as a rule to be occupied by a sample of a plant community if the normal specific assemblage will be able to develop'). The size decision in this case involves not merely the stand itself but the 'normal' composition of stands representing a phytocoenon. It is clear that before judging this in the field one should know from synthesis the characteristic species combination. Nevertheless many authors, including BRAUN-BLANQUET (1964), describe the normal species-area determination, applied analytically to the stand, as the way to find the minimal area.

TÜXEN (1970c) presented numerous examples of species-area curves ending in a horizontal part, implying that at the sample size where the horizontal part begins, full species representation or saturation has been reached. Tüxen consequently spoke of 'saturated communities,' which he found among homogeneous communities of various organization levels. He cited various authors who found curves of the same type, including DU RIETZ (1921).

Most authors, e.g. BRAUN-BLANQUET (1964), CAIN & CASTRO (1959), HOPKINS (1955), and MAAREL (1966b) have found curves that never reach an asymptote. It is true that many of their curves refer to either relatively small or relatively large areas. It remains unclear in which situation which type of species relation can be expected — the ARRHENIUS-type of a log-log relation (cf. PRESTON 1962), the ROMELL-type of a log-linear relation (cf. HOPKINS 1955, DAHL 1957, WILLIAMS 1964, MAAREL 1966b), or the KYLIN-type of saturation relation. BALOGH (1958), and MAAREL (1966b), quoting FREY (1928), who presented the three types, interpreted them as referring to very uniform and species-poor environments (limes convergens type!), highly variegated and

TABLE I
Minimal area values in square meters for various communities

Epiphytic communities	0.1–0.4
Terrestrial moss communities	1–4
Hygrophylous pioneer communities (Isoeto-Nanojuncetea)	1–4
Dune grasslands (Koelerio-Corynephoretea)	1–10
Salt marshes (Asteretea tripolii)	2–10
Pastures (Lolio-Cynosuretum)	5–10
Mobile coastal dune communities (Ammophiletea)	10–20
Hay meadows (Arrhenatheretalia)	10–25
Heathlands (Nardo-Callunetea)	10–50
Alpine meadow and dwarfshrubs (Elyno-Seslerietea)	10–50
Calcareous grasslands (Festuco-Brometea)	10–50
Chaparral, temperate sclerophyll shrubland	10–100
Weed communities (Secalietea)	25–100
Scrub communities (Rhamno-Prunetea)	25–100
Steppe communities	50–100
Temperate deciduous forest (Querco-Fagetea)	100–500
Mixed deciduous forest (North America)	200–800
Tropical secondary rainforest	200–1000
Tropical swamp forest	2000–4000

species-rich situations (limes divergens type), and intermediate types or (BALOGH) 'normal homogeneous stands', respectively.

Table I presents some minimal area values for community types, taken from various sources. (See TÜXEN, 1970 Exc. for a bibliography). They range from a few dm^2 for certain epiphytic communities to one hectare or more for climax tropical rain forests. In general the plot size is taken somewhat larger than the minimal area. It will be clear that plot size should not vary too much within one vegetation type. The shape of the plot is not standardized and may depend on the situation. If possible, a quadrat of rectangular shape is to be preferred.

20.4.4 DESCRIPTION OF STRUCTURE AND STRATA

Vegetation layering is an important structural character. Mostly only four principal layers are distinguished as the tree, shrub, herb, and moss layers. The latter is also named field layer or thallophyte layer (which is less adequate, because many of the larger fungi do not belong to it). These principal layers may be further subdivided. In the relevé as many layers are distinguished as is considered appropriate or necessary. For each layer height and coverage degree in per cent, mostly with 5–10% intervals, are estimated. For tree and shrub layers the age is estimated and additional data on stem diameter, number of dead or fallen trees, and occurrence of epiphytic communities are gathered.

Stratification can be more accurately described with the help of diagrams in which the total or combined cover of each layer is indicated (HULT 1881, BRAUN-BLANQUET 1928, cf. CAIN & CASTRO 1959, KNAPP 1971). These diagrams are related to vertical profiles, as have frequently been presented for tropical rain forest (see article 13.4.2) and by ZONNEVELD (1960) for the freshwater tidal delta of the Rhine. Both however have their own special uses. The profile can show exactly the structure of a given phytocoenose; in the coverage-stratification diagram a pattern typical for a phytocoenon can be generalized (CAIN & CASTRO 1959). A more formal method is that of DANSEREAU (1957, DANSEREAU & ARROS 1959) for the description and recording of vegetation on a structural basis (see article 13.3.3).

It is possible to classify different strata into separate synusial units. This approach is treated in detail by BARKMAN (article 16). The BRAUN-BLANQUET approach however considers the main strata of a given stand as a single phytocoenose which has to be analysed as a whole, at least as far as terrestrial communities are concerned. The main arguments are: (i) The different strata are rooting in a common substratum; (ii) The layers are ecologically closely interrelated; (iii) The plants of all other layers have originally been part of the field layer and they must pass through one or more layers as they develop from seedlings to their mature life-form stature. The layered community is, then, a dynamic totality (see WEBB et al. 1967b on the vertical integration of the rain forest).

The situation in aquatic communities of higher plants is less clear however. Here the coherence of layers is much looser, or absent. DU RIETZ (1930a) has proposed a stratification scheme, which was modified by HARTOG & SEGAL (1964) (see also article 16.4.3). Layers are distinguished according to rooting or non-rooting of plants and the positions of leaves in relation to the water surface. The layers are named by growth-forms after representative genera, e.g. the isoetid layer.

An additional analysis can be made of subterranean layers. For a treatment of root stratification see BRAUN-BLANQUET (1964: 59–62). Bibliographies of root studies were given by WILMANNS (1959 Exc., 1966 Exc.) and TÜXEN & WILMANNS (1973 Exc).

20.4.5 FLORISTIC ANALYSIS

20.4.5.1 *Cover and Abundance*

The description of structure is followed by an inventory of taxa, at least of phanerogams, pteridophytes, bryophytes and lichens.

TABLE II

The combined estimation (cover-abundance) scale of BRAUN-BLANQUET, compared with the cover-abundance scale of DOMIN (See EVANS & DAHL 1955). The subdivisions 2m, 2a, and 2b are proposed by BARKMAN et al. (1964). The ordinal transformation is discussed in 20.5.4.

BRAUN-BLANQUET, cover-abundance		ordinal transform	DOMIN	
			+	one individual, reduced vigor
r	one or few individuals	1	1	rare
+	occasional and less than 5 % of total plot area	2	2	sparse
1	abundant and with very low cover, or less abundant but with higher cover; in any case less than 5 % cover of total plot area	3	3	< 4 %, frequent
2	very abundant and less than 5 % cover, or 5–25 % cover of total plot area			
	2m very abundant	4		
	2a 5–12.5 % cover, irrespective of number of individuals	5	4	5–10 %
	2b 12.5–25 % cover, irrespective of number of individuals	6	5	11–25 %
3	25–50 % cover of total plot area, irrespective of number of individuals	7	6	26–33 %
			7	34–50 %
4	50–75 % cover of total plot area, irrespective of number of individuals	8	8	51–75 %
5	75–100 % cover of total plot area, irrespective of number of individuals	9	9	76–90 %
			10	91–100 %

The taxa are listed according to the layer in which they grow. Plants which appear to be structurally transgressive, i.e. occurring in more than one layer, have to be recorded in each of these layers separately.

Taxa occurring only outside the sample plot (but within the stand) are noted in parentheses. Next, the quantitative occurrence of each taxon is estimated. In the BRAUN-BLANQUET approach two criteria are considered most useful: abundance and coverage degree. Abundance relates to the density of the individuals of a given species in a plot. Cover degree is measured as the vertical projection of all aerial parts of plants of a given species as a percentage of the total plot area. The term 'dominance,' often used as a synonym for coverage, is less appropriate.

Abundance and cover degree are usually estimated together in a single 'combined estimation' or 'cover-abundance' scale. The five-point scale in Table II, from BRAUN-BLANQUET (1928, see also article 18.4.1) is in general use. Several authors (e.g. TUOMIKOSKI 1942, DOING 1954, EVANS & DAHL 1955, BARKMAN et al. 1964) used more detailed scales of combined estimation; the BRAUN-BLANQUET symbol 2 especially was refined. The more elaborate scales are useful for special purposes, e.g. for an accurate record of change of abundance and cover by succession on permanent sample plots in the course of years (DOING 1954). However, they often suggest more accuracy than can really be justified. Only values obtained by one investigator should then be compared (cf. CAIN & CASTRO 1959: 142–143, MAAREL 1966b). We have recently introduced as a refinement of scale interval 2 (taken from BARKMAN et al. 1964) the scale subdivisions 2m, 2a, and 2b given in Table II. This elaboration, which brings the total number of scale values to 9, has proved to be useful and reliable.

20.4.5.2 *Sociability*

Sociability or gregariousness is an expression of horizontal pattern of species. It is a measure of the degree of clustering (contagion) of the plant units of a species. A plant unit (WILLIAMS 1964) may be an individual or a shoot or a sprout-forming part of an individual. Measurement of sociability goes back (again) to HEER (1835).

In the floristic-sociological approach sociability is estimated with the following scale (BRAUN-BLANQUET 1928, 1932, 1951, 1964).
1. growing solitary, singly.
2. growing in small groups of a few individuals, or in small tussocks (caespitose), e.g. *Corynephorus canescens* in shifting sands.
3. growing in small patches, cushions or large tussocks, e.g. *Carex hudsonii* as a hummock builder in eutrophic swamps; *Silene acaulis* and *Saxifraga oppositifolia* in alpine swards.
4. growing in extensive patches, in carpets or broken mats, e.g. stands of *Hedera helix, Lamium galeobdolon, Asperula odorata* etc. in deciduous temperate forest.
5. growing in great crowds or extensive mats completely covering the whole plot area; mostly pure populations, e.g. *Erica tetralix* in *Erica*-heath; *Sphagnum rubellum* or *S. pulchrum* in raised bogs.

A variant of scale value 5 (5, loose or open 5) was proposed for populations with a cover degree of over 75% consisting, however, of plants which are sufficiently separated as to leave space for other species. Similar variants can be used for loose cover of 51-75% (4) and 26-50% (3) (MELTZER & WESTHOFF 1942, BRAUN-BLANQUET 1964).

In the relevé sociability values are written immediately behind the combined estimation values, e.g. *Scirpus maritimus* 1.1.

During the last decade several investigators (see FUKAREK 1964) expressed the opinion that the diagnostic value of sociability has been overestimated and that a certain sociability degree is a specific character of most taxa. Other authors (BRAUN-BLANQUET 1964, SCAMONI & PASSARGE 1963, WESTHOFF 1965) do not agree with this view. Only a few species have a fixed degree of aggregation based upon their innate manner of growth. The degree of gregariousness of most species is much influenced by habitat conditions and competition, and therefore is of major phytosociological importance. Sociability may also change considerably during the course of a succession; many examples are given by the authors quoted above. On the other hand, the variation in sociability will be correlated with variation in cover degree to some extent.

Sociability is commonly considered as an expression of vitality. However in various situations this is not so. JAKUCS (1970, 1972) remarked that character-taxa of the Trifolio-Geranietea (thermoxerophilous woodland fringe communities or 'Saumgesellschaften') tend to grow with sociability 1 in their optimal habitat, whilst they form polycorms with reduced vitality in suboptimal habitats. Facies of species are often connected with extreme or disturbed habitats (see 20.6.7).

20.4.5.3 *Vitality and Fertility*

Further variables in the performance of a species are its vitality and its fertility, representing vegetative and generative development respectively. BRAUN-BLANQUET (e.g. 1932, 1964) developed a scale of relative 'thriving' (Gedeihen) with four categories indicated by symbols (BRAUN-BLANQUET 1932).

●, 1 Well developed, regularly completing the life cycle (an extraordinary vitality is indicated with 'lux', luxurious).

○, 2 With vegetative propagation but not completing the life cycle.

○, 3 Feeble with low vegetative propagation, not completing the life cycle.

○○, 4 Occasionally germinating but not vegetatively propagating.

BRAUN-BLANQUET (1964) slightly altered the symbols and gave them numbers from 1 to 4, as above.

BARKMAN et al. (1964), following VARESCHI (1931) and ZOLLER (1954), stated that for many species vitality and fertility are independent or even negatively correlated parameters. They proposed separate scales for these.

20.4.5.4 *Periodicity*

In addition to vitality the seasonal phase in the life cycle of each plant, its 'phenological state' is recorded. Various scales are in use, the first one being that of GAMS (1918). BRAUN-BLANQUET (1964: 67 and 510) presented two scales (which are very similar). As was stated by GAMS (1918), ELLENBERG (1939, 1954a) and others (see also the bibliography by BALÁTOVÁ-TULAČKOVÁ, 1970 Exc.) one incidental record during the relevé is really not sufficient, a complete phenological diagram should be desired for each community. For a relevé at a given time, however, appropriate abbreviations may be used — v. (vegetative), fl. (flowering), fr. (fruiting), etc.

20.4.6 RELEVÉ PROTOCOLS

The various structural and floristic data discussed so far are written in standardized form either in field note books or on special protocol forms. These notes are preceded by some notes on the following items:
1. Date; running number; topographic locality (as detailed as possible); altitude; exposure and inclination; geology of substrate. If possible the location should be indicated on a detailed map.
2. Size and shape both of the plot and of the entire uniform stand; character of adjacent vegetation; soil profile; phreatic level; depth and differentiation of root system.
3. Character and intensity of human and animal influence, e.g. pasturing, burning, mowing, treading, manuring, irrigation.
Table III presents an example of a relevé, taken from MELTZER & WESTHOFF (1942). For each species are recorded the scale numbers for cover-abundance (before the period) and sociability (after the period), the phenological abbreviation, and the vitality symbol.

TABLE III
Protocol of a relevé (translated from MELTZER & WESTHOFF 1942)

Nr. 39462. 1st August 1939. Terschelling, Bessenplak S of beacon near beach mark 6. Gridnr. G5.61.43 in IVON-system (Institute for Vegetation Research in the Netherlands). Stand very uniform, Empetrum heath on slope of 6 m tall parabolic dune, exposition NNE, inclination 30°.

Habitat: shadowed, moist soil, by day not strongly heated and rarely desiccating. Slight shifting of sand. Little human and animal influence.
Profile: A_0: 2 cm semi-decayed material.
A_1: 5 cm dark humus containing sand.
C: bright, white dune sand.
Sample plot   100 sq m.

Herb layer cover 100 %, 20–40 cm			
Polypodium vulgare	2.3	v.	●
Empetrum nigrum	4.4	fr.	●
Hieracium umbellatum	1.1–2	fl.	●
Festuca rubra subvar. arenaria	+.1	fr.	●
Hypochoeris radicata	+.1	fr.	●
Calamagrostis epigeios	+.1	fl.	●
Jasione montana	+.1	fl.	●
Carex arenaria	1.1	v.	○
Ammophila arenaria	2.2	v.	○
Salix repens	+.2	fr.	●
Viola canina var. dunensis	r	v.	○
Moss layer cover 100 %, 2–5 cm			
Hypnum cupressiforme var. ericetorum	3.3	v.	○
Pleurozium schreberi	3.3	v.	○
Dicranum scoparium	2.3	v.	○
Mnium hornum	2.3	v.	●
Lophocolea bidentata	2.2	fr.	●
Eurhynchium stokesii	+.3	v.	○
Plagiothecium denticulatum	+.2	fr.	●
Polytrichum juniperinum	+.3	v.	○
Peltigera canina	+.2	v.	●
Parmelia physodes	+.1	v.	○
Cladonia alcicornis	+.2	v.	○

20.5 Synthetical Research Phase

20.5.1 STEPS IN SYNTHESIS

The analysis of stands is only the first step in the description of vegetation units. After relevés have been collected, they must be compared. This is the start of the synthetical phase which leads to the distinction of coena and, if wished for, the final classification of syntaxa. To this purpose, a number of relevés are tabulated in a matrix, which is usually named a relevé table or community table.

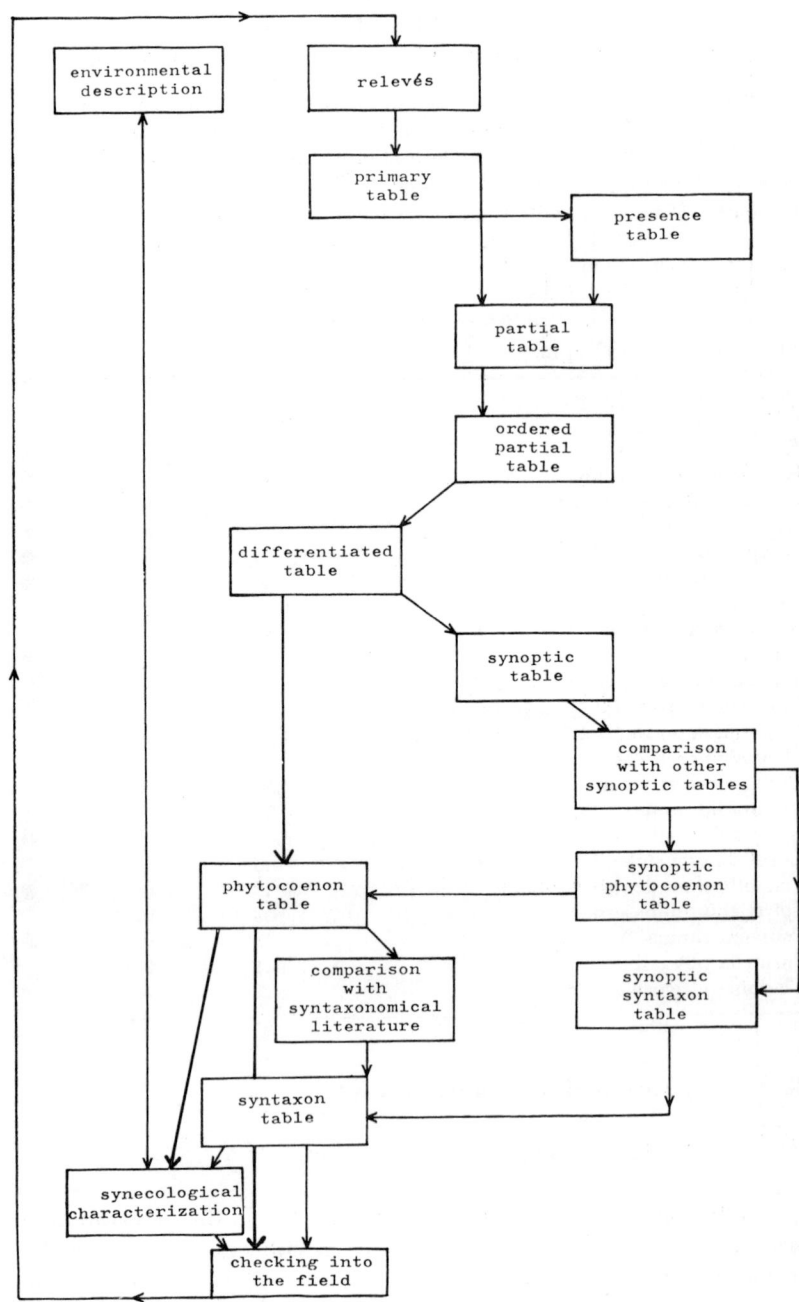

Fig. 3. Scheme of steps in the synthetic procedure.

Such a table is also called a primary or raw table (Rohtabelle). Each primary table is then rearranged into a structured table in which one or more uniform phytocoena are distinguished and characterized. After consultation of the relevant syntaxonomical literature (and possible further revision), this phytocoenon table or community table may then be presented as a syntaxon table, often for an association.

When environmental data such as soil analyses are available the vegetation types arrived at can be characterized synecologically. Finally the coena, or syntaxa, are checked in the field, e.g. whilst mapping the vegetation of the area under investigation or during the reconnaissance of a neighbouring area. This may lead to the collection of new relevés. Thus vegetation classification in the BRAUN-BLANQUET approach is essentially an iterative process, or one of successive approximation in the sense of POORE (1956). Fig. 3 shows the various steps in synthesis. SHIMWELL (1972) also outlines in English (cf. KNAPP 1958 and ELLENBERG 1956) the steps of the procedure.

The checking of units is an essential part of the method; this point seems not to be recognized by all critics of the BRAUN-BLANQUET approach (e.g. POORE 1955–56). MOORE (1962) explained that the misapprehension that the fidelity concept rests on a circular argument (e.g. POORE 1955a) is based on the failure to recognize the checking step: 'In present practice, associations are not distinguished in the field at all, but only when editing the tables of relevés. The first step involves describing uniform tracts of vegetation, *not* representative stands of a presumed association. Only when sufficient relevés have been accumulated and analysed one can discern the associations. Of course, in observing continental phytosociologists at work in their own homeland whose vegetation they know, one may be misled as to their methodology. They will now have reached the second stage of checking the reality of units already distinguished. They will not necessarily make this clear to a visiting inquirer.'

20.5.2 PRIMARY TABLE

In the primary table all taxa are listed at the left hand side of a sheet of squared paper; to each relevé a single vertical column is assigned. Each item in any column should contain either a dot or dash (in the case that a taxon is absent in the corresponding relevé) or else a combined estimation value (or an actual coverage per cent) and preferably a sociability value too. It may be convenient at this preliminary stage to group the species under separate headings according to strata and to separate phanerogams and cryptogams. The addition of new species from the later relevés produces a characteristic tailing off to the right of the table. Each row of the table represents a species (in a given stratum). Comparison of the rows of the table enables us to judge the distribution of any taxon over the relevés; comparison of the columns may lead to preliminary conclusions on relative similarities of the relevés.

The primary table should be rearranged several times, in order to establish groupings of relevés by rearrangement of columns as well as to group taxa with similar distributions in the table by rearrangement of rows. Before we discuss this rearrangement procedure (20.5.5) some concepts and techniques should be discussed.

20.5.3 Presence, Constancy, Homotoneity

Presence is the occurrence of a taxon in a vegetation table. It is usually measured as a degree by the number of relevés in which the taxon occurs (regardless of abundance and cover) expressed as a percentage of the total number of relevés compared. *Presence degree* (Stetigkeit) can thus be calculated from any number of relevés, regardless of the difference in size of the plots.

When plots of equal size are compared the corresponding per cent occurrence for a taxon is named 'constancy' (Konstanz). (Frequency, in contrast with these, is an analytical concept dealing with the distribution of a taxon within subsamples from a given stand.) Presence degree and constancy are determined from stands of different and often widely distributed localities. Constancy determinations on plots of equal size were especially favored by the Uppsala school (article 18), starting with Du Rietz et al. (1920), although the first such measurements, by Brockmann-Jerosch (1907) had used plots of various sizes. In the Braun-Blanquet approach presence degree has been termed 'Stetigkeit' and distinguished from 'Konstanz'; but confusion has resulted from the translation of both concepts as 'constancy' in some statements in English (e.g. Braun-Blanquet 1932, Moravec 1971).

Constancy and presence degree can be given in exact percentages, or in percentage classes. It is usual to distinguish five classes, noted in Roman figures:

Class:	Percentage of plots in which taxon occurs:
I	1–20
II	21–40
III	41–60
IV	61–80
V	81–100

The numbers of taxa falling into these classes are often presented in a 'constancy diagram.' Such diagrams are important characteristics of vegetation units and provide useful tools to check the uniformity or homogeneity of a table. This synthetic homogeneity concept, however, should be distinguished from analytical homogeneity. BARKMAN (1958a: 316–317) designated these concepts as 'intensive homogeneity' and 'extensive homogeneity.' However, as DAHL (1970, apud TÜXEN, 1970c: 101) pointed out, the latter term has been designated as *homotoneity* (NORDHAGEN 1943, DAHL 1957, 1960). This term was long overlooked in the BRAUN-BLANQUET approach, but now receives common use (e.g. TÜXEN 1970c, MORAVEC 1971). Homogeneity is an analytic concept, based on comparing different plots of the same size taken from an individual stand, whereas homotoneity is a synthetic concept, based on comparing similar plots from different stands of the same community-type or phytocoenon.

Homotoneity (see also 20.8.4) has been judged with the help of constancy diagrams; especially in North-European approaches (see further article 18). The classical interpretation, based on RAUNKIAER's (e.g. 1934) 'law of frequency,' of a constancy diagram is that the following relation between the constancy classes exists: $S_I > S_{II} > S_{III} \gtreqqless S_{IV} < S_V$ (the reversed J-shape). In very homotoneous tables class V may equal class I (U-shape). When classes III and IV include more species than class V, the table is considered heterotoneous (cf. GUINOCHET 1955, CAIN & CASTRO 1959, MORAVEC 1971). According to MAAREL (1972, in MAAREL & TÜXEN 1972: 209), the reasoning of WILLIAMS (1964) and observations of BARKMAN (1958a) showed that constancy class I is more or less dependent on the total number of relevés. Between the other classes the following relation exists in homotoneous tables: $S_{IV}+S_V/S_{II}+S_{III}$ is slightly over 1, whilst $S_{III}+S_{IV}+S_V/S_{II}$ is mostly about 2.

A second feature with which the homotoneity of a table can be easily checked is the variation in the number of taxa within the relevés of the table. In most cases well-developed stands of a phytocoenon do not differ much in number of species. A high variance of that number is reason to suppose that the table is heterotoneous, which may mean that more than one phytocoenon is represented in it (HOFMANN & PASSARGE 1964). After the relevés have been classed into groups according to their number of taxa, we may obtain the frequency distributions of these classes, which will indicate this variation. A table comprising relevés belonging to two different coena may produce a curve with two or more peaks (KNAPP 1971).

Apart from superficial inspection of homotoneity some simple measures have been proposed on the basis of constancy figures and species numbers (see MORAVEC 1971 for a survey). The following symbols, all corresponding to one vegetation table, are used: M is the number of relevés: S_m is the total number of species and $\bar{S}$ the mean number of species per relevé; S_v, etc. is the number of species falling into constancy class V, etc. C_i refers to the constancy per cents for species of the constancy class(es) indicated by the subscript to the summation sign, or for the prevalent species P — those belonging to the group of species with the highest constancy values, the number of which (S_p) approaches the average number of species $(\bar{S})$.

Amongst the measures based on constancy figures are:

1. The ratio S_V/S_{IV} (DAHL 1957).

2. Basic homotoneity coefficient $\dfrac{1}{S_m} \sum_{IV+V} C_i$ (MORAVEC 1971).

3. 'Homogeneity value' $\dfrac{1}{S_p} \sum_p C_i$ (RAABE e.g. 1952).

4. Index of homogeneity (CURTIS 1959) $\dfrac{\sum_p C_i}{\sum_{Sm} C_i S_m}$

5. Mean constancy for all species $\sum_{S_c} C_i / S_m$

6. Corrected homotoneity coefficient (MORAVEC 1971). This is coefficient 2 with an 'oscillation factor,' determined by the difference in species number between the richest and the poorest relevé.

TÜXEN (1970c) presented curves for various communities ending in a horizontal line (section 20.4.3); one may derive a 'minimal relevé number' from such curves. In fact TÜXEN based his saturation quotient on tables containing about this minimal number of relevés. Other observations suggest that the number of species grows continuously with the number of relevés (e.g. ETTER 1949, BARKMAN 1958a, DAHL 1957, 1960, WILLIAMS 1964). DAHL (1961) based an 'index of uniformity' on this relation (see 20.8.4)

and BARKMAN adapted KLEMENT's heterogeneity index by omitting species with a constancy of less than 10 %.
7. 'Saturation quotient' $100\bar{S}/S_m$ (TÜXEN 1970c).
As Tüxen mentioned this measure is the reciprocal of the homogeneity (i.e. heterogeneity) value of DAHL & HADAČ (1941). WHITTAKER (1972) suggested use of $BD = (S_m/\bar{S}) - 1$ as a measure of beta diversity (article 3.6). PFEIFFER's (1957) homogeneity value was reduced to TÜXEN's measure by MORAVEC. KLEMENT (1941) used the same measure under the same name as DAHL & HADAČ did, as was mentioned by BARKMAN (1958a), who concluded that this index is largely dependent on the total number of relevés in the table.

In conclusion we think it would be more realistic to base a saturation quotient explicitly on species with a minimum constancy value, e.g. species from constancy classes II–V, and to measure the rate of increase in total species number separately. For measurement of homotoneity we suggest the mean similarity coefficient for the relevés (see 20.8.5).

20.5.4 SPECIES WEIGHTS

Species may be weighted in synthetical procedures such as fidelity determination, assignment of syntaxa to higher units and calculation of spectra. Weighting is usually with the combined estimation value or some transformation of this. SCHWICKERATH (1931) used 'Artmächtigkeit' — and when species groups were concerned 'Gruppenmächtigkeit' (species and group importance value, MAAREL 1972b) — based on BRAUN-BLANQUET figures and arbitrary numerical values for symbols + and r. TÜXEN & ELLENBERG (1937) and BRAUN-BLANQUET (1946) used the Deckungswert (cover value) and Gruppenwert (group value) by taking average coverage percentages for BRAUN-BLANQUET values 2–5 and arbitrary values for 1, + and r. The latter procedure has received more application than that of SCHWICKERATH, although TÜXEN & ELLENBERG stressed its limited applicability, which was also criticized by WESTHOFF (1947), MEIJER DREES (1949), SISSINGH (1950).

DAGNELIE (1960) suggested the use of an arc-sine transformation as is usual with percentage scores. His rounded figures appear to be identical with the original BRAUN-BLANQUET values. Various other transformations have been applied, including those of ETTER (1949), BARKMAN (1958a), BARKMAN et al. (1964), MAAREL (1966b), SCHMID & KÜHN (1970), MOORE (MOORE & O'SULLIVAN 1970) and LONDO (1971). (See MAAREL, 1972b for a review.)

Without discussing this problem here (see MAAREL 1972b) we are in favour of a simple 'ordinal transformation' of the extended nine-point BRAUN-BLANQUET scale (Table II). This produces also partial accordance with the DOMIN scale in its modification by EVANS & DAHL (1955), which is much used in Scandinavia and Britain.

20.5.5 TABLE REARRANGEMENTS

The development of phytosociological tables has been stressed by the leaders of the floristic-sociological approach as a crucial procedure. E.g. BRAUN-BLANQUET (1951a, b) wrote (transl.): 'Appropriately elaborated association tables are comparable with thorough species diagnoses. By means of the tables the considerable detailed work from the minute floristic analyses of the lower vegetation units becomes accessible and evaluable. From the table it also appears whether one has worked seriously and reliably; the tables are the proper touch-stone of the concerned plant sociologist.'

At the same time the table method is said to be difficult to learn, at least by written instructions. TÜXEN & PREISING (1951) spoke of 'special aptitudes, a view of the interrelations in the tabular picture, and broad experience in the sociology, synecology and syngenetics of plant communities,' that are all needed for mastering the ordering of tables. Although personal instruction in the technique is important, as numerous pupils of BRAUN-BLANQUET and TÜXEN can confirm, it may be admitted that early BRAUN-BLANQUET sociology paid insufficient attention to explicit statements of the table techniques.

KNAPP (1948, 1958) and ELLENBERG (1956) presented detailed accounts of the table technique as it had been developed by TÜXEN in the 1930's at the Zentralstelle für Vegetationskartierung. ELLENBERG's scheme (1956: 46, slightly altered) is as follows (see Fig. 3):

(1) rearrangement of the first matrix to a 'presence table';
(2) tracing of table-differentiating species, mostly with the aid of 'partial tables';
(3) rearrangement of taxa and relevés according to the empirically established occurrence of groups of differentiating species; the result is a 'differentiated table';
(4) compilation of every relevant differentiated vegetation table into a 'synoptic table' ('Übersichtstabelle');
(5) determination of fidelity by comparing an adequate number of

synoptic tables ('zusammenfassende Übersichtstabellen' in the sense of ELLENBERG l.c.);
(6) rearrangement of the differentiated table into a final syntaxon table ('charakterisierte Tabelle' in the sense of ELLENBERG l.c.).

Similar schemes were presented by MOORE (1962), SCAMONI & PASSARGE (1963), FUKAREK (1964), KÜCHLER (1967), GOUNOT (1969), KNAPP (1971) and SHIMWELL (1972). The steps 1, 2 and 3 are the basic ones; they are essential for any local classification of relevés into phytocoena. Step 4 is preliminary to 5. Steps 5 and 6 have to do with the next level of investigation: the establishment of a (perhaps provisional) syntaxonomical classification.

TÜXEN (1970b) recommended to start with a preliminary procedure aiming at the exclusion of relevés with deviant numbers of species that suggest fragmentary or heterogeneous stands. Such relevés are thus excluded from successive rearrangements and fidelity calculations and are not considered in an association diagnosis.

The first rearrangement of a primary table aims at listing the species after their degree of presence. In the presence table it may become apparent, whether the relevés vary irregularly, or whether certain combinations of species are found recurrently. In the latter case such groups may be more or less mutually exclusive and serve as groups of differentiating species. Those are not likely to be found in the presence classes V and I; they will mostly belong to the intermediate classes II, III and IV. The supposed differentiating species are underlined or boxed in the presence table. These differentiating species, which are provisional sets of diagnostic species (see 20.3.4), are used to rearrange the table.

The next step is to set up a partial or extracted table in which only the species thought to have diagnostic value are given and used to rearrange both columns and rows. During the rearrangement of a provisional partial table, the differentiating species are written on the left hand side of the paper and the rows arranged in their corresponding groups, and the relevés (columns) are rearranged to show a diagonal order of species groups from the left to the right of the table. To simplify the rearrangement of the relevés dictation or transfer strips are used.

When the partial table has been worked out, the complete table (thus containing again the temporarily omitted species) is rearranged accordingly so that the differentiating groups appear clearly. Now we have the 'differentiated table.' The species with a high presence values indicate the degree of coherence or homo-

321

toneity of the table. The species with intermediate and lower presence values used in differentiating groups represent the heterotoneity of the table. After comparing the sizes of the two groups, one may decide how many phytocoena the table includes and whether the table should be divided, or combined with others for further treatment.

Step 4 from ELLENBERG's scheme comprises the replacement of each uniform table or partial table by a column in which for each participating species the presence degree is indicated (either as a percentage or a class value). Such a table (which may consist of only one column) is called Übersichtstabelle by ELLENBERG but is also known as Sammeltabelle, synthetische Tabelle and Stetigkeitstabelle. Following a recent denomination by DOING (1969a) and TÜXEN (1970b) we propose the neutral name 'synoptic table.'

After comparison of the synoptic table with those from other types of vegetation from the same region an idea can be formed about the local diagnostic species groups in the table under study (step 6). The differentiating groups used may represent recognized diagnostic groups, or may represent approximate groupings that should be revised on the basis of broader knowledge of diagnostic groups, or may represent diagnostic groups for newly recognized phytocoena. With those species, as well as the constant taxa, a phytocoenon may then be identified or described. Here we could speak of a synoptic phytocoenon table, from which the corresponding relevé-table or 'phytocoenon table' (Gesellschaftstabelle) can be rewritten. When moreover the synoptic table is compared with the literature and with tables from similar vegetation types from other regions, character-taxa may be found and a syntaxonomic interpretation may follow. The synoptic table now serves as a synoptic-syntaxon-table and the corresponding (rewritten) relevé-table is now a syntaxon-table. The latter will often be an association table. The fidelity research implied in steps 5 and 6 will be discussed in 20.5.7-8.

The rearrangement method might be criticized with the objection that the table pattern may be 'an artefact imposed by the clever shuffling of data and not mirroring any reality in nature,' as it has been expressed by MOORE (1962) in his refutation of this objection. In reality, the characterized table is, first, an effective organization of information, an ordering of species and relevés in relation to one another (and biotope properties). Second, it is considered a working hypothesis to be tested by further observation. It should receive confirmation: (i) from further floristic data, critically and impartially collected; (ii) from mapping the units distinguished; and (iii) from the evidence of relationships between the vegetational units and environment. The groups of differentiating taxa may be presumed to indicate ecological differences in one or more specific major biotope factors or correlated sets of factors, such as phreatic level, humus, carbonate content, amount of soil hardening; salt content, or frequency and duration of periodical inundation by sea water; exposure; shade; rate of pasturing or mowing; etc. The resulting hypotheses have to be checked by continued ecological research. If the work has been well done, the differentiating groups should have indicator value in relation to environmental factors (see 20.7.6), and the phytocoena as kinds of communities should bear consistent, reliable relations to kinds of biotopes.

20.5.6 MECHANICAL MEANS FOR REARRANGEMENT

The 'classical' rearrangement technique by manual ordering of taxa and relevés, as it has been rendered above, is a time-consuming procedure and a potential source of errors. Attempts to simplify and mechanize the procedure have been made by several investigators. The most simple means is the use of cardboard strips, usually for the relevés and thus for the columns of the table.

Various systems of blocks (wooden, plastic, or aluminium) have been developed (WILMANNS 1959, MARGL 1967, MÜLLER et al. 1972). TRENTEPOHL (1968) showed a magnetic device which seemed to be less vulnerable to manual disordering of the blocks.

For tables with a large number of relevés or a large number of species the use of punched cards is obvious. Three types of cards are in use — the edge-punched card, the visual punched card, and the machine punched card. COTTAM & CURTIS (1948) initiated use of cards with the third and most advanced type. Examples of the use of edge-punched cards are found with EMBERGER et al. (1957), GOUNOT (1957) and ZONNEVELD (early sixties, internal undated reports Netherlands Soil Survey Institute).

ELLENBERG & CRISTOFOLINI (1964) intensively used visual punch cards, mainly for sorting relevés. The corresponding cards are laid on a light-table and then compared with a standard card — e.g. one containing the characteristic species combination of a certain syntaxon. ELLENBERG (1968) published a version of this technique where combined estimation values of species are taken into consideration, however at the expense of space on the card. According to ELLENBERG, this apparatus is only efficient if the number of relevés is large, at least 200 to every table. As MÜLLER et al. (1972) stated, this technique may be useful where the vegetation has already largely been classified and the vegetation units described, and where additional relevés need only to be arranged and included in existing tables. Further use of machine punched cards in phytosociology has been described by DAGNELIE (1960), BECKING (1961), GROENEWOUD (1965) and many others; whilst recently a number of computer-based table rearrangement programmes have been developed; this will be discussed in 20.8.1 and 20.8.6.

20.5.7 DETERMINATION OF FIDELITY

The next step in the synthetic procedure is the determination of fidelity. By comparing a number of tables — if possible all available tables covering the vegetation of a given area — it is possible to discern character-species in the sense discussed in 20.3.3. As indicated there, the criteria are the distribution of the species in different coena; fidelity is the degree to which a species is concentrated in one coenon, vs. dispersed with more even occurrence in several coena. Species distributions in the tables are observed on the basis of measures already discussed — presence degree, combined estimation value, vitality, and sociability. It should be emphasized that these measures are relevant to a species' fidelity only as they differ between different coena. Character-species can as well be minor species as major ones; there is no necessary relation between fidelity and dominance. In the classification of European forests, more significance is attached to the fidelity of shrubs, herbs, and mosses, both because tree species are often widely distributed in different coena, and because disturbance has brought extensive replacement of natural tree species.

Five degrees of fidelity have been distinguished (BRAUN-BLANQUET 1921, 1928, 1932, 1951, 1964) in the relation of a species to the given coenon.

A. Character-taxa.
Fidelity degree 5: *Exclusive taxa* (treue): taxa completely or almost completely confined to one phytocoenon (vegetation unit);
Fidelity degree 4: *Selective taxa* (feste): taxa occurring with clear preference for one phytocoenon but also, though with a considerably lower presence degree, in other coena;
Fidelity degree 3: *Preferential taxa* (holde): taxa present in several coena, perhaps with about equal presence degree, but with a higher combined estimation value and (or) with a higher vitality degree in one particular coenon;

B. Companions.
Fidelity degree 2: *Indifferent taxa* (vage): taxa without pronounced preference for any coenon.

C. Accidentals.
Fidelity degree 1: *Strange taxa* (fremde): taxa having a definite presence degree optimum and mostly also a cover-abundance optimum outside the considered coenon. These are often accidental intruders from neighbouring coena or relics from a coenon that preceded in succession.

TABLE IV

Determination of fidelity according to Szafer and Pawłowski
(Braun-Blanquet 1932)

F = fidelity degree;
A = cover-abundance combined estimation;
C = presence or constancy class;
V = vitality.

F	in phytocoenon under consideration		in comparable phytocoena	
	C	A	C	A
5	IV–V	3–5	I–II	+–2(1)
	IV–V	+–2	I	+–2
	I–III	+–5	absent or very rare	
4	IV–V	3–5	II–III (IV) (relic or pioneer)	+–2(1)
	IV–V	+–2	II–III	+–1(2)
	III–IV	+–2	I–II (III)	+–1(2)
	I–III	+–2	I (rare)	+
3	I–V	3–5	I–V	+–2
	C, A various	V normal	C, A lower	V reduced
2	C, A, V various		similar	
1	I	+–1	higher	
	V reduced			
	– outskirts and disturbed parts of the stand(s) –			

As early as 1927, Szafer & Pawłowski gave a quantitative outline for the determination of the fidelity class. It has been taken over, with slight modifications, by Braun-Blanquet (1928, 1932, 1951, 1964) and it is presented here with slight alterations (Table IV). In this table the comparison is between a phytocoenon that is being studied and comparable phytocoena. It seems logically incorrect to specify in such a table 'the given association' versus 'other associations,' as Braun-Blanquet (l.c.) has done in all editions of his manual.

BECKING (1957) has modified the scheme of SZAFER & PAW-
ŁOWSKI (1927) considerably, especially in omitting the possibility
of obtaining fidelity degree 5 with a low presence degree. However,
it has been crucial in the BRAUN-BLANQUET approach that faithful
species need not be constants, and apart from diagnostic implications (see 20.6.1) one cannot simply change this principle.

20.5.8 GEOGRAPHICAL EFFECTS ON FIDELITY

Since the beginning of the BRAUN-BLANQUET approach (BRAUN-
BLANQUET 1921) it has been recognized that the diagnostic value of
character-taxa is geographically limited. Naturally enough this
recognition has become more and more general with the growing
geographical extension of phytosociological knowledge. The vegetation of a region with a uniform climate and a uniform geological
history usually shows great uniformity in the ecological amplitudes
and therefore in the sociological positions of its species. Most
species however occur in larger, climatically and geologically
heterogeneous areas and present different ecological amplitudes
or habitat relations in different parts of their areas. In many cases
these differences in local habitat response within the species are
correlated with genetic differences: we are dealing then with
ecotypic variation. In other cases genetic differences, if present, are
not evident. Geographic variation in habitat response may involve
such shifts in topographic or other local distribution, as tend to
compensate for difference in climate (law of relative habitat constancy, cf. WALTER & STRAKA 1970, article 3.2).

A number of species which behave as xerophytes in Central
Europe, e.g. *Bromus erectus, Koeleria gracilis, Carex humilis, Peucedanum
cervaria*, are bound to humid soils of cool northern exposures in the
Mediterranean. Species which are real woodland plants in a
continental climate may present much wider amplitudes in oceanic
climates: e.g. *Osmunda regalis, Listera ovata, Dryopteris filix-mas*.
On the contrary a number of thermoxerophilous species are indifferent to soil conditions in Central Europe, whereas in the cooler
and more humid climate of NW-Europe they are strictly confined
to calcareous soils. *Iris pseudacorus*, a species of wet eutrophic swamps
in Central and Western Europe, grows in moderately damp grassland in the perhumid Western Irish climate. *Schoenus nigricans*, in
Central Europe faithful to calcareous marshes or fens, occurs in
Ireland both in fens and in ombrotrophic blanket bogs. Even within
a much more limited geographical range the differences in ecological amplitude may be considerable. Within The Netherlands,

Orchis morio is faithful to the Mesobromion of dry chalk in Southern Limburg, whereas it is bound to moorland (Calluno-Genistetum molinietosum) in the Pleistocene part of the country, to mesohalinic peat soil (community of *Orchis morio* and *Ophioglossum vulgatum*) in the brackish Holocene parts of Holland (sensu stricto) and to a narrow ecotone at the margin of wet heather dune slacks on the West Frisian islands (cf. WESTHOFF & HELD 1969). *Hymenophyllum peltatum* is in the major part of Ireland bound to damp oakwood, but on the extremely oceanic cliff of Slievemore on Achill Island (Co. Mayo) it thrives freely in the open air in between *Calluna vulgaris* and *Empetrum nigrum* (PRAEGER 1934).

Consequently, one and the same species can be a character-taxon for association *A* in one area and for association *B* in another one. BRAUN-BLANQUET (1964: 100) concluded that the validity and diagnostic value of character-taxa for a given association are mostly restricted to a climatically uniform area. It follows that a typology of character-taxa with respect to their geographical extension of their validity may be useful.

BRAUN-BLANQUET (1928, 1932) gave a first distinction between general and regional character-taxa. Later (1951a) he mentioned absolute, territorial and local character-taxa (cf. ELLENBERG 1956). The distinction between territorial and local was unclear, however; apparently it refers to the size of the range of the association in which the taxon shows fidelity. As BARKMAN (1958a) pointed out the replacement of general by absolute suggests a change of criterion, i.e. from range of fidelity to degree of fidelity, which was in fact realized in BRAUN-BLANQUET's definition of 'Absolute Treue.' With BARKMAN we think that these two criteria should be used independently. A third criterion is the relation between the range of the character-taxon and the range of the corresponding syntaxon.

MEIJER DREES (1951) elaborated this criterion and came to four possibilities: 1) ranges of taxon and syntaxon coincide, 2) taxon range forms part of syntaxon range, 3) syntaxon range forms part of taxon range and 4) ranges differ but overlap. MEIJER DREES maintained BRAUN-BLANQUET & MOOR's term 'regional' which he applied to all cases in which the taxon is a character-taxon throughout the common range of taxon and syntaxon, contrasted with the term 'local' for cases in which the taxon is a character-taxon only in part of the common area.

The corresponding scheme was taken over by BECKING (1957, 1961) who incorporated it moreover in a complex scheme in which also the degrees of fidelity were considered. BARKMAN (1958), who wished to maintain the more obvious meaning of the terms local and regional, adopted the eight types of MEIJER DREES (1951), but largely renamed them. Finally KNAPP (1971), who only mentioned MEIJER DREES and BECKING, refined the division into four types of geographical character-taxa in the taxon-bound sense of MELTZER & WESTHOFF (1942): viz. 'local', 'territorial or regional,' 'region-' and 'absolute.' The following scheme may be given as a summary:

i) Local character-species, with fidelity restricted to an area that forms only part of the total area of the syntaxon.

ii) Regional character-species, with fidelity for the total area of the syntaxon, but with the species area exceeding that of the syntaxon. Subcategories could be distinguished in relation to BRAUN-BLANQUET's phytogeographical typology including provinces and sectors.

iii) General character-species, with fidelity for the total area of the syntaxon, and the species area coinciding with the syntaxon area.

20.5.9 PHYTOCOENON CHARACTERIZATION

After a primary table has been rearranged into a phytocoenon table, presence values have been calculated, and the fidelity and differentiating value of the taxa towards neighbouring communities have been established, the corresponding phytocoenon can be characterized. First the general structure can be described. Mostly the presence class of each species is indicated at the right hand side of the table, in some cases the presence per cent. To the right of the presence figure the range of the combined cover-abundance estimation values may be given.

Then, according to procedures described above the faithful, differentiating and constant taxa may be indicated. The diagnostic value may be only provisional, when the relevés come from a limited area or when insufficient other material could be compared. We are dealing here with the first of three phases of systematic phytosociology, viz. the local phase. The taxa that are of lowest presence degree and lack diagnostic value (mostly of presence class I) are often presented in an addendum to the table. In many cases the description of phytocoena does not immediately lead to a formal classification. Without this, the phytocoenon tables may summarize much of what phytosociology seeks to understand of relations amongst species, kinds of communities, and environments. When data are thus organized into phytocoenon tables, however, the stage is set for the syntaxonomic treatment next to be discussed.

Table V

Association table of the Artemisietum maritimae in the SW Netherlands by W.G. Beeftink

	Artemisietum maritimae typicum									Artemisietum maritimae armerietosum				Artemisietum maritimae agrostidetosum							
	Initial phase with Artemisia maritima					Typical form								Typical form				Fragment			
			4)		5)																
Number of relevé	53136	53134	53133	52390	52371	52338	54081	53022	55058	52343	54092	54089	52051	53043	53054	52207	53071	54130	54133	54139	54152
Locality[1]	Ka	Ka	Ka	Ka	Ka	He	Ko	Ba	Ka	He	Sp	Sp	Kats	Wa	Wa	Bath	Bath	Os	Os	Os	Li
Habitat[2]	cr	cr	cr	dike	dike	cr	cr	ab	cr	ac	cr	cr	cr	ab	ab	cr	cr	cr	ac	cr	ac
Salinity flood-water[3]	p	p	p	p	p	p	e	p	p	p	e	e	e	m	m	m	m	m	m	m	m
Date	2-9-53	2-9-53	2-9-53	21-8-53	20-8-52	5-8-52	13-8-54	5-8-53	5-8-55	5-8-52	14-8-54	14-8-54	4-9-52	6-8-53	6-8-53	3-7-52	6-8-53	2-9-54	2-9-54	2-9-54	2-9-54
Surface in m²	2x12	2x12	3x10	2½x20	4x12	2x25	3x20	3x15	3x12	100	3x25	4x20	2½x12	5x10	6x8	1½x15	2x10	5x50	10x20	2x15	4x20
Coverage in %	98	95	95	95	100	100	100	100	100	95	100	100	90	100	100	100	100	100	100	100	100
Character-taxon of the association																					
Artemisia maritima	5.4	4-5.4	5.4	4.5.5	4.3-4	2.1	2-3.1-2	2.1-2	2.1-2	1-2.1	2.1-2	2.1-2	2.1	3.1-3	3.1-2	2.1-2	3.1-2	-	-	-	-
Differential-taxa of the subassociations																					
Armeria maritima	-	-	-	-	-	-	-	-	-	1-2.2	1.2	1.2	+.2	-	-	-	-	-	-	-	-
Agrostis stolonifera var. compacta subvar. salina	-	-	-	-	-	-	-	-	-	-	-	-	-	4.5	5.5	2.2	2.2-3	3.2-3	3.5	1.2	3.5
Atriplex hastata	-	-	-	2-3.2	3.3	-	-	-	-	-	-	-	-	1.1	1.1	1.1	+.1	1.1	2.1-2	1.1-2	-
Character-taxa Armerion maritimae																					
Festuca rubra f. litoralis	+.2	+.2	1.2-3	-	+.2	5.5	5.5	5.5	5.5	5.5	5.5	5.5	5.5	3.5	2.5	5.5	5.5	4.5	4.5	5.5	4.5
Glaux maritima	-	-	-	-	+.1	-	+.1-2	-	-	1.1	+.1	-	1.1	2.1-2	1.1-2	-	-	-	-	-	+.1
Juncus gerardii	-	-	-	-	-	-	-	-	-	-	-	-	+.2	-	-	-	-	-	-	-	+-1.1
Parapholis strigosa	-	-	-	-	-	-	-	-	-	1.1-2	-	-	-	-	+.1-2	-	-	-	-	-	-
Character-taxa Puccinellion maritimae																					
Puccinellia maritima	+.2	+.2	-	-	+.1-2	-	-	-	+.2	-	+.2	-	-	-	-	-	-	-	-	-	1.2
Halimione portulacoides	2.2-3	2-3.3	2.2-3	+.1-2	1.1-2	+.1-2	1.1-2	+.2	1-2.2	+.1-2	1.1-2	+.1-2	1.2.1	-	-	-	-	-	-	-	-
Bostrychia scorpioides	-	-	+.2	-	-	-	-	-	-	-	-	-	-	-	-	-	-	-	-	-	-
Character-taxa Glauco-Puccinellietalia																					
Spergularia marginata	-	+.1	-	-	-	1.1	1.2	+.1-2	-	1.1	1.1	+.1	+.1-2	+.1-2	-	+.1	-	-	-	-	-
Limonium vulgare ssp. vulgare	-	-	+.2	-	+.1-2	+.1-2	1.1-2	+.2	1.1-2	1.1-2	2-3.2	2.2	+.1-2	+.2	-	2.1-2	+.2	-	-	-	-
Character-taxa Asteretea tripolii																					
Aster tripolium	+.1	+.1-2	+.1	-	+.1	+.1	1.1-2	2.1-2	1-2.2	-	2.1-2	+.1-2	+.1-2	+.1-2	+.1	1.1-2	+.1	+-1.1	+.1	2.1-2	-
Triglochin maritima	-	-	-	-	-	-	+.2	-	+.2	-	+.2	+.2	1.2	-	-	+.1-2	-	-	+.2	-	+.2
Plantago maritima	+.2	+.2	+.2	-	+.1	+.1-2	-	+.2	+.2	2.2	+.2	+.2	1.2	2.2	1-2.2	+.2	+.1-2	-	-	+.2	+.1-2
Character-taxa Thero-Salicornion, Thera-Suaedion and Spartinion																					
Salicornia europaea	-	-	-	-	-	-	-	-	-	-	+.1°	-	+.1°	-	-	-	-	-	-	-	-
Suaeda maritima	-	+.1	-	+.1	+.1-2	-	+.1	-	+.1	-	1.1	-	-	-	-	-	-	-	-	-	-
Spartina townsendii	-	+.2	-	-	-	-	-	-	-	-	-	-	-	-	-	-	+.1	-	-	-	-
Other taxa																					
Elytrigia pungens	-	-	-	+.1-2	1.1-2	1-2.1	-	-	+.1	-	+.1-2	+.1	+.2	-	+.1-2	+.1-2	+.2	+.2	+.2	2.1-2	-
Atriples littoralis	-	-	-	+.1	-	-	-	-	-	-	-	-	-	-	-	-	-	-	-	-	-
Lolium perenne	-	-	-	-	-	-	-	-	-	-	-	-	-	-	-	-	+.2	-	-	-	-

[1] Localities : Ka = Kaloot near Borssele (South Beveland);
He = "Slikken van de Heene" (St. Philipsland);
Ko = Marshes near Kortgene (North Beveland);
Ba = Marshes near Baarland (South Beveland);
Sp = Spieringschor near Kamperland (North Beveland);
Kats = Marsh near Kats (North Beveland);
Wa = Marshes east of Waarde (South Beveland);
Bath = Marshes west of Bath (South Beveland);
Os = Marshes west of Ossendrecht and
Li = "Galgeschoor" north of Lilloo (Belgium).

[2] Habitats : cr = on elevated creek banks;
dike = at the foot of the dike, where tidal drift is washed ashore;
ab = on elevated parts with an abrasion edge;
ac = on highly accreted parts in the marsh.

[3] Salinity flood water : e = euhaline;
p = polyhaline and
m = mesohaline.

[4] Variant with Halimione portulacoides

[5] Variant with Atriplex hastata

20.6 Syntaxonomical Research Phase

20.6.1 SYNTAXON TABLE CHARACTERIZATION

The syntaxonomical research phase starts when a phytocoenon is to be fitted into the hierarchic system of syntaxa. The phytocoenon table is then interpreted by consulting the relevant syntaxonomic literature, especially synoptic syntaxon tables. Questions in this are: (i) Which already described association can be recognized in the characteristic taxon combination of the phytocoenon; (ii) which lower units could be recognized on the basis of the established differentiating taxa; (iii) which taxa can be recognized as character- or differential-taxa from higher units already distinguished?

Careful treatment of these questions may lead to the conclusion that the phytocoenon under study is one already recognized, or that it must be described as a new syntaxon, or that part of the existing hierarchy must be redefined. Here, as ELLENBERG (1956: 57) said 'begins the range of the 'tact' of which BRAUN-BLANQUET, 1951: 18 spoke.' The expression 'sociological view' ('soziologischer Blick') is relevant. Though some of the decisions on phytocoenon similarity are objective, the essential role of judgement and experience in a typological discipline should be recognized.

The results of these considerations are expressed in the structuring and labeling of the phytocoenon table, which now becomes a syntaxon table. Usually the highest unit treated in a given syntaxon table is the association. In such an association table the subordinate units distinguished are presented as well. Table V presents an example of an association table of an already recognized association, viz. the Artemisietum maritimae, as it occurs in a region not systematically described before, the southwestern Netherlands.

When a number of related associations have been described, their regional descriptions might be presented together in a synoptic syntaxon table. An example of such a table is presented in Table VI. Both tables, composed by W. G. BEEFTINK (cf. BEEFTINK 1962, 1965, 1968), whose cooperation is gratefully acknowledged, will be discussed in the subsequent paragraphs.

20.6.2 CHARACTERIZATION OF THE ASSOCIATION

As was discussed in 20.3.2 the association concept of the BRAUN-BLANQUET approach was always based on the presence of a characteristic taxon combination, including character-species. However, as the number of described associations grew, the role of character-taxa gradually diminished whilst the diagnostic importance of differential-taxa increased. In some association descriptions character-taxa are not mentioned separately but are given together with the differential-taxa ('Kenn- und Trennarten'). This led to the character-combination concept which in fact recognized the possibility that none of the participating taxa is a character-taxon in a strict sense, but rather the correlated occurrence of taxa is the essential 'character' of the phytocoenon.

BRAUN-BLANQUET (1964: 122) seemed to confirm this development when he stated that 'the essential association features' should be found in a stand in order to assign it to that association and then defined this as follows: 'primarily the normal characteristic species combination should be present, that is a minimum number of character- and differential-species and some of the more important companions.' However, during recent colloquia and symposia of the International Society for Plant Geography and Ecology the presence of at least one character-taxon was demanded by others, especially by OBERDORFER (1968); see also the following discussions, during the 1964 symposium at Stolzenau (TÜXEN 1968: 132–141) and DIERSCHKE (1971). DIERSCHKE (1971) even stated that 'the basic unit for the Prodromus is the association, the rank of which is determined by the constant occurrence of at least one character-species and which is defined by its characteristic species combination.' This definition may be considered the standard from which, however, vegetational circumstance may force some departures.

20.6.3 GEOGRAPHIC PROBLEMS WITH ASSOCIATIONS

Many of the problems with character-taxa result from complexities in the geographic behaviour of species in relation to coena, as discussed above (20.5.8). The difficulties led ELLENBERG (1954b) to speak of 'the crisis of the character species doctrine' and to propose a much more local delimitation of associations without general character-species. As ELLENBERG (1956) added, the use of geographically restricted 'local character-species' for associations is then appropriate. The local establishment of associations emphasized by ELLENBERG (1954b) had, in fact, been common practice by then for some 20 years, as TÜXEN (1955) stated in a reply to ELLENBERG, whilst referring to BRAUN-BLANQUET & MOOR (1938) and TÜXEN (1937). In fact OBERDORFER (1957, but also

	Asteretea tripolii						
	Glauco-Puccinellietalia						
	Armerion maritimae			Puccinellio-Spergularion salinae			Halo-Scirpion
netum	Artemisietum	Juncetum	Junco-Caricetum	Puccinellietum	Puccinellietum	Puccinellietum	Halo-Scirpetum
coidis	maritimae	gerardii	extensae	distantis	fasciculatae	retroflexae	maritimi
	6	7	8	9	10	11	12
	61	64	7	37	10	12	19
	30(+-1)°	42(+-1)°	14(+)°	35(+-2)°	70(r-2)°	100(1-2)°	-
	-	-	-	-	-	-	-
	-	-	-	-	-	-	-
	13(+)	20(+-2)	-	8(+)°	-	17(r)	37(+-2)
	43(+-2)	31(+-2)	14(+)	62(+-1)	90(r-3)	25(r-1)	26(1-3)
	98(+-3)	62(+-1)	-	5(+)	-	-	-
	93(+-3)	59(+-1)	-	8(+-1)	-	-	-
	2(+)	90(+-3)	14(r)	-	-	-	-
	-	-	100(1-4)	-	-	-	-
	-	-	-	100(1-5)	20(r-2)°	17(r-+)	-
	-	-	-	-	100(2-4)	17(r)	-
	-	-	-	-	-	100(1-4)	-
	-	-	-	32(+-2)°	10(+)°	8(r)°	100(3-5)
	5(+-2)	-	-	-	-	-	-
	-	86(+-5)	100(2-4)	14(+-2)	30(+)	-	5(+)
	100(3-5)	97(+-5)	86(+-3)	35(+-3)	10(r)	-	-
	33(+-1)	98(+-3)	100(2-3)	35(+-3)	30(r-2)	-	11(+)
	5(+-1)	64(+-3)	43(+-2)	19(+-2)	10(+)	8(+)	-
	-	11(+-1)	100(+-3)	65(+-2)	30(+-2)	-	42(+-4)
				97(+-3)	80(+-2)	100(r-2)	-
	72(+-2)	47(+-1)	14(+)	19(+-1)	20(1-2)	8(+)	-
	66(+-2)	91(+-3)	57(r-2)	3(+)°	-	-	-
	100(+-2)	72(+-2)	71(r-1)	65(+-3)	100(r-3)	100(+-4)	74(+-2)
	29(+-1)	77(+-2)	29(+)	5(+-1)	50(r-2)	8(r)	16(+-1)
	72(+-3)	98(+-3)	100(+-2)	11(+)	-	8(+)	-
	31(+-1)	20(+-1)°	-	38(+-1)°	30(r-1)°	8(1)°	-
	11(+)	5(+)°	-	62(+-2)°	40(r)°	-	63(1-3)
	31(+-2)	28(+-2)	-	30(+-2)	-	-	32(+-2)
	-	2(+)	-	30(+-3)	10(r)	-	-
	-	2(+)	29(+)	11(+-1)	10(1)	25(+)°	-
	-	2(+)	43(r-+)°	30(+-3)	10(r)°	67(r-2)°	11(+-2)

Carex distans 5(+), Sagina maritima 2(+), Solanum dulcamara 2(+); Column 8: (r-1), Lotus tenuis 29(2), Hippophae rhamnoides 29(r-+), Trifolium fragiferorale 14(+)°; Column 9: Polygonum aviculare 41(+-2), Elytrigia repens umnalis 5(+), Trifolium repens 11(+)°, Coronopus squamatus 8(+-2), Matri- Poa annua 11(+-2), Cochlearia officinalis 11(+), Festuca arundinacea 8(+), cum sp. 5(+), Sonchus arvensis 3(+), Poa pratensis 3(+), Solanum nigrum 3(+); Column 10: Centaurium pulchellum 10(+), Plantago major 20(r)°, Ma- cus bufonius 40(r-2), Hordeum marinum 10(r), Samolus valerandi 10(r); learia officinalis 5(+), Atriplex littoralis 5(+).

of the alliance Puccinellion maritimae and exclusive character-taxon of

Table VI

Classification of salt-marsh communities in the SW Netherlands
according to the Braun-Blanquet method by W.G. Beeftink

Classes	Thero-Salicornietea		Spartinetea		
Orders	Thero-Salicornietalia		Spartinetalia		
Alliances	Thero-Salicornion		Spartinion		Puccinellion maritim
Associations	Salicornietum strictae		Spartinetum maritimae	Spartinetum townsendii	Puccinellietum maritimae	Puccinellietum Halim portu
Column	1		2	3	4	5
Number of relevés	14		24	30	124	40

Character-taxa of the associations					
Salicornia europaea coll.[1]	100(1-3)	33(+-2)	14(+-2)	78(+-2)	40(+-
Spartina maritima	14(+)	100(2-4)	-	6(+)	-
Fucus vesiculosus var. lutarius	-	79(+-5)	7(2-3)	4(+-2)	-
Spartina townsendii agg.	86(+-1)	79(+-2)	100(3-5)	50(+-2)	35(+-
Puccinellia maritima [2]	25(+-1)	33(+-1)	30(+-1)	100(3-5)	92(+-
Halimione portulacoides	8(+)	-	33(r-1)	83(+-2)	100(3-
Artemisia maritima	-	-	-	5(+-1)	-
Armeria maritima	-	-	-	6(+-1)	-
Carex extensa	-	-	-	-	-
Puccinellia distans	-	-	-	-	-
Puccinellia fasciculata	-	-	-	-	-
Puccinellia retroflexa	-	-	-	-	-
Scirpus maritimus var. compactus[3]	-	-	27(+-2)	-	-

Faithful taxa Puccinellion maritimae					
Bostrychia scorpioides	-	17(+-4)	14(+-2)	21(+-4)	37(+-

Character-taxa Armerion maritimae					
Juncus gerardii	-	-	-	2(+)	-
Festuca rubra f litoralis	-	-	-	14(+-2)	62(+-
Glaux maritima	-	-	10(r-1)	50(+-2)	15(+-
Parapholis strigosa	-	-	-	-	-
Agrostis stolonifera var. compacta subvar. salina	-	-	-	-	-

Character-taxon Puccinellio-Spergularion salinae					
Spergularia salina	-	-	-	-	-

Character-taxa Glauco-Puccinellietalia					
Spergularia marginata	-	-	7(r-+)	73(+-2)	42(+-
Limonium vulgare ssp vulgare	8(+)	8(+)	30(r-+)	73(+-2)	60(+-

Character-taxa Asteretea tripolii					
Aster tripolium	50(+-2)	33(+-2)	77(r-2)	98(+-2)	97(+-
Triglochin maritima	8(+)	-	27(r-2)	86(+-4)	62(+-
Plantago maritima	-	-	20(r-1)	65(+-4)	65(+-

Other taxa					
Suaeda maritima	50(+-1)	12(+-2)	37(+-1)	64(+-2)	60(+-
Atriplex hastata	-	-	77(+-2)	25(+-2)	5(+)
Elytrigia pungens	-	-	7(r-+)	1(+)	22(+)
Lolium perenne	-	-	-	-	-
Plantago coronopus	-	-	-	-	-
Phragmites communis	-	-	-	-	-

Addenda: Column 1: Zostera noltii 29(+-2); Column 7: Centaurium pulchellum 17(+-
Centaurium pulchellum 57(r-2), Carex distans 29(r-2), Juncus maritimus
rum 14(r), Sonchus arvensis 14(r), Trifolium repens 14(r), Centaurium l
22(+-1), Potentilla anserina 8(+-1), Plantago major (24+-2), Leontodon
caria inodora 11(+)°, Bromus mollis 8(+), Ranunculus sceleratus 16(+-1)
Cirsium arvense 5(+-1), Poa trivialis 3(3), Hordeum secalinum 3(1), Tax
3(+), Senecio vulgaris 3(+), Anagallis arvensis 3(+), Leontodon nudicau
tricaria inodora 10(r)°, Bromus mollis 10(+), Sagina maritima 40(+-2),
Column 11: Bromus mollis 8(r); Column 12: Ranunculus sceleratus 5(+), C

1. In the alliances Thero-Salicornion and Spartinion represented by S. stricta L
2. Preferential character-taxon of the association; also selective character tax
   the order Glauco-Puccinellietalia.
3. Also character-taxon of the alliance Halo-Scirpion.

1968 and OBERDORFER et al. 1967) also worked with local character species, though always in close connexion with alliance character-species.

This local association approach as such has been criticized from another point-of-view by SCHWICKERATH (1942, 1954a, 1963). This author emphasized that geographically differentiating taxa should be used in the delimitation of syntaxa, but whenever a series of geographically different syntaxa should have one or more character-taxa in common they should be grouped together into one association. SCHWICKERATH especially criticized the splitting up of the Xerobrometum by BRAUN-BLANQUET & MOOR (1938).

To solve the problem use of different terms for more localized, vs. more widely distributed, phytocoena has been suggested. KNAPP (1942, 1948, 1958) introduced the concept of chief-association (Hauptassoziation) for the type of association SCHWICKERATH was referring to. MEIJER DREES (1951) defined it as follows: the smallest unit possessing absolute or regional character-species. Absolute = general or regional is meant here in the sense of BRAUN-BLANQUET & MOOR (1938) and MEIJER DREES, i.e. not occurring outside the association's range. KNAPP (1971) later mentioned the concept rather incidentally as referring to character-species that are valid 'at least within an entire flora-region.' The more local units comprised in a Hauptassoziation were called Gebietsassoziationen, or 'regional associations'. PASSARGE (1968) and PASSARGE & HOFMANN (1968) elaborated the suggestion by KNAPP towards a separate system of historic-geographical units besides the usual edaphic-ecological units. 'Regional' (territorial or provincial) associations could then be distinguished as parallel but subordinated units to the 'elementary association,' with which the chief-association is identical. PASSARGE & HOFMANN (1968) subjected the regional association to a trinary nomenclature as is applied in zoological systematics for geographical races.

This brings us back to OBERDORFER (1957, 1968) who sharply distinguished between 'association,' 'regional-association' and 'geographical race.' (See also MÜLLER 1968). A regional association ('Gebietsassoziation') has a limited distribution-area; it is characterized by a combination of regional (or local) character-taxa, corresponding with SCHWICKERATH's geographically differentiating taxa, and general alliance character-taxa. A geographical race is a regional expression of a larger, regional or general association, without such a typical combination of territorial and alliance character-taxa. As OBERDORFER (1968) remarked, it will often be very difficult to distinguish between the two levels. Geographical races have not yet found a place in the formal hierarchy but have nonetheless frequently been described, particularly in Germany (e.g. OBERDORFER 1957), Czechoslovakia (e.g. NEUHÄUSL & NEUHÄUSLOVÁ-NOVOTNÁ 1972) and Poland (e.g. MATUSZKIEWICZ 1962).

We have thus a surfeit of terms and concepts. To resolve this matter we refer to the remarks on character-taxa (20.5.8) and adopt the distinctions made by OBERDORFER and others under three terms for units of expanding geographic extent: the local association, regional association, and (general) association. Clearly these intergrade, but in principle they should possess local, regional and general character-taxa, respectively. Groups of related regional associations, which are in fact vicariant associations (cf. PASSARGE 1968) could be called 'vicariant association groups'; this term seems more appropriate than either chief or elementary association. Another acceptable name for the vicariant association-group is collective association (cf. MEIJER DREES 1951).

We shall now briefly discuss the Artemisietum maritimae association as presented in Table V. It is characterized by one character-taxon, *Artemisia maritima* which has, according to Table VI, a high fidelity degree (at least in the investigated area) and which is, according to the synchorological data presented by BEEFTINK (1965) as well as chorological data on the species, a territorial character-taxon. It follows from Table VI that most of the salt marsh associations in the southwestern Netherlands are typified by few but generally 'good' character-taxa (fidelity degrees 4 or 5). The normal characteristic taxon combination of the Artemisietum maritimae in the southwestern Netherlands consists of *Artemisia maritima, Festuca rubra* f. *litoralis, Limonium vulgare* ssp. *vulgare, Aster tripolium* and *Plantago maritima*. BEEFTINK has listed most of the taxa under a specific head so that the companions are few and simply called 'other taxa'. Note the four relevés at the right end of the table, that lack the character-taxon and are considered a fragmentary syntaxon.

20.6.4 HIGHER UNITS (ALLIANCE, ORDER, CLASS)

When a new association is recognized, it must be placed in the system of higher units. The assignment of an association to an alliance (and other higher units) is primarily based on comparison of floristic relationships. As in plant taxonomy (cf. DAVIS & HEYWOOD 1963) the terms relation(ship) and affinity refer only to the possession of common characters. They do not refer to syngenetical (successional) links; we may compare with SOKAL & SNEATH's

(1963) emphasis of phenetic relationships. Group importance values of the character- (and differential-) taxa of the relevant alliances are calculated, either on a presence-absence or on a quantitative basis, and compared (cf. WESTHOFF 1947, RAABE 1957). It may occur that the affinity towards two alliances (or orders or classes) is almost equal. An example of close relationship towards two alliances is presented by the Artemisietum maritimae. Generally the Armerion maritimae alliance is the obvious higher unit, as follows from tables V and VI, but the subassociation Artemisietum maritimae typicum is also clearly related to the Puccinellion maritimae. In fact the association was once assigned to the Puccinellio-Salicornion alliance (BRAUN-BLANQUET & LEEUW 1936) from which the Puccinellion maritimae was split off, but later (e.g. TÜXEN 1937, BEEFTINK 1965, WESTHOFF & HELD 1969) it was considered part of the Armerion maritimae. In cases of doubt equal use is made of structural, physiognomical and synecological considerations (cf. OBERDORFER 1957, WESTHOFF & HELD 1969).

It may also be the case that in a little-known area a number of new associations are described; and some of these may bear so little relation to known alliances (etc.) that they should be classified into new higher units. As was introduced in 20.3.2, various higher units above the association level, viz. the alliance, the order, and the class are characterized in the same way as the association. Species equally occurring in a group of related associations but faithful to that group, become character-taxa for higher units to which they are faithful. Contrary to most associations the higher units generally do possess a number of regional or general character-taxa of high fidelity. When an alliance character-taxon shows a preference for one association within an alliance, it may be used both as alliance character-taxon and as differential-taxon for an association within the alliance. The species is then termed a 'transgressive character-taxon.' Species may of course bear similar character- and differential- relations to orders vs. alliances, and classes vs. orders. Interpolated units like the suballiance and suborder are characterized by character-taxa and/or differential-taxa. Their use is to be avoided; but the suballiance has had some use.

Formal definitions for higher syntaxa and instructions for their ranking have not been agreed upon (cf. PIGNATTI 1968b, DIERSCHKE 1971). In practice the number of related associations tends to determine the number of the higher units classifying these. Growth in number of associations expresses its influence upward, in growth in the number of alliances, etc. For example, TÜXEN (1937) described 94 associations and 41 alliances for northwest Germany;

333

in TÜXEN (1955) the numbers were 189 and 76. In the Netherlands the number of described alliances grew from 39 (WESTHOFF et al. 1946) to 85 (WESTHOFF & HELD 1969). PIGNATTI (1968b) called the attention to the 'inflationary' character of this phenomenon. His suggestive title stimulated much discussion (TÜXEN 1968b: 89—97), which produced few solutions apart from plans for cooperation and standardization now being realized in the European Prodromus of plant communities.

20.6.5 UNITS ABOVE THE CLASS

The growing number of classes, as well as the extension of the BRAUN-BLANQUET approach to various, and floristically different parts of the world, lead to the introduction of units above the class. The grouping above the class originally proposed by BRAUN-BLANQUET, the 'vegetation circle' was a phytogeographical rather than a sociological unit. BRAUN-BLANQUET (1959) and TUXEN (1970a) on a suggestion of SCHMITHÜSEN sought to establish a new unit, the *class group* (cf. TÜXEN 1963: 213—218).

The class group (TÜXEN 1970a) is a set of territorially defined, largely vicariant classes analogous with the 'vicariant association group' (20.6.3). The characterizing taxa are in this case *genera* with vicariant species in different floristic provinces or regions; for example the classes Querco-Fagetea silvaticae (Europe), Querco-Fagetea grandifoliae (North America), and Querco-Fagetea crenatae (East Asia) form a class group: Querco-Fagea, with the suffix -ea probably taken from JAKUCS (e.g 1972). Curiously, BRAUN-BLANQUET (1964) presented the class group under a different name 'community kingdom' (Gesellschaftsreich) and apparently as a somewhat different phytogeographical concept. Although he first (p. 140) stated that this unit 'takes together classes floristically related by vicariant species and identical higher categories', he concluded that community kingdoms should coincide with SCHMITHÜSEN's (1961) vegetation kingdoms established by means of vicariant classes.

TÜXEN's class group should not be confused with the geographically determined class-group of SCAMONI et al. (1965). However, according to the elaboration given by PASSARGE (1968) and PASSARGE & HOFMANN (1968), these class groups correspond actually to the usual European classes. It is questionable whether PASSARGE & HOFMANN's (1968) complicated system of two hierarchies can be integrated in the BRAUN-BLANQUET system. TÜXEN (1971) strongly rejected such an integration.

Three other treatments of higher units deserve mention. (i)

CHAPMAN (1952, 1959) used terms like Coeno-Salicornietalia for groups of higher units dominated by species of one widely distributed genus. For such a unit the name Coeno-Salicornietum should be preferred. (ii) The concept of 'vegetation type' was developed by HADAČ (e.g. 1962, 1967) for groups of more or less similar classes. The vegetation type is based on the occurrence of common species (indicating or 'bezeichnende' species). However, the floristic basis of these units seems to be rather loose, and as HADAČ admits the vegetation types are 'practically formations'. (iii) JAKUCS (1961) also used the class group concept, which he later (e.g. 1972) named 'division' and for which he introduced the suffix -ea. HADAČ (1967) concluded that the division is largely identical with his vegetation type, which term should then have priority. The identity seems to be real, but for it the term 'vegetation type' is least descriptive and most confusing; *division* should be preferred. BOLOS (1968) presented a system of divisions for classes occurring in Spain. As term and concept the division is consonant with the hierarchy.

In conclusion, a division may be defined as a syntaxon above the class level that unites related classes within a floristic region (or province) on the basis of common division character-taxa. The character-taxa may be species, or genera, or both. Let us call this a 'vertical' unit in the sense that it unites lower units sympatric in a given region. The class-group concept, in contrast, is comparable with the vicariant association group in that it joins units on the same level from different floristic territories. It is in this sense a 'horizontal' unit. Its character-taxa will more usually be genera than species. Therefore the name 'vicariant class group' may be proposed, defined as a group of classes allopatric in geographic occurrence, but linked by the occurrence of vicariant species in one or more genera. Since two units would be distinguished in this way, it may be questioned whether the suffix -ea should be used for both. It seems logical to reserve this suffix for the division, for which it was proposed.

The definition of the division is not physiognomic. We suggest that, the difference in definition notwithstanding, divisions may converge in practice with the formations of Anglo-American ecology (article 13), as broad physiognomic units limited to a given region or continent. Vicariant class groups, in contrast, would link floristically related phytocoena of different continents. Thus the Querco-Fagetea-group would comprise forest phytocoena with genera in common in Europe, eastern North America, and western Asia, whilst a Spartinetea-group would comprise a set of more narrowly defined phytocoena (*Spartina* associations) in the salt marshes of the world.

20.6.6 HORIZONTAL AND VERTICAL CLASSIFICATION

Thus we follow the suggestion of KNAPP (1948), to classify independently by two principles, the one on an edaphic-ecological ('vertical') and the other on a historic-geographical ('horizontal') basis. This suggestion has been amply discussed (WESTHOFF 1950, MEIJER DREES 1951, BARKMAN 1958a, BEEFTINK 1965, SCAMONI et al. 1965, PASSARGE & HOFMANN 1968, PASSARGE 1968). All authors agree with the incorporation of both viewpoints in syntaxonomy. MEIJER DREES emphasized the vertical approach, because syntaxonomy should build upward from locally classified associations. BEEFTINK suggested that the choice of the criterion would depend on the peculiarities of the vegetation concerned: for some phytocoena with disjunct distributions the stronger floristic relations and preferred grouping would be horizontal whereas for other phytocoena the more natural grouping would be vertical. PASSARGE & HOFMANN use both principles on each level, with the geographical units subordinated to the corresponding edaphic units, which makes the system very complicated (cf. TÜXEN 1971, in his critical review of PASSARGE & HOFMANN 1968).

We would modify BEEFTINK'S suggestion in recommending that the main-axis of the classification should be 'vertical' on all levels of the hierarchy. Given the main direction of the classification as vertical, secondary to this, 'horizontal' vicariant groupings may be recognized on any syntaxonomic level. We have mentioned such groupings for associations (20.6.3) and higher syntaxa (20.6.5); we may now trace the problem and the principle into the lower syntaxa.

20.6.7 LOWER UNITS (SUBASSOCIATION, VARIANT, FACIES)

Associations are on the one hand joined into higher units, and on the other divided into subordinate units. As in idiotaxonomy the former procedure is compulsory, the latter facultative. The lower syntaxa are mostly characterized by differential-taxa. Usually, also, the syntaxa below the association level are described as deviations from average or typical situations. Thus to complete the syntaxonomy on a certain level a 'typical' subassociation (variant, etc.) and one or more 'deviant' syntaxa are described and characterized in parallel.

Variations within an association have been approached in three directions; to the edaphic-ecological ('vertical') and historic-geographical ('horizontal') may be added a syndynamical (successional) point-of-view. It has been proposed repeatedly that different syntaxa should correspond to these approaches each of which could form a hierarchy of subordinate units. Thus the subassociation might be based on local edaphic or micrometeorological differences within the association, the variant on geographical or climatological differences (with the terms vicariant and geographical race as synonyms), and the phase on differences in successional status. This distinction is in accordance with the original approach of BRAUN-BLANQUET (1928, 1932, BRAUN-BLANQUET & PAVILLARD 1928). SCHWICKERATH (1942) and DUVIGNEAUD (1946, 1949) distinguished five types of differential-taxa on the basis of differences in nutrient status, moisture conditions, successional status, geographical area, and human influences.

The system was further extended by the incorporation of the sociation, a unit defined by the dominant species of its strata (see article 18). This suggestion of NORDHAGEN (e.g. 1937) was acknowledged by the 1935 Botanical Congress (see DU RIETZ, 1936). In the Netherlands it has become common practice to describe species-poor communities without character-species but with one or more dominants as sociations or consociations (WESTHOFF 1947, 1949, BEEFTINK 1965, WESTHOFF & HELD 1969). Contrary to DU RIETZ (1930b, 1936, 1965) the sociation and consociation are considered of association-rank (BRAUN-BLANQUET 1955, MAAREL et al. 1964). They may be assigned to an alliance and higher syntaxa (WESTHOFF & HELD 1969).

From the beginning many communities have been described without receiving a definite syntaxonomic position. They are often called 'community of ......' ('Gesellschaft von ......') (cf. ELLENBERG 1956, NEUHÄUSL 1963, FUKAREK et al. 1964). These units can be either provisional local coena, which might later be ranked in the system when more evidence is available, or coena without character-species or dominants. BARKMAN et al. (1958) and WESTHOFF et al. (1959), proposed the term 'consortium' for the latter units.

An additional term for subordinate units, 'form,' is used in different ways for syntaxa of lower rank. It may mean a unit of no specified rank (as seems to be the case with SCHUBERT 1960, for example). It is also used in relation to altitude (OBERDORFER 1957, 1968, BRAUN-BLANQUET 1964) or moisture conditions (DUVIGNEAUD 1946). The 'Ausbildungsform' is considered the lowest unit in the hierarchy apart from the facies by TÜXEN (1970a, cf. SCHUBERT 1960, PASSARGE 1968).

The distinction of local-edaphic and geographic units has been maintained by BRAUN-BLANQUET (1951, 1964), MEIJER DREES (1951), OBERDORFER (1957, 1968), BARKMAN (1958b) and others.

Under the influence of Tüxen (1937) the variant became a subunit under the subassociation (Braun-Blanquet 1951a, 1964) rather than a differently defined (geographic) unit. In line with this practice we suggest as the downward extension of the primary hierarchy, for units based on local variations: association, subassociation, variant, subvariant, and facies. The last two may be often unneeded, but facies may be ecologically significant and useful as indicators in some cases (Krause 1954). Geographic races of lower syntaxa would be separately indicated by the territories in which they occur (Oberdorfer 1957, Müller 1968).

The *subassociation* is indicated with the suffix -etosum. Of the various ecological possibilities and corresponding symbols of Duvigneaud (1946, 1949) d and δ may be used for subassociations differentiated by nutrient and moisture status, respectively. Schwickerath (e.g. 1942), Duvigneaud (1946) and Meijer Drees (1951) showed how subassociations of an association may in fact indicate the transitions of that association towards related associations. The subassociation 'typicum' may then be considered as the 'nucleus' of the association (cf. Fig. 1); in fact this term is commonly used for the subassociation that lacks differential-species. Westhoff (1965, Westhoff & Held 1969) observed that many subassociations of the latter type were relatively poor in species, and particularly poor in character-species of the association. Such units were indicated as 'inops' (= having shortage of); the 'typical' Querco-Betuletum is an example of such an 'inops'-subassociation.

The next lower rank is called the *variant*. According to Braun-Blanquet (1964: 124) a variant would not be characterized by differential-species, but 'either by a strong prominence of certain species, which cannot be considered as differential-taxa, or by a slightly deviating species assemblage.' This criterion is, however, diagnostically too vague. Most phytosociologists share the opinion that a variant also must be characterized by differential-species. The same holds for the next lower rank, the subvariant. The variant-differentiating taxon (taxa) may be differentiating only because of its 'strong prominence,' i.e. its higher combined estimation values in one variant compared with others and not because it is clearly present resp. absent. In most cases the variant is distinguished within the subassociation; but it is also possible to divide an association directly into variants, if the mutual differences are considered too small to justify the rank of subassociation. Below the variant the subvariant may be used if needed.

The lowest unit is the *facies*. It is not even characterized by differential-taxa; it thus does not employ diagnostic-species and is not part of the formal hierarchy. A facies is usually characterized by the dominance, in a high cover degree (scale 4 or 5), of one of the species belonging to the normal floristic assemblage of an association. It is not usual to consider a stand as a facies if this dominance is a normal feature of that association. Since, for instance, dominance of *Fagus sylvatica* is normal for any association assigned to the alliance Fagion sylvaticae, it is not appropriate to construct such a coenon as 'a *Fagus* facies of the Fagetum.' A facies, therefore, is a deviation phenomenon. It may be brought about by special, and sometimes by extreme conditions of abiotic factors, but in many cases it is the result of human disturbance (e.g. MELTZER & WESTHOFF 1942, KNAPP 1971). Examples are a facies of *Rubus* sect. *Sylvatici* or *Rubus* sect. *Heteracanthi* in a stand of Querco-Carpinetum, or a facies of *Acorus calamus* in a stand of Scirpo-Phragmitetum. It is often possible to deduce an effect of human disturbance from the appearance of a certain facies.

20.6.8 NOMENCLATURE

Formal nomenclature of syntaxa started with the naming of associations by BRAUN-BLANQUET (1913), following precedents of CAJANDER (1903), BROCKMANN-JEROSCH (1907) and RÜBEL (1912). The first general proposals were discussed at the Brussels Botanical Congress (FLAHAULT & SCHRÖTER 1910); the nomenclature of associations and other syntaxa developed gradually thereafter. With the growth of the number of described syntaxa the need for standardization and rules increased. A start toward systematic nomenclature was made in 1933 (BRAUN-BLANQUET 1933b), and DAHL & HADAČ (1941) proposed a coherent set of rules based on the code of botanical nomenclature. During the Stockholm Botanical Congress BARKMAN (1953a) proposed a general set of rules and MEIJER DREES (1951) a complete system of nomenclature. These were discussed in a number of papers in *Vegetatio* vol. 4. Contributions by BACH et al. (1962) and RAUSCHERT (1963) appeared later, and BRAUN-BLANQUET (1964) reported on the discussion and presented 7 rules. The next phase started during the Symposium on Plant Sociological Systematics at Stolzenau, 1964, where MORAVEC (1968) introduced a set of 26 articles and urged a general acceptance of these amongst active plant sociologists. A special Working Group is now preparing a code for the new Prodromus project (see TÜXEN 1968b: 152, MORAVEC 1968, 1969, 1971, NEUHÄUSL 1968).

Some instructions for the construction of names and some proposed general rules for nomenclature of syntaxa follow.

According to MORAVEC (1971) five necessary principles are:
1. Each syntaxon (with definite rank, position and delimitation) has only one correct name.
2. Each name can be correctly used for one syntaxon only.
3. The correct name is established according to rules based on the priority principle.
4. The association is the fundamental nomenclatural unit (syntaxon).
5. The validity of nomenclatural rules is retroactive.

20.6.8.1 *Construction of Names*

For the standardization of syntaxon names, the following procedure is in general use.

To the generic part of the names of one or two (not more) characteristic (not necessarily faithful) taxa of a syntaxon a suffix is added. These suffixes are specific for the different syntaxonomic ranks, see Table VII. The suffix '-etum' for the association goes back to classical Latin and has been in use since HUMBOLDT (1805). The suffix '-ion' for the alliance was proposed by Moss (1910) and taken over by BRAUN-BLANQUET (1921: 347). The suffix '-etalia' for the order was presented in BRAUN-BLANQUET (1928). For the class, up to 1932 (BRAUN-BLANQUET, FULLER & CONARD) no specific suffix was used, since the classes were designated by circumscriptions such as 'communities of maritime dunes' (later the class Ammophiletea). In 1934 MEIER & BRAUN-BLANQUET proposed the suffix '-etales' for the class, but (BRAUN-BLANQUET et al. 1939) this suffix was later changed into '-etea'. The suffix '-etosum' was introduced by BEGER (1922).

If we are dealing with an association or alliance that is sufficiently characterized by one taxon (e.g. by a faithful dominant character-taxon), the genus name of the taxon is used with the appropriate suffix (-etum, -ion), followed by the species name in the genitive. For example the association Ericetum tetralicis is named after *Erica tetralix*, and the alliance Alnion glutinosae after *Alnus glutinosa*.

In some cases the specific epithet has been replaced by a geographical adjective indicating the area characteristic of the syntaxon, e.g. Agropyretum boreoatlanticum for the association of *Agropyron junceum* (syn. *Elytrigia juncea*) of the North Atlantic Coast. In this case this has been done to distinguish it from a different association of *Agropyron junceum* occurring in the Mediterranean. In recent decades, however, there is a tendency to avoid such geographical names as far as possible and to replace them by bigeneric names. In the case of orders and classes and, though less frequently, alliances the specific epithet is omitted when the context leaves no doubt – for example, 'Fagion' instead of 'Fagion sylvaticae' in a European context.

When a syntaxon was to be named after two characteristic (not necessarily faithful) taxa, the name of the second taxon was provided with the appropriate suffix; whilst the name of the first taxon was joined with the second one by the suffix -eto. Examples are the association Querceto-Betuletum, the alliance Alneto-Ulmion, the order Glauceto-Puccinellietalia, and the class Querceto-Fagetea. However in many cases this suffix was abbreviated to a connecting vowel after the stem of the first name, e.g. Alneto-Padion became Alno-Padion. BACH et al. (1962) proposed to do this systematically, mainly on the argument that, for example, 'Querceto-Betuletum' could mean a mixture of, or transition between, a *Quercus* stand and a *Betula* stand, which is of course something quite different. This proposal has been widely accepted. However, the correct choice of the connecting vowel, especially in the case of the third Latin declension and with names derived from Greek words, gives difficulties. BACH et al. (1962) and RAUSCHERT (1963) gave directions for a large number of cases.

There are two reasons for choosing a bigeneric name. The first and more obvious is that a combination of two names should characterize the syntaxon better. The first name may be a dominant, generally the second name represents the character-taxon considered diagnostically most important. The other reason was already mentioned, viz. monogeneric names with a geographical adjective are replaced by bigeneric names, in which the first name comes from a species characteristic for the geographical region originally indicated. The example given above, the Agropyretum boreo-atlanticum, has been renamed in fact the Minuartio-Agropyretum (TÜXEN 1955).

In some cases the nomenclature is at variance with the standard procedure. The following deviations should be mentioned:
(1) Contractions. It has become general use to follow the proposal by KOCH (1925) to replace clumsy terms like 'Potamogetonion' (and, later on, also 'Potamogetonetalia' and 'Potamogetonetea') by 'Potamion', 'Potametalia' et 'Potametea'. However, OBERDORFER et al. (1967) proposed to return to the complete and cumbersome names.
(2) In special cases descriptive adjectives or substantives are in-

cluded in the name. 'Magnocaricion' indicates an alliance characterized by tall sedges; 'Macrophorbio-Alnetum' is meant to characterize a moist woodland (carr) association in which 'macrophorbiae' (tall forbs, Hochstauden) form a group of differentiating species.

(3) Finally, sometimes specific epithets have been used instead of generic ones. This was common practice in the early days of alpine phytosociology; BROCKMANN-JEROSCH (1907) and RÜBEL (1912), for example, used the names 'Curvuletum' and 'Firmetum' instead of 'Caricetum curvulae' and 'Caricetum firmae.' Later this practice has been applied only very rarely; a rather recent example is 'Alno-Padion' (KNAPP 1942, MATUSZKIEWICZ & BOROWIK 1957), to designate an alliance characterized by *Alnus glutinosa* and *Prunus padus*.

20.6.8.2 *Validity, Changes and Authors' Names*

Various rules have been proposed for the settlement of the validity of syntaxon names. A name should have been published after 1900 in printed form, available at least in generally accessible libraries. Publication of a name should be combined with an adequate diagnosis.

For associations and lower-rank units this would include the presentation of a table with the complete floristic composition of at least three relevés and the assignment of a type-relevé. For higher units the listing of character- and differential-taxa and the assignment of a type-unit of the next lower level would be required. Such types are called nomenclatural types.

When two or more correctly published names appear to refer to one and the same syntaxon, the oldest name has priority. When a syntaxon is divided into two or more units of the same rank, the original name goes to that new unit that contains the nomenclatural type.

When the valid name of a taxon used in a syntaxon name changes, corresponding change in the syntaxon name should be considered. In any case a syntaxon name should be changed when one of the name-giving taxa appears to be a younger homonym of a different legitimate name. For example, the Isoëtetum setacei should become the Isoëtetum delilei, since *Isoetes setacea* (BOX ex) DELILE appeared to be a homonym of *Isoetes setacea* LAM. = *I. echinospora* DUM. and was then renamed *I. delilei* ROTHMALER.

No agreement has been yet reached on the – still very frequent – simple nomenclatural changes. E.g. *Carex nigra* (L.) REICH-

ARD appears to be the correct name for *C. fusca* ALL. Should the order Caricetalia fuscae now be renamed the Caricetalia nigrae? MORAVEC (1968) would say not; WESTHOFF would say yes and in fact (WESTHOFF & HELD 1969) changed this name.

Rules for the citation of authors' names have also been proposed, more or less parallel to those of idiobotanical nomenclature. The name of the author(s) who published a valid name is added to that name together with the year of publication. When this name refers to a previously published but invalid name, the earlier author's name is added in the form, Thero-Airion TÜXEN ex OBERDORFER 1957. When a name is changed, e.g. when the syntaxonomic level is changed, the author of the changed name is added between brackets. When a syntaxon is given a new contents or delimitation, both old and new authors are mentioned connected with the abbreviation em. (= emendavit).

These proposals were advanced by MORAVEC (1968). PIGNATTI (1968b: 87) recommended dropping authors' names to avoid the description of new syntaxa by authors who primarily want to have their names immortalized. In the discussion following, p. 89—97, there was agreement on omitting author's names when no confusion could occur, especially with higher units and when a prodromus can be referred to.

20.6.9 SCHEME OF SYNTAXA

All in all phytosociological nomenclature may be considered a necessary evil. Contrary to the opinion of DOING (1962, 1966, 1970) that a rigid system of nomenclature should be abandoned for practical reasons, we would emphasize that the value of the system for ordering information and communicating among phytosociologists outweighs its difficulties. We would further state that descriptions of formal phytosociological syntaxa should either be published in strict accordance with nomenclatural rules, or they should not be published as such.

Table VII presents the current syntaxa and their nomenclature, as discussed so far. Examples together with their denominating taxa are added. The variant (and the facies) are given here without suffixes. The geographically determined syntaxa (20.6.3, 20.6.5), which have not yet been agreed upon, are not included.

TABLE VII

Scheme of Syntaxa

Levels and units of the formal hierarchy from highest (Division) to lowest subvariant, with suffixes and examples of the construction of names after denominating taxa.

Syntaxon	Suffix	Examples	Denominating taxa
Division	-ea	Querco-Fagea	genus Fagus
Class	-etea	Phragmitetea	Phragmites australis
		Querco-Fagetea silvaticae	Quercus robur Fagus sylvatica
Order	-etalia	Littorelletalia	Littorella uniflora
		Festuco-Sedetalia	Festuca and Sedum L. div. spp.
Alliance	-ion	Agropyro-Rumicion crispi	Agropyron repens, (syn.: Elytrigia repens) and Rumex crispus
		Alnion glutinosae	Alnus glutinosa
Suballiance	-ion (-esion)	Ulmion carpinifoliae (Ulmesion)	Ulmus carpinifolia
Association	-etum	Ericetum tetralicis	Erica tetralix
		Elymo-Ammophiletum	Elymus arenarius, Ammophila arenaria
Subassociation	-etosum	Arrhenatheretum elatioris brizetosum	Briza media
Variant		ibid., Salvia variant	Salvia pratensis
Subvariant		ibid., Bromus sub-variant	Bromus erectus

20.7 Extension of the Approach

20.7.1 OTHER GEOGRAPHIC AREAS

From its center of origin in the Alps and the western Mediterranean the BRAUN-BLANQUET approach spread into many European countries and Japan. Larger parts of the Mediterranean and Euro-Siberian, and smaller parts of the Sino-Japanese, floral regions have been synsystematically described. Still, a considerable part of the original working area is insufficiently known, particularly the eastern Mediterranean, Russia and Scandinavia (see the map presented by DIERSCHKE 1971). Expansion of the approach's influence continues, and we shall give a short survey of more recent literature applying the approach outside its European homeland. The survey cannot be exhaustive; only regions will be mentioned in which substantial work has been done. For further references the reader is referred to *Excerpta Botanica Sectio Sociologia* (bibliographic contributions to which are indicated below as 'Exc.'). This series can also serve as a survey of phytosociological work in the three regions mentioned above, with bibliographies for almost all European countries.

Japan may deserve special attention. The phytosociological study of Japan was initiated by SUZUKI/TOKIO and developed especially by MIYAWAKI and his pupils (see MIYAWAKI 1960 and 1966a, 1971a Exc.). Of special interest is the manner in which the Japanese communities are compared with Mediterranean and Indo-Malaysian communities. Other comparative studies in Japan include OHBA (1972), in which various vicariant salt marsh associations were described, partly belonging to the 'European' class Asteretea tripolii; TÜXEN, MIYAWAKI & FUJIWARA (1972) on the class Oxycocco-Sphagnetea in which Europe and North America (work of R. KNAPP and A. DAMMAN) are included; and TÜXEN (1966a) on holarctic Honckenyo-Elymetea communities. These studies were all based on extensive relevé tables and close cooperation between European and Japanese phytosociologists. From these studies the concept of vicariant class group developed (see TÜXEN 1970a and 20.6.5). Vegetation mapping on a phytosociological basis has reached a high level in Japan (see MIYAWAKI 1966b, 1971b, Exc. for a survey of maps).

The *arctic-subarctic* region has only occasionally been described. Iceland is becoming relatively well-known; a general survey is in preparation by R. TÜXEN based on phytosociological excursions by HADAČ, BÖTTCHER, the International Society for Plant Geography and Ecology, and others. Newfoundland was partly described by DAMMAN (1964). The research on Oxycocco-Sphagnetea by DAMMAN in Canada was already mentioned under Japan as was the study of TÜXEN (1966a). See also STEINDÓRSSON (1966, Exc.) for Iceland, BÖCHER (1961, Exc.) for Greenland, and HANSON (1959, Exc.) for Alaska.

The *west and central Asiatic* (Irano-Turanian) region has been described in a fragmentary way. An extensive survey of xerophytic and summer-dry hygro- and mesophilous communities of Afghanistan was presented by GILLI (1969, 1971), who could reasonably establish associations, though not as sharply delimited as in Europe because of the large number of species in most communities. HARTMANN (1968, 1972) published accounts on mountain grassland and scrub communities.

For the *Macaronesian* region the Canaries are relatively well-known through work of RIVAS-GODAY & ESTEVE CHUECA (1965), OBERDORFER (1965), LEMS (1968), and SUNDING (1969), as well as

LOHMEYER & TRAUTMANN (1970) who reviewed this literature (see also references in SUNDING 1969-70, Exc.). Studies on the Azores and the Cape Verdes were listed by PINTO DA SILVA & TELES (1962, Exc.). OBERDORFER (1970) reported on some plant communities on the Canaries and described complex, stable, mosaics of succulent scrubs and open sclerophyllous woods, which are amongst the most complicated communities of the world and should be studied as complexes (cf 20.7.4 and the South-African Kingdom below).

Rather numerous, though still scattered phytosociological studies have been carried out in the *North-American* and *Canadian* regions, starting with the description of Long Island by CONARD (1935) including 71 associations. CONARD (1952) also published on the vegetation of Iowa and DANSEREAU (1957) on the middle St. Lawrence Valley. TOMASELLI (1958, quoted in KÜCHLER 1967) described plant communities in eastern Kansas. Monographs are concentrated on forest communities (except for the studies of LOOMAN e.g. 1969 on grasslands). They include those of MEDWECKA-KORNAŚ (1961) in the Montreal area, GRANDTNER (1966) in Quebec, KORNAŚ (1965) in North Carolina, CRISTOFOLINI (1967) in Tennessee, JANSSEN (1967) in Minnesota and KRAJINA (1969) in British Columbia. KNAPP (1957, see also 1965) presented a preliminary survey of higher phytosociological units, viz. 72 classes, some of which were divided into orders and later (KNAPP 1964) extended it to the whole holarctic Kingdom.

The *Indo-Malaysian* and the *Polynesian* subkingdoms are practically phytosociological terra incognita. Some studies have been made in India (BHARUCHA & MEHER-HOMJI 1963, Exc., GUPTA 1966, 1967, Exc., MEHER-HOMJI 1969, Exc., MEHER-HOMJI & GUPTA 1972a,b, Exc.).

The *African* territory has been studied rather intensively, here and there. French studies on the North African desert region culminate in the monograph of QUÉZEL (1965), which presents a coherent survey of over 100 desert communities largely fitted into a syntaxonomic system (see also ROUSSINE & SAUVAGE 1961, Exc.). Relatively intensive studies have been performed in the West African rain forest region by LEBRUN (e.g. 1960), LEBRUN & GILBERT (1954) and SCHNELL (1952). The associations and higher units described are all based on relevés and range from xerophytic and hydrophytic pioneer communities to climax forests. LEBRUN concludes that the BRAUN-BLANQUET approach is very well applicable. Savanna studies in this area include those of DUVIGNEAUD (1949), ADJANOHOUN (1962), and DEVRED (1956). In Moçambique a first phytosociological inventory was carried out by GOMES PEDRO & GRANDVAUX BARBOSA (1955), see also PINTO DA

SILVA & TELES (1962, Exc.). KNAPP (1965, 1966a) presented preliminary surveys of higher units for West, Central and East Africa and reviewed the literature for some countries (KNAPP 1969—70, 1971a,b, Exc.).

Phytosociological research in *South Africa* started recently and is now expanding rapidly (EDWARDS 1967, TAYLOR 1969, ZINDEREN BAKKER 1971, WERGER 1973, see WERGER for references). WERGER et al. (1972) presented the first account of plant communities in the Cape Kingdom, including vegetation types of the famous 'Fijnbos' formation. Their experiences with the BRAUN-BLANQUET approach in this extremely varied mosaic of species-rich communities are comparable with that in European xerotherm woodland-scrub-borderline communities (cf. JAKUCS 1961, 1972).

The *Caribbean* region has also been treated in general phytosociological studies, e.g. KNAPP (1965). Detailed descriptions of savanna vegetation in Surinam have been published by DONSELAAR (1965, 1969), who found the BRAUN-BLANQUET approach applicable. The North Surinam open savannas could be grouped into one class. A special problem was formed by the floristic similarities between community types of deviating structure. DONSELAAR considered the floristic relations as decisive, as did WERGER et al. (1972) for similar situations in Fijnbos communities of the Cape. Earlier Surinam work is mentioned in BOERBOOM (1970, Exc.). *Brazilian* rain forest descriptions are very limited (VELOSO 1962 in BRAUN-BLANQUET 1964, DONSELAAR 1965). The *Andean* region has received more attention; OBERDORFER (1960), SCHMITHÜSEN (1960), ESKUCHE (1968, 1969) described various scrub and forest communities, and KOHLER (1968, 1970) desert and coastal communities. See also ESKUCHE (1967, Exc.) for this, and adjacent regions in South America. KNAPP(1966b) presented a survey of higher vegetation units in *Patagonia*.

The BRAUN-BLANQUET approach has thus been applied in nearly all floristic regions, and to almost all types of vegetation. The approach should in principle be universally applicable; and in fact no difficulties have been met in its expansion out of Europe more fundamental than those encountered in Europe. The principal directions of difficulty are: (i) In species-poor vegetation in the North and the Antarctic, character-taxa may be difficult to find and use, and species dominance (and the sociation as a unit) may be more emphasized (see article 18). (ii) Rich tropical vegetation confronts the approach with such a wealth of species as to make application more difficult; physiognomy has been more emphasized in the tropics as a means of recognizing phytocoena. (iii) Delimitation of uniform stands becomes difficult in mosaic structures of open xerothermic scrubs and woodlands, such as found in the Canaries and South Africa (cf. 20.4.1).

20.7.2 BIOTIC COMMUNITIES

Approaches to the classification of animal communities are diverse (WHITTAKER 1962); applications of the Braun-Blanquet approach have mostly been concentrated on the analysis of one or more animal groups in plant communities already described. General bibliographies were presented by RABELER (1957b, 1964, Exc.). More or less complete descriptions of biotic communities are restricted to marine benthic environments, in which animals and plants were used together to construct a hierarchy according to Braun-Blanquet principles (ROGER MOLINIER 1960).

The following conclusions may be drawn from work on biotic communities (cf. BRAUN-BLANQUET 1951, 1953, 1964, RABELER 1937, 1960, 1962, 1965, WESTHOFF & WESTHOFF-JONCHEERE 1942, MÖRZER BRUYNS 1950, WHITTAKER 1962, MAAREL 1965, TÜXEN 1965a and BARKMAN article 16 in this volume):
1. Species from most animal groups form characteristic combinations, often indicated as 'communities.' Mostly they are merely taxocoenoses (cf. MAAREL 1965).
2. The local distribution pattern of such groupings may coincide with that of the plant communities in which the groupings have been established, but mostly it is of a larger, sometimes of a smaller scale.
3. Most animal groupings are primarily bound to specific abiotic factors like soil surface moisture or to vegetation structural features (or, rather, to the micrometeorological features determined by them).
4. The plant community can be characterized by animal species groups, although the number of faithful animal species is low. As long as the synusial structure of animal communities is not sufficiently known, the phytocoenoses are the most realistic framework for biocoenotic studies. This implies that attempts like that of QUÉZEL & VERDIER (1953) to create a separate hierarchy of 'communities' of members of one – or a few – animal groups is unrealistic when these communities are in fact taxocoenoses.

20.7.3 MICROCOMMUNITIES AND SYNUSIAE

Since these community-types are treated by BARKMAN (article

16), we refer only to some current literature. The BRAUN-BLAN-QUET approach has been applied to various kinds of microcommunities and synusiae. Bibliographies have been composed for: epiphytic communities (BARKMAN 1962, Exc., 1966, Exc.), marine algal communities (HARTOG 1967 Exc.), epigeic moss and lichen communities (HÜBSCHMAN & TÜXEN 1964, Exc., TÜXEN 1964, Exc., 1968/1969, Exc.) and soil inhabiting communities (APINIS 1969, Exc.) Classification systems have been developed by HARTOG (1959) and BOUDOURESQUE (1971) for marine algal communities; HARTOG distinguished associations which were grouped into formations, BOUDOURESQUE presented a complete hierarchic system. Broad classification systems of moss and lichen communities have been presented by KLEMENT (1955), KOPPE (1955), BARKMAN (1958), WILMANNS (1962). From these studies one may conclude that the BRAUN-BLANQUET system is generally applicable to microcommunities and synusiae, and that it is advantageous to approach the latter through a classification separate from that for phytocoenoses (see also article 16.5 for references).

20.7.4 COMMUNITY COMPLEXES

A 'complex' is a set of contiguous or continuous communities forming a mosaic or pattern. Complexes can thus be studied on various levels from the microtopography of bogs to whole landscapes. Small-scale topographic complexes may often be studied without direct concern for successional relationships, even though the communities may have developmental relations to one another (as in bogs) or to the potential natural vegetation (SEIBERT 1968). In situations where man has drastically changed the vegetation pattern as well as the substrate or in geologically young systems such as coastal dunes it may be more realistic to approach complexes directly in their local mosaic or zonation structure. A difficult type of complex is the 'superposition complex' (Überlagerungskomplex) within which rapidly changing community patterns occur, e.g. with aquatic or with ephemeral communities (cf. MÜLLER 1970).

For communities related by successional process the concept of community ring (Gesellschaftsring) was mentioned by KRAUSE (1952) and developed by SCHWICKERATH (1954b) and SCHMITHÜSEN (1961). The ring comprises the series of communities that are syndynamically related to a terminal community. SEIBERT (1968) translated the concept as 'circle of communities', a term that unfortunately invites confusion with BRAUN-BLANQUET's (1932, 1964) 'circle of vegetation' or 'Vegetationskreis'. The English-language terms 'climax-complex' and 'successional complex' (see article 14.4.2) may well be preferred.

Study of broader topographic and edaphic complexes in relation to climax theory was initiated by TÜXEN. (For an English account of TÜXEN's views, see KÜCHLER 1967). TÜXEN & DIEMONT (1937, see also TÜXEN, 1933) developed the parallel concepts of 'climax swarm' and 'climax group' for spatially contiguous stable communities (Dauergesellschaften) that occur in the same climate, but differ in consequence of direction of exposure in mountains (swarm), or soil parent material in lowland areas (group). TÜXEN (1933) proposed also the term 'paraclimax' for a widespread stable community the characteristics of which are determined by soil development, and not climate alone. TÜXEN furthermore recognized a series of climax-regions for Europe, each characterized by a number of character-associations bound to it in their distributions, together with companion-associations (Begleitassoziationen) of wider distribution in more than one climax-region. The latter idea is a development beyond the climax-complex of BRAUN-BLANQUET (1928, 1964) and others; it characterizes regions not by a single, somewhat hypothetical climatic climax (Schlussgesellschaft), but by associations used as diagnostic-communities.

TÜXEN (1937) also recognized the coherence between the various substitute communities (Ersatzgesellschaften), both the completely cultural and the semi-natural ones, which may occur in the space that could be (and may actually partly be) occupied by a terminal community. Such locally coherent communities were consequently called 'contact communities.' The recognition of the coherence of substitute communities in their potential development towards one terminal community led TÜXEN (1956) to the concept of *potential natural vegetation*. This concept is defined in relation to a given habitat, as the vegetation that would finally develop (terminal community) if all human influences on the site and its immediate surroundings would stop at once, and if the terminal stage would be reached at once. With the latter restriction abiotical (e.g. macroclimatological) changes during this development are meant to be excluded. Although in many cases the potential natural vegetation may be identical with the original vegetation and interpretable as such, the distinction of this concept is useful for the following reasons: (i) There may have been no 'original' vegetation at all, since man has influenced the ecosystem under consideration at least as long as the present climate (and thus the present climax complex) lasts; (ii) Man may have induced irreversible changes in the ecosystem, so that the presumed 'original' vegetation can never establish again; (iii) reconstruction of the original vegetation may theoretically be possible in some cases, but precarious in practice.

Further discussion of climax interpretation is beyond our purposes here; we refer to WHITTAKER (1953), DANSEREAU (1957), SCHMITHÜSEN (1961) and BRAUN-BLANQUET (1964). We mention as significant, however, the phytosociological study of community complexes over wide geographic areas by KRAUSE (1952). KRAUSE observes the consistency of community composition over extensive areas together with the occurrence of vicariant communities, particularly between eastern and western Europe. The community-complex of a given area is shaped both by its macroclimate, and by local factors producing mosaic-like community relationships. Regions are characterized by different complexes, and different prevailing or dominant communities. Complexes give way to other complexes through geographic distance in three ways: (i) compression of communities into smaller areas and exclaves, (ii) increasing floristic impoverishment until a given community can be traced only in infrequent fragments, and (iii) the replacement of dominant communities by vicariants suited to other climates. For further discussion of vegetation areas and regions see SCHMITHÜSEN (1961), BRAUN-BLANQUET (1964), and SEIBERT (1968).

20.7.5 APPLIED PHYTOSOCIOLOGY

The floristic-sociological approach has been applied in many neighbouring disciplines. Early indications of this application were presented by BRAUN-BLANQUET (1930) and by TÜXEN (in BRAUN-BLANQUET & TÜXEN 1932) who observed several non-botanical sciences with which phytosociology might have mutual relationships: geology, hydrology, geomorphology, climatology, soil science, animal sociology (see 20.7.2), geography, archaeology, palaeobotany (particularly palynology), nature conservation and land use planning. Applications of phytosociology were often based on vegetation maps. Above all, under TÜXEN the Zentralstelle (later Bundesanstalt) für Vegetationskartierung (Federal Institute for Vegetation Mapping) at Stolzenau/Weser, West-Germany, developed vegetation mapping and its application in land interpretation and management and thereby contributed considerably to the acceptance of phytosociology as a significant applied, as well as basic science. Many studies at the Institute, models of such research, can be found in the series *Angewandte Pflanzensoziologie* (Stolzenau/Weser) and *Mitteilungen der Floristisch-Soziologische Arbeits-*

gemeinschaft (Neue Folge, Stolzenau). See also TÜXEN (1956) for a survey of maps, BRAUN-BLANQUET (1969) for a survey of TÜXEN's work, and KÜCHLER (1967) for an English account of floristic-sociological vegetation mapping.

Since applied phytosociology will be the subject of other volumes of the *Handbook,* we present here only a survey of main applied floristic-sociological literature and bibliographies. General publications on agriculture and forestry include textbooks by KNAPP (1949) and ELLENBERG (1950, 1952, 1954a) as well as the series *Angewandte Pflanzensoziologie* (Stolzenau) and the symposia on Vegetation Mapping, Anthropogenic Vegetation, Plant Sociology and Palynology, Landscape Ecology and Experimental Phytosociology of the International Society for Plant Geography and Ecology (TÜXEN 1963, 1966b, 1967c, 1968a, 1969b). Further general references are found in *Excerpta Botanica, Sectio Sociologica.*

An important general conclusion may be drawn from applied research: detailed floristic-sociological description of vegetation combined with adequate analysis of its environment provides a basis for applied phytosociology in many fields of research which can be of great social importance, at least in the cultural landscapes of Europe.

20.7.6 STRUCTURAL CONSIDERATIONS

BRAUN-BLANQUET established the association on a floristic-sociological basis, but since the floristic-sociological characters of an association are supposed to reflect all other characters a floristicsociologically uniform association might be expected to be structurally uniform as well. The original association concept thus implied the physiognomic uniformity of the association. As WESTHOFF (1967) showed, it is not always true that floristically consistent units are also structurally uniform. Many examples are known of considerable structural differences – even on the formation level – within associations or higher syntaxa (e.g. BARKMAN, 1958b). In a number of cases the discrepancy between floristic and structural uniformity has been solved by a refined syntaxonomic treatment.

An example is the Salicornieto-Spartinetum BRAUN-BLANQUET & DE LEEUW (1936) of NW-Atlantic mud flats. This association was considered a mosaic of patches of two different life forms, hence of two different structural types. New relevés taken separately from the two kinds of patches revealed them to be floristically different, which led to separated associations (cf. BEEFTINK 1962, 1965). These associations were in fact assigned to separate

classes, viz. Spartinetea and Thero-Salicornetea (e.g. BEEF-
TINK, 1962). A second example from a very different situation is the
classification of scrub and mantle communities separately from the
woodland classes to which they used to be assigned. TÜXEN (1952)
created for these the order Prunetalia spinosae which he later
(1962) placed in a separate class Rhamno-Prunetea (after
RIVAS GODAY & BORJA CARBONELL).

Besides attempts to use structural criteria in floristic-sociological classification various suggestions for an integration of the
latter system with physiognomic systems have been put forward.
A connexion between the syntaxonomical class and the physiognomic formation has been sought by DOING (1962) and PASSARGE
(1966, 1968, SCAMONI et al. 1965). PASSARGE (1966) adopted a
formation definition that was based on the original concept of
Grisebach (See BEARD, article 13), as well as a growth-form definition from SCHMITHÜSEN (1961): a phytosociological formation is a
sociological vegetation unit dominated by related growth-forms and
hence showing a uniform physiognomy. Informal arrangements of
classes within formations have been presented by ELLENBERG (1963)
and WESTHOFF (1967), WESTHOFF & HELD (1969). One could also
think of connexions between formations and either of the floristic-
sociological units above the class, viz. the division and the vicariant
class group.

20.7.7 INDICATOR GROUPS

One of the essential bases of applied phytosociology is the use
of indicator groups of species. It is consistent with – indeed it expresses – the perspective of the BRAUN-BLANQUET approach that
species are used, on the basis of their distributions, as indicators of
biotope characteristics and other factors, and that when possible
groups of species are thus used. Groups of species may often give
much more effective indication of environmental factors than dominant species, or other single species alone, or abundance relations that may be much affected by disturbance. Discussions of
indicator use of species have been given by, among others, Du-
VIGNEAUD (1946), ELLENBERG (1950), WHITTAKER (1954a) and
TÜXEN (1970b). We may consider indicator species in four contexts –
in connexion with synsystematic, biotope or ecological, successional,
and geographic indication. In each of these, but particularly in the
biotope-ecological function, it is possible to use indicator species
in a perspective of either classification or gradient analysis.

20.7.7.1 *Indicators for Classification*

One direction of development – the synsystematic use of groups of species as indicators for syntaxa – is implicit in the BRAUN-BLANQUET approach. The concept was developed further, however, with the distinction of differentiating taxon groups (SCHWICKERATH 1942, 1954a, DUVIGNEAUD 1946), and the idea of sociological groups (SCAMONI & PASSARGE 1959, 1963, DOING 1962, 1969b). The latter are established as groups of species that are distributionally related, and that consequently occur in and characterize, particular syntaxa; the idea is related to that of 'commodal' species (WHITTAKER 1956). SCAMONI et al. (1965) constructed a sort of 'sociological profile' by arranging relevés for phytocoena and determining the distribution of sociological groups over the units compared. Treatment in terms of syntaxonomical groups has been developed by SEGAL & WESTHOFF (1959) and MAAREL (1969 et seq.).

Clearly, groups of species should indicate characteristics of biotopes or habitats, as well as the particular phytocoena that occur in those biotopes. The concept of ecological groups — sets of species that, because of their similar distributional response to environmental factors, tend to occur together and to indicate properties of biotopes — may be traced from DUVIGNEAUD (1946, 1949). Ecological groups have had most extensive application by ELLENBERG (1950, 1952, 1956), who relates them both to environmental factors and to the syntaxa of the BRAUN-BLANQUET approach. By describing species responses to various important factors — such as pH, lime, moisture, nitrogen — one may arrive at integrated ecological characterizations of species (ELLENBERG 1956, compare article 5.4.1.) Ecological groups have been used also by investigators of the Centre d'Études Phytosociologiques et Écologiques at Montpellier (GOUNOT 1969, GUILLERM 1971, DAGET et al. 1972).

A third application is recognition of geographic indicator groups that express geographic-historic influences and may characterize vicariant phytocoena. SCHWICKERATH (1942) and OBERDORFER (1957) have used geographic groups in dealing with vicariant associations. Given classification of species into geographically defined groups, or areal types, it is possible to characterize communities by geographic spectra, based on representation of these groups. The geographic perspective has been developed by MEUSEL et al. and by MØLHOLM HANSEN and BÖCHER in the Danish school (see article 18.5.3-4), and in the United States WHITTAKER (e.g. 1954b, 1960) has compared areal-type spectra of communities.

20.7.7.2 *Indicators for Gradient Relations*

Community spectra may intergrade, perhaps continuously, along major axes of biotope characteristics, geography and climate, and succession. Spectra (per cent representation of different species groups) may consequently be used to indicate position along these axes. There seems to have been little application of indicator groups to the study of succession, though the recognition of 'decreaser, increaser, and invader' groups in response to grazing disturbance (article 4.2.1) should be noted. Study of gradients of geographic spectra also has been limited, though ELLENBERG (1950) deals with climatic indicator groups, and article 3 (Fig. 8) illustrates a gradient in geographic spectra. A third possibility, use of representation of ecological groups to indicate position along environmental gradients, has had such extensive development as to be beyond review here. Use of ecological groups in direct ordination is discussed in articles 2—5 and 8; use in indirect ordination in article 9. The most extensive application in phytosociology is that of ELLENBERG (1950, 1952, 1956). ELLENBERG classified European species (weeds in agricultural fields, and plants of meadows and pastures) into ecological groups on the basis of their responses to various environmental factors. Weighted averages (see 2.2.3.1) for samples expressed their positions in relation to these factors. At the same time the relation of the species to syntaxa was known, and it was consequently possible to interpret the relations of the syntaxa to one another and environment. The technique thus created a multi-dimensional direct ordination, and belongs also to section 20.9.1.

20.8 Numerical Techniques

There remain to be discussed two (closely related) recent developments in phytosociology - numerical techniques and ordination. For review of similarity measurements and approaches to numerical classification we refer the reader to GOODALL's articles (6 and 19) in this volume. We discuss briefly, however, work in this direction employing the BRAUN-BLANQUET approach.

20.8.1 STORAGE OF RELEVÉS

Although the numbers of relevés treated in syntaxonomical studies may be very high - often several thousands - the number of storage systems is low. Most individual phytosociologists and their institutes work with an archive of relevé note books and tables. Such is the case at syntaxonomic centres like the Station Internationale de Géobotanique Méditerranéene et Alpine (S.I.G.M.A.) at Montpellier and the former Bundesanstalt für Vegetationskartierung at Stolzenau (data from the latter are now at TÜXEN's Arbeitsstelle für theoretische und angewandte Pflanzensoziologie at Todenmann über Rinteln, West Germany). Some archives have been set up with visual punched cards. The only such archive of some size seems to be that of the Geobotanical Institute Rübel at Zürich including about 5500 forest relevés from Switzerland (ELLENBERG 1968).

Besides incidental use of machine punched cards in the application of computer techniques, two archive systems based on computer hardware are known to us: the system of the C.E.P.E., Centre d'Études Phytosociologiques et Écologiques at Montpellier and the system of the Working Group for Data-Processing of the International Society for Plant Geography and Ecology. The C.E.P.E. system (see EMBERGER 1968) has the following characteristics: (i) The basic unit is the relevé, which is defined as the whole of ecological and phytosociological observations at a particular site. (ii) General vegetation structure, situation of the site, general climate and soil are described by means of measurements or estimations of numerous factors. (iii) For each factor a code is devised with, in most cases, 10 classes. (iv) All coded determinations of general features are punched in three 'parent' relevé cards, 'Cartes-Maîtresse-Relevé' (CMR) of which columns 1—71 (except for CMR 3) are used. (v) Each taxon from the relevé is treated separately for each vegetation layer in which it has been observed. Coded determinations include vegetation layer, abundance, and dominance (in separate 0—9 codes), pattern within the site, phenological state, vitality and life-form. The corresponding punched cards are called 'detailed cards', Cartes Détail (CD) and used only from column 72 onwards. (vi) Species are enumerated from 0001—4779 according to the Quatre flores de France by FOURNIER (1961). Cryptogams are provisionally enumerated from 5000 onwards. (vii) Data per species per layer are combined with CMR data, which results in three combined cards.

In this way a relatively large archive arises, which has the advantage of being rapidly accessible to all kinds of spectrum and correlation calculations.

The 'Working Group' was established in 1969 during the International Symposium at Rinteln after a proposal by S. PIGNATTI, G. CRISTOFOLINI and D. LAUSI (Trieste). One major aim (see MAAREL 1971) was the treatment of relevés in such a way that the data could be readily stored and retrieved. An outline of the system has been presented by CRISTOFOLINI et al. (1969); its main characteristics are: (i) The basic unit is a full relevé table, either published or in manuscript form. (ii) Rows of relevé tables are punched – from column 21 onwards. Cover-abundance values per species are coded either as direct BRAUN-BLANQUET symbols or in a 0—9 transformation code (Table II). (iii) In the first 20 columns coded data on publication, table number in publication, and taxon number are punched. (iv) Taxa are coded with 7 digits, 4 for the genus, 3 for the species – as yet no space is left for infraspecific taxa. The genus enumeration is according to the world survey *Genera Siphonogamarum* by DALLA TORRE & HARMS (1900—1907, reprint 1963). Species enumeration is as yet restricted to European species (Russia not included), according to *Flora Europaea* as far as possible, otherwise according to a provisional enumeration based on six large standard floras of Europe. Additional provisional codes for cryptogams are being developed. (v) A separate punched card is used for bibliographical data on the paper (or set of unpublished tables). (vi) A separate set of punched cards is used for general data on the site and its situation, for each relevé in each table. This part of the system is still in development. The Montpellier system, which is already elaborated in this respect, could be taken as a starting-point.

The Working Group decided to concentrate activities on salt marsh communities. Some 3400 relevés (including 576 Spartinetum relevés) have been stored in punched card decks and on magnetic tape.

20.8.2 SPECIES CORRELATION

The study of species correlation was introduced by VRIES (e.g. VRIES et al. 1954, DAMMAN & VRIES 1954). The plexus of grassland species he presented (article 7, Fig. 3, p. 169) was taken over in the textbooks of ELLENBERG (1956) and BRAUN-BLANQUET (1964), who underlined VRIES' conclusion 'that the resulting species groups broadly coincide with the associations of the Zürich-Montpellier school.'

Measurement of species distributional similarities may be based on either binary or quantitative data. Distributional similarity based on binary (presence and absence) data may be termed 'species association,' that based on quantitative representation in relevés, may be termed 'species correlation' (see article 6). In the BRAUN-BLANQUET approach the quantitative weightings may be either the abundance-cover scale (20.4.5.1) in individual relevés (or a transformation of that scale), or presence degrees (20.5.3) in different phytocoena. Species association seems to be more appropriate for the detection of species groups in relevés representing an extensive range of community difference, whilst correlation studies are more appropriate for intensive study within the range of broadly overlapping species (cf. GREIG-SMITH 1964). The use of species correlation in the establishment of sociological groups is obvious, but phytosociologists using such groups have hardly applied this technique. The studies of HEGG (1965; see article 7, Fig. 7, p. 174) and STOCKINGER & HOLZNER (1972) may, however, be mentioned as examples (see also article 7).

The detection of ecological groups has been particularly developed by phytosociologists using factor analysis and principal component analysis. DAGNELIE (1960, see also article 9) described the use of factor analysis for the establishment of sociological groups, and the use of joint species-environmental factor analysis for the establishment of ecological groups. GODRON (1966) used species correlations for the checking of 'imbricating' ecological groups; GUILLERM (1971) and DAGET et al. (1972) constructed ecological profiles of species against environmental factors. LACOSTE & ROUX (1971) and ROMANE (1972) used 'factor analysis of correspondences' in which floristic and environmental variables were analysed in combination to arrive at ecological-sociological groups.

Another such method was developed by FRESCO (1971) as part of his 'compound analysis.' He constructed overlapping species groups based on similarities between species with respect to their loadings on eigenvalues extracted in subsequent factor analyses. This method could be applied to very large data sets, e.g. the first selection of salt marsh data (1296 relevés, 245 species). From this salt marsh treatment (with an arbitrary similarity level) 65 such groups were derived which could be interpreted roughly as alliance character-species groups.

20.8.3 FIDELITY TESTS

Although it seems obvious to use tests on the significance of the exclusiveness or differentiating value of species (or species groups) for a given vegetation type, phytosociologists have hardly done so. MEIJER DREES (1949) suggested the use of the t-test for the significance of the difference between the group amounts of species groups in different relevés of a set. In his example the relevés were assigned to a particular association and the two species-groups examined were syntaxonomic groups for two orders; the problem was the assignment of the association to one or the other order. For such situations GOODALL (1953b) used FISHER's discriminant function, which is more powerful but also more laborious.

GOODALL (1953b) also proposed an index of fidelity for a species based on the ratio of constancy values for that species in two communities, which could be tested with the χ^2 test or by exact calculation of P. In addition he suggested the use of this index as an 'indicator value' for the comparison of the constancy of a species in one community with that in various other communities — GOODALL spoke of 'all other communities in the area.' In usual phytosociological terms GOODALL's indicator value is the real fidelity index and the application of this measure to the two-community comparison would be an 'index of differentiation.' Such tests may have limited value, particularly because of the non-random selection of relevés from which they are derived (GOODALL 1953b). Still, their use may be experimented with for syntaxon diagnoses from tabular summaries.

20.8.4 RELEVÉ SIMILARITY

Measurements of similarity between relevés have long been accepted as one basis of grouping these into phytocoena (BRAUN-BLANQUET, 1928, 1932). Various coefficients of similarity and dissimilarity are in use in phytosociology, including the formulas of JACCARD and SØRENSEN (article 6, formulas 39 and 40). Recently an anonymous 'similarity ratio,' which was introduced by WISHART (1969) in CLUSTAN, a set of classification programmes, was used by KORTEKAAS & MAAREL (1972). This formula reads

$$\sum_i x_i y_i / \sum_i x_i^2 + \sum_i y_i^2 - \sum_i x_i y_i,$$

and is thus a generalisation of JACCARD's formula. Comparatively little use has been made of direct dissimilarity measures (GROENEWOUD 1965, MAAREL 1966b). Similar formulas, most of them derived from SØRENSEN's formula, are in use for the comparison of sets of relevés or the calculation of average similarity within a set (cf. SØRENSEN 1948, RAABE 1952, ČEŠKA 1966, 1968, FRYDMAN & WHITTAKER, 1968, MAAREL, 1969). Weighting of species may use presence values or a combination of presence with an importance value. ELLENBERG (1956) as well as PIGNATTI & MENGARDA, see PIGNATTI 1964, suggested the use of a similarity coefficient in the assignment of a relevé to a type by comparing it with a 'standard' or average relevé, e.g. one consisting only of the normal characteristic species combination. HOFMANN & PASSARGE (1964) and SCAMONI et al. (1965) determined similarity between relevés and sets of relevés with reference to the sociological groups ('group affinity').

20.8.5 MEASUREMENT OF HOMOTONEITY

Homotoneity we have defined as the homogeneity of a community table, the relative consistency of the relevés the table comprises (20.5.3). In addition to the simple measures mentioned in 20.5.3, the mean similarity coefficient for the relevés, as calculated in a direct way by ČEŠKA (1966), is a convenient measure, particularly in classifications based on relevé similarity. Either the average of all interrelevé similarity values within a table, or the average relevé similarity with the average for the table (i.e. the centroid of the relevé cluster) may be used. See also article 6 and 20.8.4 and 20.8.8 on similarity measures.

Alternative techniques based on species relationships include the use of positive interspecific correlation by GOODALL (1953a), association analysis (WILLIAMS & LAMBERT 1959), and information analysis (WILLIAMS et al. 1966) (see article 19.5.1.5 and 19.5.2.5). MAAREL (1966b) derived an information measure that is related to that of WILLIAMS et al. and appeared to be a useful heterogeneity coefficient in a study of a local grassland community complex. GODRON (1966) developed an information measure, based on BRILLOUIN's information formula, for determination of the heterogeneity within a series of contiguous quadrats in the field. The frequencies of the species occurring in the set determine the value of this measure. It could also be used for heterotoneity in sets of relevés and in that case is rather similar to the formula of WILLIAMS et al. (1966).

DAHL (1957, 1960) derived an index of uniformity from the FISHER model (logarithmic series) for species-individual and species-area relations ($S = a \log_e (1+N/a)$, S is number of species, N number of individuals, and a a diversity index, see WILLIAMS 1964). DAHL considered the species-relevé relation to be similar and defined the diversity parameter for a table as $(S_m - \bar{S})/a \log M$ (see symbols and formulas in 20.5.3). The index of uniformity is accordingly defined as $\bar{S}/a$. A clear relationship with CURTIS' (1959) index was shown by DAHL (1957).

20.8.6 NUMERICAL CLASSIFICATION

Of the various numerical classification techniques which are described in article 19, only a few are in use in quantitative phytosociology. Direct use of similarity matrices was developed by Polish phytosociologists (MATUSZKIEWICZ 1948, FALIŃSKI 1960) and by GUINOCHET (1955, GUINOCHET & CASAL 1957), who spoke of the 'differential analysis of CZEKANOWSKI' (after the Polish originator of the approach). See further article 7.3.1.

Agglomerative clustering techniques (19.5.2) based on similarity coefficients like the JACCARD and SØRENSEN indices were used by SØRENSEN (1948), LOOMAN & CAMPBELL (1960), MAAREL (1966b), MOORE & O'SULLIVAN (1970), MOORE et al. (1970) and KORTEKAAS & MAAREL (1972). Agglomerative techniques can be profitably combined with re-allocation techniques (cf. WISHART, 1969). The relevé-groups arrived at in these classifications could generally be typified with characteristic species combinations and connected with syntaxonomical units.

IVIMEY-COOK & PROCTOR (1966) applied nodal analysis (association analysis followed by an inverse species clustering) to salt marsh, fen and woodland data. They obtained tables with clear blocks of relevés and species and concluded that the results confirmed the arrangement of the data arrived at by traditional phytosociological methods. The technique was considered particularly useful for the detection of differentiating species. Moreover they found minor lines of division which had not been obvious in previous studies.

20.8.7 TABLE REARRANGEMENT

A successful nodal analysis should produce an ordered table. A number of special techniques for table rearrangement have been developed, from the early work of BENNINGHOFF & SOUTHWORTH (1964) to the more recent of MOORE (1972, MOORE et al. 1970, MOORE & O'SULLIVAN 1970). The following computer programmes are known to us:

i) BENNINGHOFF & SOUTHWORTH (1964), G. W. MOORE et al. (1967), LIETH & MOORE (1971). This programme finds species clusters by application of LIN's algorithm for the 'travelling salesman problem.' It is applied to species with intermediate presence in the table. Additionally, relevés are sorted. Final table ordering is by hand.

ii) J.J. MOORE (1971, last version). This programme 'PHYTO' rearranges both species and relevés. It finds pairs of species with intermediate presence values and high co-occurrence values. Opposing pairs are used as differentiators. Further arrangements are

361

dictated by the user, either through suspected mutually exclusive species or through preferred order of species and relevés.

iii) SPATZ (1969, 1972 last version). This programme forms groups of relevés on a similarity basis. Differentiating species are defined as species with presence of $> 50\%$ in one group and $< 10\%$ in all others. Output includes a synoptic table with synthetic figures.

iv) SCHMID & KÜHN (1970). This programme calculates D^2 values and forms groups on lowest D^2 basis. Differentiated tables can be dictated.

v) ČEŠKA & ROEMER (1971). This programme finds species-relevé groups through iteration. Diagnostic species are defined as in programme 3 (values vary from 50 vs. 10 to 66 vs. 33). A relevé is considered a member of a group if it contains at least 50 % (or 66 %) of the diagnostic species of that group. Groups are arranged in order of size and within-group similarity. Final arrangement is usually by hand.

vi) JANSSEN & MAAREL (1972, JANSSEN 1972). This programme is based on programmes HIERAR and RELOC of the CLUSTAN set (WISHART 1969) with a choice of 10 similarity coefficients. It starts with an allocation of relevés to groups, either at random or as determined by previous classifications and phytosociological experience. Group arrangement is based on group properties, particularly the relation between relevé number and species number, or alternatively on position along the first component of a principal components analysis of groups. Species are arranged so as to form a diagonal structure in the table. Significance of positive differentiating value of species in one or more groups is determined by a χ^2 test. Final arrangement is usually through MOORE'S PHYTO-programme.

vii) STOCKINGER & HOLZNER (1972). This programme finds species groupings through similarity analysis. Relevés are sorted according to occurrence of species groups.

It follows from this description that all programmes require the personal finishing touch of the user; this is felt to be not an inconsistency but rather a matter of efficiency. Most designers claim a considerable gain in accuracy, since re-writing tables by hand is no longer necessary. The speed of the programmes will depend on type of computer and size of the table. Programmes i, ii, v, and vii are most adapted to tables with comparatively many relevés and few species, programmes iii and iv to tables of the reverse form. Programme vi has a larger capacity, but requires preliminary experience to be fully efficient.

20.8.8 NUMERICAL SYNTAXONOMY

Numerical syntaxonomy has been mainly restricted to the characterizing of various syntaxon levels by average similarity. Various coefficients of similarity and dissimilarity are in use in phytosociology. 'Coefficients of community' comparing samples by the per cents of their species shared have been used most, with about equal interest in the formulas of JACCARD and SØRENSEN (article 6, formulas 39 and 40). Formulas using importance values for species (cover-abundance in relevés, especially) include the 'percentage similarities,' notably the CZEKANOWSKI index (article 6, formula 50). Similar formulas are in use for comparing sets of relevés (representing different phytocoena), or calculating average similarity within sets (see SØRENSEN 1948, RAABE 1952, ČEŠKA 1966, 1968, FRYDMAN & WHITTAKER 1968, MAAREL 1969, 1972a). Comparison of sets of relevés may use presence per cents, or mean cover-abundance values, for the species in these sets. Numerical classification is further discussed in article 19.

SØRENSEN (1948) concluded that grassland groupings with similarity levels of 0.40 roughly corresponded with syntaxa on the alliance level. ELLENBERG (1956) stated that the average similarity of relevés within one association is between 0.25 and 0.50, whereas subunits may be distinguished by levels above 0.50 (JACCARD values, which are 10% lower than SØRENSEN values on the average). LOOMAN & CAMPBELL (1960) calculated SØRENSEN values of > 0.70 within subunits belonging to one grassland association, whereas values between subunits were all < 0.50. HOFMANN & PASSARGE (1964) presented group affinity values between various woodland associations and subassociations. Within associations values were mostly > 0.60, between typical subassociations of related associations values were mostly between 0.30 and 0.50. RAABE (1952) obtained affinity values (KULCZIŃSKI coefficient) between associations and alliances of weed, salt marsh and alpine communities. Within-alliance values were between 0.40 and 0.50, between-alliance values were 0.20 to 0.40. Within salt marsh associations values varied more widely, from 0.30 to 0.80, and here the influence of geographical distance between local representatives of associations was evident.

NEUHÄUSL & NEUHÄUSLOVÁ-NOVOTNÁ (1972) studied within- and between- group similarities (SØRENSEN values) of many woodland associations of the alliance Carpinion. Within-group similarities of locally established associations and lower units were generally over 0.55. When larger areas were involved, these values were lower but still over 0.45. Similarities between associations and between lower units were always lower, mostly 0.10 to 0.20. These calculations made possible some improvement in the delimitation of Carpinion associations (which, as the authors stated, must be checked by renewed phytosociological table studies). KORTEKAAS & MAAREL (1972) carried out a numerical analysis of European *Spartina* communities involving an agglomerative classification with relocation of 576 relevés. They arrived at four separate dendrograms each comprising relevés dominated by one of the *Spartina* species. On various levels of the hierarchy groups were inspected for character-species and compared with the existing syntaxonomy (BEEFTINK & GÉHU, in prep). For these communities values of the similarity indices between 0.40 and 0.60 could be connected with the association level, 0.61—0.70 with the subassociation level and 0.71—0.80 with the variant level. Figure 4 presents the dendrogram for the *Spartina maritima* relevés. Each 'subassociation' had exactly one 'good' differential-taxon. The examples suggest that in related communities a reasonable parallel between syntaxonomical level and similarity level can be established. Furthermore, the

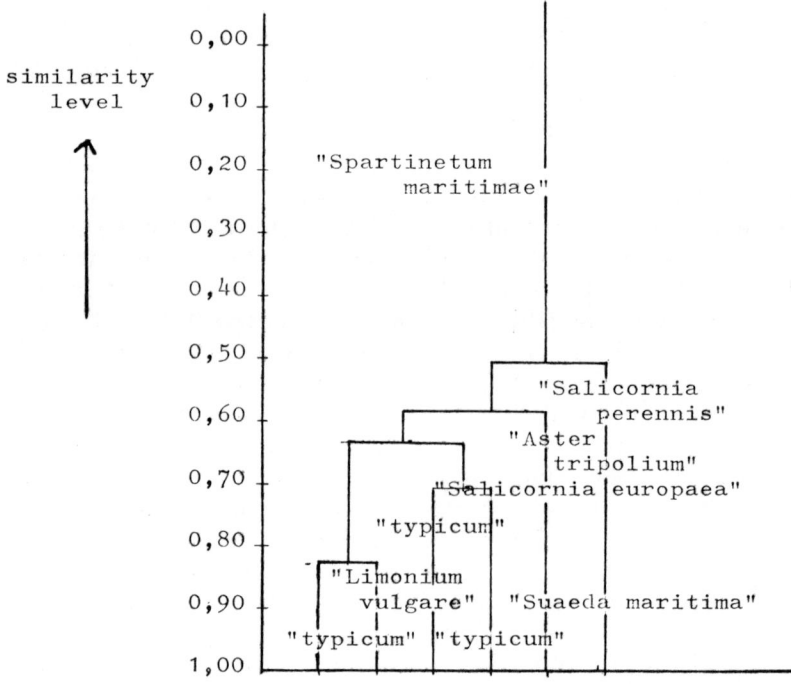

Fig. 4. Dendrogram of Spartinetum maritimae relevés. (KORTEKAAS & MAAREL, 1972).

similarity values for a given syntaxonomic level in different vegetation types are similar, suggesting that various authors have made comparable intuitive judgments of relative similarity for various kinds of communities. Similarity measurements should thus be thought a useful aid to synsystematics.

20.9 Ordination

'Ordination' refers to the arrangement of entities (generally samples, or species) in a uni- or a multidimensional order (RAMENSKY 1930, GOODALL 1954, articles 1 and 2.) Ordination is often considered to stem from the concept of vegetation as a continuum, whereas the floristic-sociological classification may seem to assume discontinuity. It may consequently be argued whether ordination is appropriate in phytosociology. Even if we assume for the moment that associations are generally discontinuous with one another, there is no reason ordination cannot be used as an aid to understanding the relations of relevés and lower syntaxa to one another and environment, within the association. On a different level, ordination and the perspective of gradient analysis can well be applied to the relations of associations and higher syntaxa as wholes, to one another and environment. We feel consequently that ordination is a fully acceptable supplementary approach in phytosociology.

There has been sòme dispute, mainly between BRAUN-BLANQUET (1939, 1951b, 1955) and GAMS (1918, 1941, 1954) on the question 'linear or multidimensional system in plant sociology.' GAMS emphasized the multidimensional relationships of plant communities, without indicating the way of constructing an appropriate system, as BRAUN-BLANQUET remarked. WAGNER (1954, 1968) observed that many syntaxa show multidimensional relationships to various syntaxa of the next higher level. The 'flexibility' ELLENBERG (1954b) demanded at lower syntaxonomic levels could be thought of in this perspective. Surely it is true both that linear arrangements into hierarchies are possible and valuable, and that study of multidirectional relationships may be rewarding.

20.9.1 INFORMAL ORDINATION

A number of approaches that represent ordination but do not employ similarity measurements have been applied in phytosociology. In many ordered phytocoenon tables a suspected underlying environmental gradient is chosen as a basis for ordering that may be completed by inspection of ecological groups (MOORE et al. 1970). Often a moisture gradient is chosen and a 'dry' a 'typical' and a 'moist' subassociation are arranged within an association table. The unidimensional approach known as compositional ordination (article 2) has a phytosociological precursor in the approach of SCHWICKERATH (1931 et seq.), in which relevés of one association were arranged on the basis of representation of diagnostic species groups. The development of weighted-average ordinations by ELLENBERG (1950, 1952) has been referred to (20.7.7.2).

More abstract two- or many-dimensional schemes including Russian examples (discussed in article 5) were recommended by GAMS (1941, 1961). The approach of DUVIGNEAUD (1946) was an early floristic-sociological ordination without numerical basis. A number of phytosociologists have presented patterns ('mosaic charts', 2.3.3) of phytocoena in relation to environmental gradients (see DUVIGNEAUD 1946, WAGNER 1954, 1968, ELLENBERG 1952, Fig. 5, 1963, ZONNEVELD 1960, HEGG 1965, and article 18, Fig. 1, p. 555).

20.9.2 FORMAL ORDINATION

The first ordination of relevés on the basis of their similarities is found in the Polish approach of MATUSZKIEWICZ & TRACZYK (1958), and FALIŃSKI (1960). They constructed 'dendrites' as arrangements that can be ecologically interpreted (cf. GAMS 1961, 1967, and article 7).

Ordination in the sense of axis construction was first applied in phytosociology by DAGNELIE (1960) on beechwoods (see article 9). A comparable factor analysis on heathlands was presented by FRESCO (1969). GROENEWOUD (1965), MOORE et al. (1970) and MAAREL (1972a) applied principal component analysis to similarity matrices. The Wisconsin polar ordination of BRAY & CURTIS (1957), or slightly modified forms, was applied in Dutch work (unpublished MSc theses, Universities of Utrecht and Nijmegen, MAAREL 1966b, 1969, MAAREL & LEERTOUWER 1967, LONDO 1971) and further by ROGERS (1970). LACOSTE & ROUX (1971) applied factor analysis of correspondences (see 20.8.2) to relevés assigned to various subalpine associations and subassociations.

Ordination can be applied also to phytocoena (or syntaxa), with each phytocoenon treated as a composite sample with its species composition summarized as presence per cents or mean importance values. One may then conceive of environmental gradients

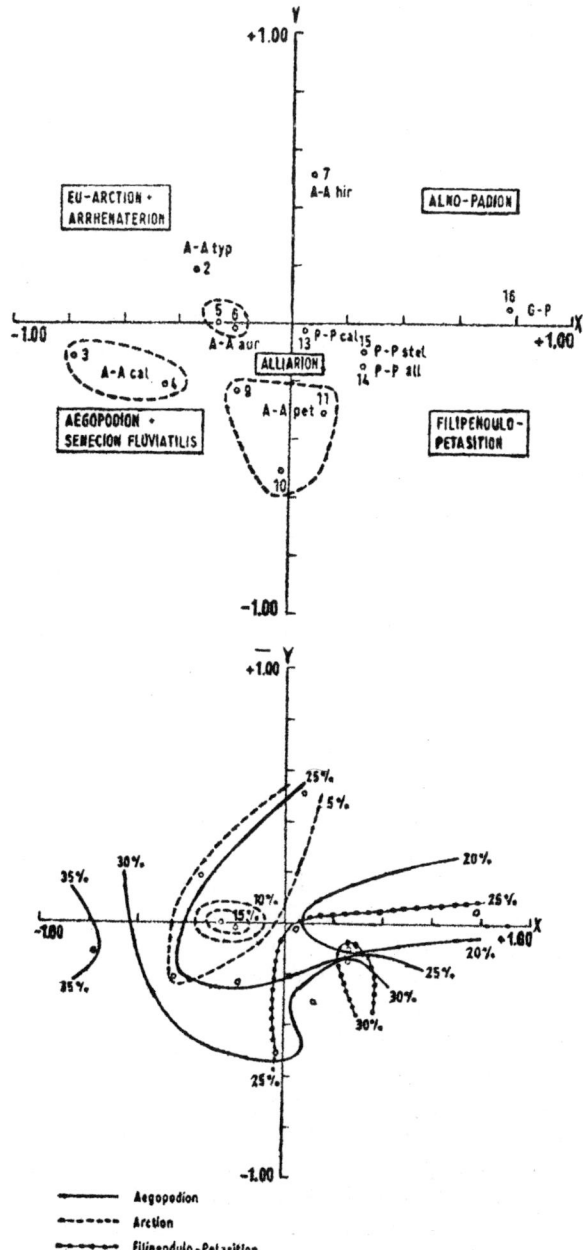

Fig. 5. Ordination of nitrophilous edge communities described by Tüxen (1967a), from Maarel (1969).

relating these phytocoena as a multidimensional habitat space (GOODALL 1963, WHITTAKER 1967) in which the centroids of the phytocoena (as clusters of relevés) are ordinated. ORLOCI (1966) carried out such an ordination by principal component analysis; GITTINS (1965), CRAWFORD & WISHART (1966, 1967), FRYDMAN & WHITTAKER (1968), MAAREL (1969), and WHITTAKER (1972) applied polar ordination to community-types. Mostly the similarities between types were measured by presence or constancy values. All the resulting ordinations were effective and easily interpretable. An effective approach to interpretation is the plotting of contour lines for biotope measurements (e.g. FRYDMAN & WHITTAKER 1968), to determine the relationship between axes and environmental factors, and the plotting of community-types and representation of diagnostic groups (FRYDMAN & WHITTAKER 1968, MAAREL 1969, WHITTAKER 1972) to show the relation of these to environment and one another. Fig. 5 illustrates the latter technique (see further articles 2, 8, and 9).

When no appropriate environmental data are available it is possible to interpret the ordination by plotting the distribution of syntaxonomical groups. Figure 5 presents an example. Nitrophilous edge communities, bordering woodlands and scrubs, as described by TÜXEN (1967a) are ordinated and the ordination space is characterized by various alliances. The associations are reasonably separated in the ordination space, except for the Agropyro repentis-Aegopodietum. The subassociation with *Chaerophyllum hirsutum*, in particular, is distinct and has a considerable representation of Alno-Padion species. Patterns of three alliances are presented. Lines with equal percentage amounts of one alliance are called 'isocenes' and outlined. The greatest difference in the set of types is between the relatively dry, open nitrophilous Agropyro repentis-Aegopodietum calystegietosum sepium from anthropogenic edges at low altitudes, and the relatively moist and shaded Geranio phaei-Petasitetum along streamlets in the montane region. WHITTAKER (1972) and FRYDMAN & WHITTAKER (1968) presented a similar approach with similar results. Besides species populations, species groups, and environmental measures, various spectra can be plotted in an ordination field. RAUNKIAER life-forms, for example, showed clear distribution patterns when thus plotted by FRYDMAN & WHITTAKER (1968).

Other uses of ordination include applications to succession (MAAREL 1969, LONDO 1971) and ordinations based not on species composition but on structural characters, life-forms or sociological (syntaxonomic) groups (KNIGHT 1965, GOFF & COTTAM 1967, MAAREL 1969, WHITTAKER 1972). MAAREL (1972a) used

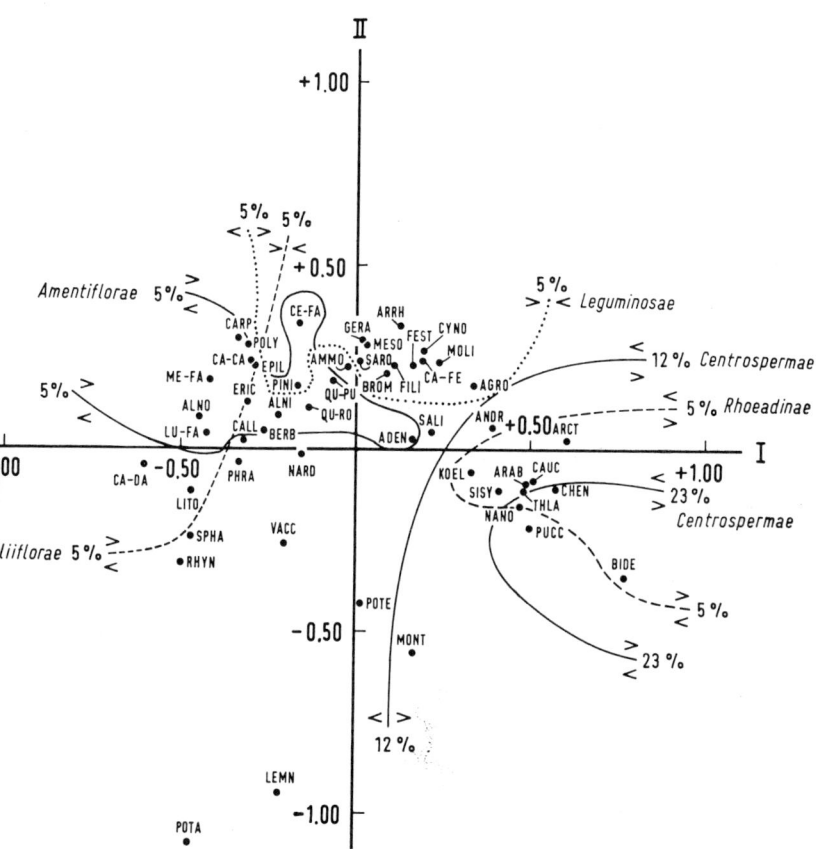

Fig. 6a. Ordination of 51 European associations and lower syntaxa on the basis of their plant order spectra, dimensions 1 and 2. (MAAREL, 1972a).

higher idiotaxonomical units as attributes to compare and ordinate vegetation types. Figure 6 presents some results of an ordination of 51 associations and subassociations belonging to 49 alliances, and thus covering almost the entire variation in central and western Europe. The corresponding similarity matrix was based on representations of plant orders in synoptic tables, mainly from OBERDORFER (1957). The first line of variation shows the sociological progression from pioneer comminities, with alliances such as the Bidention (BIDE) and Nanocyperion (NANO) on the right, to mature communities like the Fagion (—FA) and Carpinion (CARP) on the left. Most grasslands, marshes, and heathlands are intermediate. This variation is illustrated by isolines of various idiotaxonomic orders, e.g. Centrospermae vs. Liliiflorae. The second axis involves a moisture gradient from wet (Eu-Potamion,

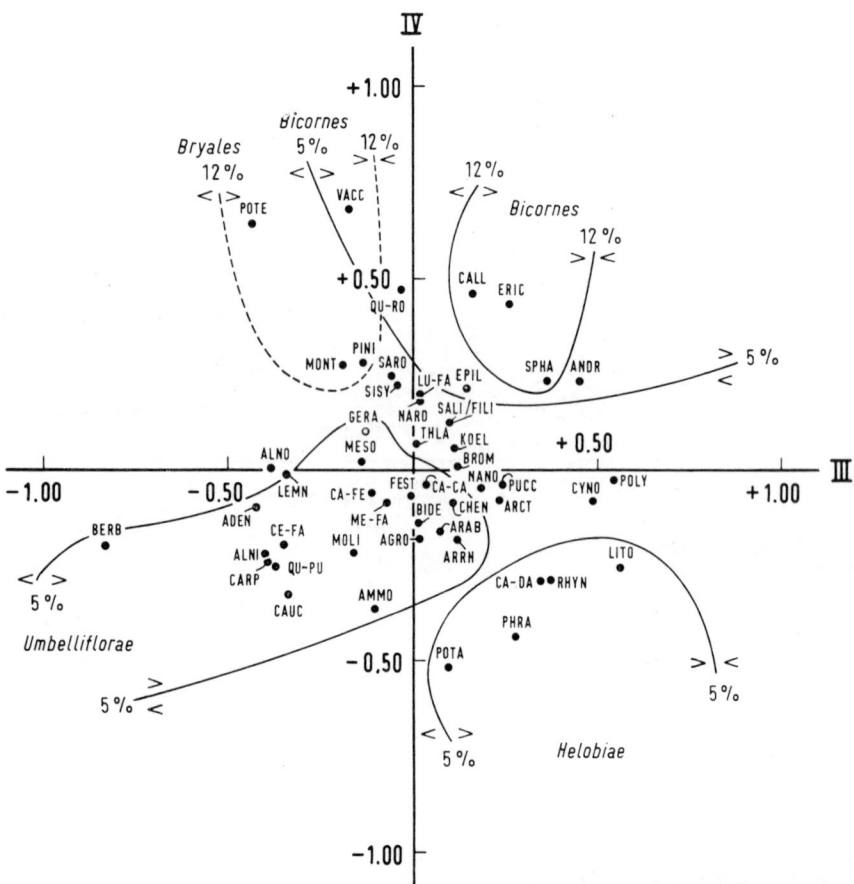

Fig. 6b. Ordination of 51 European associations and lower syntaxa on the basis of their plant order spectra, dimensions 3 and 4. (MAAREL, 1972a)

POTA, below) to dry, (Mesobromion, MESO, upper middle). Between axes III and IV (Fig. 6b) two more or less oblique lines of variation can be discerned. The third axis may be tentatively interpreted as a shading gradient from Polygonion avicularis (POLY) and Phragmition (PHRA) below, to Cardamino-Montion (MONT) and Vaccio-Piceion (VACC) above, from prevalence of Helobiae to that of Bryales. The fourth axis runs from raw humus soils with a relatively high C/N ratio to soils with a low C/N ratio and high microbiological activity (Ericion tetralicis, ERIC, with Bicornes, upper right, to Caucalion, CAUC, with Umbelliflorae, lower left). We regard Fig. 6 as representing only a first venture in broad-scale ordination of phytocoena; its significance may be in its demonstration of possibility.

20.10 Conclusion

In perspective, these numerical techniques and ordinations should be seen as ancillaries to the BRAUN-BLANQUET approach. We re-emphasize the essential ideas with which we began: the floristic approach to understanding, the functional value of diagnostic species for classification and environmental indication, and the utility of the hierarchical classification for ordering knowledge and expressing understanding. Given this as the approach, then the numerical techniques serve to reduce labour, and ordinations to enhance interpretation. Ordination without prior classification has accomplished significant research in American ecology, but ordinations in different areas are not easily coordinated with one another. It is the European experience that the BRAUN-BLANQUET approach provides such coordination of the work of different investigators through its classification, whilst the classification provides a context that increases effectiveness of ordination. On the one hand prior classification makes possible the ordination of phytocoena, rather than an often unmanageable number of relevés; on the other hand the known relationships of species groups and phytocoena to one another and environment give basis for interpreting an ordination. The ordination should, in turn, express and further clarify those relationships.

We return to another theme – the complexity of the BRAUN-BLANQUET approach as a technical system. The complexity has causes; these include the numerousness of species and phytocoena, the multifarious distributional relations of species to one another and phytocoena, and the many directions of environmental, geographic, and developmental relationships amongst species and phytocoena. The complexity of the BRAUN-BLANQUET system is thus a response of science to the complexity of vegetation. It is no real statement of preference that the approaches through physiognomy and species dominance are 'simpler'; it is not necessarily true that they are, and to the extent that they are, they are only because they do not treat and coordinate so wide a range of information. We thus express our judgement (or our bias) that in the study of vegetation one approach has most fully faced the demands for a method that is both detailed and generalizing, both locally intensive and integrative of the results from different local areas – and that that one approach is BRAUN-BLANQUET'S.

This judgement we must balance with another. The BRAUN-BLANQUET approach is very demanding of research effort by many investigators, if the vegetation of extensive areas is to be known through it. The effort has been possible in Europe and some other areas but is not yet possible everywhere — there are not phytosociologists enough. The approach may be particularly difficult to initiate in an area lacking prior studies to offer suggestions toward classification and diagnostic species groupings, unless the initiator is experienced in the approach. This difficulty is one reason for the importance of approaches through physiognomy and dominance in many areas; by these approaches knowledge of vegetation may be gained that might not otherwise be obtained at all. It is true also that the physiognomic and other approaches may reveal vegetational relationships that are important, but are outside the main concerns of the BRAUN-BLANQUET approach. The statement that BRAUN-BLANQUET's is the most fully developed approach to vegetation is thus no rejection of the contributions of others amongst the many ways for the ordination and classification of communities.

20.11 SUMMARY

(20.1) Amongst the central ideas of the approach are that: (i) classification and interpretation of communities should be based on their full floristic composition, (ii) with emphasis on diagnostic species, whose relative restriction to samples characterizes communities and indicates their environments and (iii) which may be used to organize the communities into a formal, hierarchical classification. (20.2) This floristic-sociological approach to vegetation had its origin in southern European phytosociology centered in the cities of Zürich and Montpellier, but especially in the ideas of BRAUN-BLANQUET.

(20.3) It is important to distinguish particular, concrete plant communities or phytocoenoses, and abstract classes of plant communities; the term phytocoenon is recommended for the latter. The phytocoena of the formal BRAUN-BLANQUET hierarchy are termed syntaxa, in analogy with the taxa into which individual organisms are classified. Syntaxa are characterized by diagnostic species of three types: character-species are centered in or relatively restricted to a particular syntaxon compared with all others, and therefore characterize it and indicate its environment; differential-species distinguish two closely related syntaxa by presence in most samples of the one and absence in most samples of the other; constant companions are not restricted to a given syntaxon but help to characterize it and indicate its relationships to higher units. In addition to the hierarchy, syntaxa may be arranged along a sociological progression from simple and poorly organized to complex and highly differentiated communities. The BRAUN-BLANQUET classification is considered natural in the sense that its syntaxa (though their limits may be arbitrary) are consistent with a large number of relationships between environment, species populations, and biocoenoses.

(20.4) The research procedures include three phases — analytical, synthetical, and syntaxonomical. In the analytical phase samples or relevés representing kinds of phytocoena to be studied are taken in the field. The relevés should be chosen to represent the phytocoena to be studied, and each should sample a uniform area of sufficient size to represent that phytocoenose adequately. A relevé includes information on environment and location and a list of all plant species in the sample area, with species recorded by strata and rating scales (for combined cover and abundance, and if possible sociability, vitality, and periodicity).

(20.5) In the synthetical phase the relevés are compared with one another to derive a preliminary classification. A set of relevés are first listed in their full species composition in a primary table. Species of intermediate presence values are emphasized in seeking groups of differentiating species that characterize some of the relevés but not others. Partial tables are prepared using these species to rearrange the relevés into groups characterized by the groups of differentiating species. When the rearrangement is satisfactory, the data from all species are recopied into a phytocoenon table in which boundaries may be drawn to mark off differential-species groups and phytocoena. A synoptic table may now be prepared in which each column summarizes data for species occurrence in the relevés of a phytocoenon. By use of synoptic tables the phytocoena under study may be compared with others from the same area in a search for character-species. Problems in the recognition of character-species, including geographic differences in fidelity of species to phytocoena, are discussed.

(20.6) In the syntaxonomical phase the studied phytocoena may enter the formal hierarchy. The phytocoenon table is revised into a formal syntaxonomic table indicating the character-species and differential-species groups represented. If the phytocoenon represents a new association, this is named and placed in the hierarchy in which associations are grouped into alliances, alliances into orders, orders into classes. Syntaxa on all these levels may be defined by character-species; in some cases character-genera are used for higher units. Apart from the hierarchy, associations (and other syntaxa) from different geographic areas may be grouped by floristic affinities, especially as marked by vicariant species. Associations are divided into subordinate units: subassociations and variants characterized by differential-species, and facies as the lowest-level units, characterized by quantitative differences in species representation. Associations are named on the basis of one or two characteristic species and the suffix -etum. Further rules of nomenclature for associations and other syntaxa are discussed, and an outline of the syntaxa is given.

(20.7) Various extensions of the approach are described. Applications to vegetation outside the European homeland are summarized. The approach is applied also to biotic and animal communities, and to stratal and life-form subcommunities to produce a separate classification of these into synusiae. Complexes (mosaics or patterns of contiguous or continuous communities) can be studied on various levels from the microrelief of bogs to whole landscapes. Mapping of phytocoena is valuable in applied phytosociology and land management. For applied work, diagnostic species groups can be used to indicate environment. Species can also be classed into ecological groups by their distributional responses to environmental gradients, and representation of ecological groups can be used to indicate position of a phytocoenose along an environmental gradient.

(20.8) Numerical techniques are being developed to store relevés, to measure homogeneity of phytocoenon tables, and to aid in sample rearrangement in tables and the earlier stages of classification. (20.9) Ordinations are possible on either a less formal basis, using relevés or phytocoena along known environmental gradients, or on a more formal basis using quantitative comparisons amongst samples to arrange them along abstract axes. Ordinations can be applied to phytocoena (each summarizing data from several relevés) as well as to individual relevés. Results to date indicate that this is a promising means of clarifying relationships amongst species and syntaxa.

(20.10) Numerical techniques and ordination are aids to the essential floristic-sociological procedures of the BRAUN-BLANQUET approach. Despite the complexity of the procedures, the approach is felt to be flexible to different kinds of vegetation and research purposes, productive of understanding of the relations of species and samples to one another and environment, useful for applied purposes, and valuable for its coordination of the results from different areas into a single classification. It is judged the most fully developed and most widely useful approach to the classification and interpretation of vegetation.

20.12 Additions to the Second Edition

No essential changes affect the BRAUN-BLANQUET approach as treated in the first edition. Our additions are mainly references to a (selected) number of publications that came to our notice after the manuscript for the first edition was completed. They will be given by sections.

(20.2-3) History and General Concepts

The history of the BRAUN-BLANQUET approach was studied in further detail by MAAREL (1975) on the occasion of BRAUN-BLANQUET's 90th birthday. VAN DER MAAREL emphasizes the crucial position of BRAUN's work in the development of phytosociology, with the conclusion that it is a synthesis of ideas and approaches from the entire 19th century in a framework that has incorporated many developments of the 20th century.

(20.4) Analytical Research Phase

The book by MUELLER-DOMBOIS & ELLENBERG (1974) presents full information on the relevé method and compares various methods of measuring species quantities, without giving any preference for one particular method. An important new French contribution was published by GUINOCHET (1973), who paid special attention to the choice and delimitation of sample plots.

(20.4.3) MINIMAL AREA AND PLOT SIZE

WERGER (1972) discussed minimal area and emphasized that no analytical minimal area can be found through a species-area curve. He suggested (cf. MAAREL, 1966b*) an optimal relevé-size representing a given fraction of the total information in the stand. MORAVEC (1973a) tried to establish a minimal area through analysis of similarity between multiple plots of increasing size (as was done earlier by MAAREL, 1966b* following GOUNOT & CALLÉJA, 1962).

(20.5) Synthetical research phase

Again the book by MUELLER-DOMBOIS & ELLENBERG (1974) may be mentioned as a general reference. For homotoneity MORAVEC (1973b) proposed an empirical correction for small sets of relevés as well as a simplified calculation for large sets.

(20.6) Syntaxonomical Research Phase

Further evidence is presented by MUELLER-DOMBOIS & ELLEN-

BERG (1974) on the geographical variation in habitat preference of many species and the consequently limited significance of character-species. The authors regret the increasing splitting of associations and emphasise alliances as general units, recurring in similar forms in many areas, which largely fulfil the requirements once set up for the association.

An important approach to plant communities with an insufficient number of character-species was developed by KOPECKÝ & HEJNÝ (1974, see also 1973 and KOPECKÝ 1974). They distinguish 'basal' and 'derivate' communities in addition to the cenologically saturated communities that can be fully characterised as syntaxa. Basal communities are composed of species with relatively broad habitat amplitudes. They may arise either after disturbance of saturated communities, or during a succession on newly formed anthropogenic sites. They lack character and differential species on the association level, but may show them on the alliance, the order or even only the class level. KOPECKÝ & HEJNÝ propose a special nomenclature for those communities. E.g. a basal community (BC) characterised by dominance of *Urtica dioica* and *Aegopodium podagraria* in which only class character-species occur is indicated BC Urtica-Aegopodium (Galio-Urticetea).

A derivate community is characterised by a dominating species showing, at least regionally, a narrow amplitude and a rapid spread from diaspores. It may develop as a special form of a basal community during primary succession, or, again, by disturbance of a saturated community, or during the disintegration of a basal community due to disturbance. Derivate communities (DC) are indicated as are basal communities; for example, within the range of the BC Urtica-Aegopodium (Galio-Urticetea) may occur a DC Chaerophyllum aromaticum (Galio-Urticetea). This approach may be applied to all kinds of anthropogenic plant communities (cf. BRAAKHEKKE & BRAAKHEKKE-ILSINK 1976) as well as to communities in naturally disturbed or otherwise extreme environments.

Nomenclature of syntaxa has reached a milestone through the publication of the Code of Phytosociological Nomenclature in *Vegetatio* (BARKMAN et al. 1976).

(20.7) **Extension of the approach**

The number of applications of the Braun-Blanquet approach in non-European areas has further increased. The reader is referred to numerous contributions in *Vegetatio*, the new periodical *Phytocoenologia*, and Symposium volumes (notably GÉHU 1975 on coastal dunes).

Many references are found in new issues of *Excerpta Botanica Sociologica* and also in *Fortschritte der Botanik* (Progress in Botany) volumes 36 and 37.

With respect to community complexes a new development can be mentioned: the distinction of 'sigmassociations' or association complexes, coherent series of associations characterizing certain landscapes (TÜXEN 1973, GÉHU 1974). Also syntaxa on higher levels can be described in such a framework, with the general term sigmasyntaxon. The approach is similar to the description of vegetation complexes in recent Dutch landscape ecological descriptions (cf. MAAREL & STUMPEL 1975).

As to indicator groups, a recent synthesis of ecological knowledge on European plant species by ELLENBERG (1974) can be mentioned.

(20.8) **Numerical techniques**

The Working Group for Data-Processing in Phytosociology ended its research. The results are now to be published in *Vegetatio*, starting with a general survey of activities and perspectives (MAAREL et al. 1976, see also MAAREL 1974).

The use of "analyse factorielle des correspondences' (reciprocal averaging or correspondence analysis, article 11, HILL 1973, 1974) has shown further successes in the delimitation of plant communities and the establishment of characterising species groups, particularly through the work of LACOSTE (1975, 1976). The Montpellier group proceeded in establishing ecological groups by applying correspondence analysis to floristic and environmental data (BOTTLIKOVÁ et al. 1975) (see also GUINOCHET 1973).

With respect to numerical classification COETZEE & WERGER (1975) considered the BRAUN-BLANQUET approach in its essence superior to association analysis, whilst STANEK (1973) found it as effective as a sum-of-squares agglomeration technique. This is a divisive polythetic classification and an approach in that direction through the method of indicator species analysis as suggested by HILL et al. (1975) seems very promising.

Table rearrangement procedures have been treated at length by MUELLER-DOMBOIS & ELLENBERG (1974) mainly on ELLENBERG's (1956*) original test data, with applications of numerical methods divised by SPATZ (1972*, SPATZ & SIEGMUND 1973). STOCKINGER & HOLZNER (1973) proceeded with their approach through the erection of correlated species groups. DALE & QUADRACCIA (1973) used display on monitors in an interactive table sorting. DALE & WEBB

(1975) applied a two-parameter analysis, i.e. an analysis which is symmetric in its use of species and reléves, in such a way that relevé groups are established during the divisive procedure at a point where discontinuities between relevé groups change to continuities within them.

The table rearrangement program TABORD (JANSSEN 1972★) has been further extended and implemented in various centres; JANSSEN (1975) devised a simple clustering procedure for the division of very large data-sets (up to 6000 relevés) into smaller groups manageable with TABORD.

The attempts towards a numerical syntaxonomy of *Spartinetum* communities by KORTEKAAS & MAAREL (1973) proceeded. A comparison with the classical syntaxonomy became possible through the publication of the *Spartinetea* issue of the *Prodrome* of European plant communities (BEEFTINK & GÉHU 1973), the first issue of this series, a new milestone in traditional European phytosociology. The comparison showed substantial agreement whilst some new lower rank syntaxa were suggested by the numerical output.

(9.9) Ordination

For the rapidly growing literature on phytosociological ordination we may refer to ORLÓCI (1975), various contributions in *Vegetatio* and, of course, the companion volume of this book. Like LACOSTE (1974, 1976) with correspondence analysis, many phytosociologists have used principal component analysis to obtain effective boundaries between community types (e.g. BOUXIN 1975, FEOLI 1973, GILS et al. 1975, and PIGNATTI & PIGNATTI 1975). DALE & CLIFFORD (1976) further explored the possibilities of using higher taxonomic ranks in the classification of vegetation (cf. MAAREL 1972★).

(9.10) Conclusion

The general effectiveness and efficiency of the BRAUN-BLANQUET approach was emphasised once more, by WERGER (1974). PIGNATTI (1975), contemplating the future of phytosociology, concluded that neither a one-sided if sophisticated mathematical phytosociology, nor a sterile perfecting of the hierarchic system of syntaxa should be desired. The approach should instead seek first summarization and fundamental understanding of relations of plant communities to one another and environment, and second increasing application of this knowledge to the conservation and management of our natural heritage.

★References to publications already listed in the first edition are marked with ★; new publications are in a separate list at the end of the References.

REFERENCES

This bibliography does not intend to be complete. In particular, those papers have been omitted dealing with a similar topic as the first and (or) main paper of the same author referred to.

ADJANOHOUN, E. J., – 1962 – Étude phytosociologique des savanes de Basse Côte d'Ivoire (Savanes lagunaires). *Vegetatio* 11: 1—38.
AICHINGER, E., – 1951 – Soziationen, Assoziationen und Waldentwicklungstypen. *Angew. Pfl. Soziol.*, Wien 1: 21—68.
AICHINGER, E., – 1954 – Statische und dynamische Betrachtung in der pflanzensoziologischen Forschung. *Veröff. geobot. Inst. Rübel* 29: 9—28.
ALECHIN, W. W., – 1926 – Was ist eine Pflanzengesellschaft? *Repert. spec. nov. Regni veg.*, Beih. 37: 1—50.
BACH, R., R. KUOCH & M. MOOR, – 1962 – Die Nomenklatur der Pflanzengesellschaften. *Mitt. flor.-soz. Arbeitsgem. N. F.* 9: 301—308.
BALOGH, J., – 1958 – Lebensgemeinschaften der Landtiere. Ihre Erforschung unter besonderer Berücksichtigung der zoozönologischen Arbeitsmethoden. Akademie Verlag GmbH, Berlin-Budapest. 560 pp.
BARKMAN, J. J., – 1953b – Some proposals for a phytosociological nomenclature. *Proc. 7th int. bot. Congr. Stockholm 1950*: 662—663; discussion: 663—666.
BARKMAN, J. J., – 1958a – Phytosociology and Ecology of Cryptogamic Epiphytes. Van Gorcum, Assen. 628 pp.
BARKMAN, J. J., – 1958b – La structure du Rosmarineto-Lithospermetum helianthemetosum en Bas-Languedoc. *Blumea suppl.* 6: 113—136.
BARKMAN, J. J., H. DOING, C. G. VAN LEEUWEN & V. WESTHOFF, – 1958 – Enige opmerkingen over de terminologie in de vegetatiekunde. *Corr. Bl. Rijksherbarium* 8: 87—93.
BARKMAN, J. J., H. DOING & S. SEGAL, – 1964 – Kritische Bemerkungen und Vorschläge zur quantitativen Vegetationsanalyse. *Acta bot. neerl.* 13: 394—419.
BECKING, R. W., – 1957 – The Zürich-Montpellier school of phytosociology. *Bot. Rev.* 23: 411—488.
BECKING, R. W., – 1961 – Mathematical analysis of plant communities. *Rec. Adv. Botany*: 1346—1350.
BEEFTINK, W. G., – 1962 – Conspectus of the phanerogamic salt plant communities in the Netherlands. *Biol. Jb. Dodonaea* 30: 325—362.
BEEFTINK, W. G., – 1965 – De zoutvegetaties van ZW-Nederland beschouwd in Europees verband. *Meded. Landbouwhogeschool Wageningen* 65(1): 1—167.
BEEFTINK, W. G., – 1968 – Die Systematik der europäischen Salzpflanzengesellschaften. In, 'Pflanzensoziologische Systematik', ed. R. TÜXEN, *Ber. Symp. int. Ver. Vegetationskunde*, Stolzenau 1964: 239—263. Junk, The Hague.
BENNINGHOFF, W. S. & W. C. SOUTHWORTH, – 1964 – Ordering of tabular arrays of phytosociological data by digital computer. *Abstr. 10. Int. bot. Congr. Edinburgh*: 331—332.
BHARUCHA, F. R., – 1956 – Methods for the study of tropical vegetation. *Proc. Kandy Symp. Study Tropical Vegetation*, UNESCO, Paris. p. 89—90.
BOLOS, O., – 1968 – Tabula vegetationis Europae occidentalis. *Acta geobot. Barcinonensa* 3: 5—8.
BOUDOURESQUE, C. F., – 1971 – Contribution à l'étude phytosociologique des peuplements algaux des côtes varoises. *Vegetatio* 22: 83—184.
BRAUN-BLANQUET, J., – 1913 – Die Vegetationsverhältnisse der Schneestufe in den Rätisch-Lepontischen Alpen. *Schweiz. Naturf. Gesellsch. Neue Denkschr.* 48: 1—347.
BRAUN-BLANQUET, J., – 1915 – Les Cévennes méridionales. Étude phytogéographique. *Biblioth. Univ. Arch. Sci. Phys. Nat. Ser.* 4: 39: 72—81 a.f.

BRAUN-BLANQUET, J., – 1918 – Eine pflanzengeographische Excursion durchs Unterengadin und in den Schweizerischen Nationalpark. *Beitr. geobot. Landesaufn. Schweiz* 4: 1–80.
BRAUN-BLANQUET, J., – 1921 – Prinzipien einer Systematik der Pflanzengesellschaften auf floristischer Grundlage. *Jahrb. St. Gallen Naturw. Ges.* 57: 305—351.
BRAUN-BLANQUET, J., – 1923 – Origine et Développement des Flores dans le Massif Central de France. Zürich-Paris.
BRAUN-BLANQUET, J., – 1925 – Zur Wertung der Gesellschaftstreue in der Pflanzensoziologie. *Vierteljahrsschr. Naturf. Ges. Zürich* 70: 122—149.
BRAUN-BLANQUET, J., – 1928 – Pflanzensoziologie. Grundzüge der Vegetationskunde. Biologische Studienbücher 7. 1. Ed. Berlin. x + 330 pp.
BRAUN-BLANQUET, J. – 1930 – L'importance pratique de la sociologie végétale. *Comm. S.I.G.M.A.* 4: 1—8.
BRAUN-BLANQUET, J., – 1932 – Plant Sociology. (Transl. by G. D. FULLER and H. S. CONARD). New York, xviii + 439 pp. Reprint 1966.
BRAUN-BLANQUET, J., – 1933a – Phytosociological nomenclature. *Ecology* 14: 315—317.
BRAUN-BLANQUET, J., – 1933b – Prodrome des Groupements végétaux, Fasc. 1. Ammophiletalia et Salicornietalia medit. Com. Int. Prodrome Phytosociol. Montpellier. 23 pp.
BRAUN-BLANQUET, J., – 1939 – Lineares oder vieldimensionales System in der Pflanzensoziologie. *Chronica bot.* 5: 391—395.
BRAUN-BLANQUET, J., – 1946 – Über den Deckungswert der Arten in den Pflanzengesellschaften der Ordnung Vaccinio-Piceetalia. *Jahresber. Naturf. Gesell. Graubündens* 130: 115—119.
BRAUN-BLANQUET, J., – 1950 – Sociologica vegetal. Versión espan. A. P. L. Gigilio & M. M. Grassi. Buenos Aires.
BRAUN-BLANQUET, J., – 1951a – Pflanzensoziologie. Grundzüge der Vegetationskunde, 2nd ed. Springer, Wien. 631 pp.
BRAUN-BLANQUET, J., – 1951b – Pflanzensoziologische Einheiten und ihre Klassifizierung. *Vegetatio* 3: 126—133.
BRAUN-BLANQUET, J., – 1953 – Essai sur le classement des Biocénoses. *Comm. S.I.G.M.A.* 118: 8—12.
BRAUN-BLANQUET, J., – 1955 – Zur Systematik der Pflanzengesellschaften. *Mitt. flor.-soz. Arbeitsgem. NF* 5: 151—154.
BRAUN-BLANQUET, J., – 1959 – Grundfragen und Aufgaben der Pflanzensoziologie. *Vistas in Botany* 1: 145—171.
BRAUN-BLANQUET, J., – 1962 – Zur pflanzensoziologischen Systematik, Erinnerungen und Ausblick. *Comm. S.I.G.M.A.* 159: 1—11.
BRAUN-BLANQUET, J., – 1964 – Pflanzensoziologie, Grundzüge der Vegetationskunde, 3rd ed. Springer, Wien-New York. 865 pp.
BRAUN-BLANQUET, J., – 1968 – L'école phytosociologique Zuricho-Montpelliéraine et la S.I.G.M.A. *Vegetatio* 16: 1—78.
BRAUN-BLANQUET, J., – 1969 – Reinhold Tüxen, Meister-Pflanzensoziologe. *Vegetatio* 17: 1—25.
BRAUN-BLANQUET, J. & E. FURRER, – 1913 – Remarques sur l'étude des groupements de plantes. *Bull. Soc. languedoc. Géogr.* 36: 20—41.
BRAUN-BLANQUET, J. & H. JENNY, – 1926 – Vegetations-Entwicklung und Bodenbildung in der alpinen Stufe der Zentralalpen. *Schweiz. Naturf. Ges., Denkschr.* 63(2): 181—349.
BRAUN-BLANQUET, J. & W. C. DE LEEUW, – 1936 – Vegetationsskizze von Ameland. *Ned. kruidk. Arch.* 46: 359—393.

BRAUN-BLANQUET, J. & M. MOOR, – 1938 – Verband des Bromion erecti. Prodromus der Pflanzengesellschaften. *Comm. S.I.G.M.A.* 5: 1—64.
BRAUN-BLANQUET, J. & J. PAVILLARD, – 1922/28 – Vocabulaire de sociologie végétale. 1. Ed. Montpellier, 16 pp. 3. Ed. Montpellier, 23 pp.
BRAUN-BLANQUET, J. & J. PAVILLARD, – 1930 – Vocabulary of Plant Sociology. (Transl. by Bharucha, F. R.). Cambridge. 23 pp.
BRAUN-BLANQUET, J., G. SISSINGH & J. VLIEGER, – 1939 – Prodrome des Groupements Végétaux. Fasc. 6 Klasse der Vaccinio-Piceetea. Montpellier 123 pp.
BRAUN-BLANQUET, J. & R. TÜXEN, – 1932 – Die Pflanzensoziologie in Forschung und Lehre. *Biologe* 1931/32: 175—186.
BRAY, J. R. & J. T. CURTIS, – 1957 – An ordination of the upland forest communities of southern Wisconsin. *Ecol. Monogr.* 27: 325—349.
BROCKMANN-JEROSCH, H., – 1907 – Die Pflanzengesellschaften der Schweizeralpen. I. Die Flora des Puschlav und ihre Pflanzengesellschaften. Engelmann, Leipzig. 438 pp.
CAIN, S. A. & G. M. DE OLIVEIRA CASTRO, – 1959 – Manual of Vegetation Analysis. Harper and Brothers, New York. xvi+325 pp.
CAJANDER, A. K., – 1903 – Beiträge zur Kenntniss der Vegetation der Alluvionen des nördlichen Eurasiens. I. Die Alluvionen des unteren Lena-Thales. *Acta soc. sci. Fenn.* 32 (1): 1—182.
ČEŠKA, A., – 1966 – Estimation of the mean floristic similarity between and within sets of vegetational relevés. *Folia geobot. phytotax.* 1: 92—100.
ČEŠKA, A., – 1968 – Application of association coefficients for estimating the mean similarity between sets of vegetational relevés. *Folia geobot. phytotax.*, 3: 57—64.
ČEŠKA, A. & H. ROEMER, – 1971 – A computer program for identifying species-relevé groups in vegetation studies. *Vegetatio* 23: 255—276.
CHAPMAN, V. J., – 1952 – Problems in ecological terminology. *Rep. Austr.-Nw Zealand Ass. Adv. Science* 29th Meeting 259—279.
CHAPMAN, V. J,, – 1959 – Salt marshes and ecological terminology. *Vegetatio* 8: 215—234.
CLEMENTS, F. E., – 1936 – Nature and structure of the climax. *J. Ecol.* 24: 252—284.
CONARD, H., – 1935 – The plant associations of Central Long Island. *Amer. Midl. Nat.* 16: 433—516.
CONARD, H., – 1952 – The vegetation of Iowa. An approach toward a phytosociologic account. *Stud. nat. Hist. Iowa Univ.* 19 (4): 1—166.
COTTAM, G. & J. T. CURTIS, – 1948 – The use of the punched card method in phytosociological research. *Ecology* 29: 516—519.
CRAWFORD, R. M. M. & D. WISHART, – 1966 – A multivariate analysis of the development of dune slack vegetation in relation to coastal accretion at Tentsmuir, Fife. *J. Ecol.* 54: 729—743.
CRAWFORD, R. M. M. & D. WISHART, – 1967 – A rapid multivariate method for the detection and classification of groups of ecologically related species. *J. Ecol.* 55: 505—524.
CRISTOFOLINI, G. and S. M., – 1967 – Phytosociological study on some Liriodendron tulipifera L. forests of Tennessee, U.S.A. *N. Gior. bot. ital.* 101—6: 317—346.
CRISTOFOLINI, G., D. LAUSI & S. PIGNATTI, – 1969 – Survey of the system for coding of plantsociological records used by the 'Trieste Group'. Report Working Group Data-Processing Phytosoc. (mimeographed) 13 pp.
CURTIS, J. T., – 1959 – The vegetation of Wisconsin. An Ordination of Plant Communities. Univ. of Wisconsin Press. Madison, Wisconsin. xi + 657 pp.

CURTIS, J. T. & R. P. MCINTOSH, – 1951 – An upland forest continuum in the prairie-forest border region of Wisconsin. *Ecology* 32: 476—496.
DAGET, PH. et al., – 1972 – Profils écologiques et information mutuelle entre espèces et facteurs écologiques. In, 'Grundfragen und Methoden in der Pflanzensoziologie', ed. E. VAN DER MAAREL & R. TÜXEN, *Ber. Symp. int. Ver. Vegetationskunde*, Rinteln 1970: 121—148. Junk, The Hague.
DAGNELIE, P., – 1960 – Contribution à l'étude des communautés végétales par l'analyse factorielle. *Bull. Serv. Carte Phytogéogr. CNRS* 5: 7—71, 93—195.
DAHL, E., – 1957 – Rondane; Mountain vegetation in South Norway and its relation to the environment. *Skr. norske Vidensk-Akad.; mat.-naturv. Kl.* 1956(3): 1—374.
DAHL, E., – 1960 – Some measures of uniformity in vegetation analysis. *Ecology* 41: 785—790.
DAHL, E. & E. HADAČ, – 1941 – Strandgesellschaften der Insel Ostoy im Oslofjord. *Nytt Mag. Naturv.* 82: 251—312.
DAHL, E. & E. HADAČ, – 1949 – Homogeneity of plant communities. *Studia Bot. Cechosl.* 10: 159—176.
DAMMAN, A. W. H., – 1964 – Some forest types of central Newfoundland and their relation to environmental factors. *Forest Sci. Monogr.* 8: 1—62.
DAMMAN, A. W. H. & D. M. DE VRIES, – 1954 – Testing of grassland associations by combinations of species. *Biol. Jb. Dodonaea* 21: 35—46.
DANSEREAU, R., – 1957 – Biogeography; an Ecological Perspective. Ronald Press Comp., New York. xiii + 394 pp.
DANSEREAU, P. & J. ARROS, – 1959 – Essais d'application de la dimension structurale en phytosociologie. I. Quelques exemplaires européens. *Vegetatio* 9: 48—99.
DAVIS, P. H. & V. H. HEYWOOD, – 1963 – Principles of Angiosperm Taxonomy. Oliver & Boyd, Edinburgh. xx + 556 pp.
DEVRED, R., – 1956 – Les savanes herbeuses de la région de Mvuazi (Bas Congo). *Publ. INEAC Bruxelles Sér. Scient.* 65: 1—115.
DIERSCHKE, H., – 1971 – Stand und Aufgaben der pflanzensoziologischen Systematik in Europa. *Vegetatio* 22: 255—264.
DOING KRAFT, H., – 1954 – L'analyse des carrés permanents. *Acta bot. neerl.* 3: 421—425.
DOING, H., – 1962 – Systematische Ordnung und floristische Zusammensetzung niederländischer Wald- und Gebüschgesellschaften. *Wentia* 8: 1—85.
DOING, H., – 1969a – Sociological species groups. *Acta bot. neerl.* 18: 398—400.
DOING, H., – 1969b – Assoziationstabellen von niederländischen Wäldern und Gebüschen. Rep. Lab. Plantensyst. Geogr. Wageningen 29 pp. + tab.
DONSELAAR, J. VAN, – 1965 – An ecological and phytogeographic study of northern Surinam savannas. *Wentia* 14: 1—163.
DONSELAAR, J. VAN, – 1969 – Observations on savanna vegetation types in the Guianas. *Vegetatio* 17: 271—312.
DRUDE, O., – 1890 – Handbuch der Pflanzengeographie, Stuttgart, 582 pp.
DU RIETZ, G. E., – 1921 – Zur methodologischen Grundlage der modernen Pflanzensoziologie. Akad. Abhandl. Uppsala. 272 pp.
DU RIETZ, G. E., – 1930a – Vegetationsforschung auf soziationsanalytischer Grundlage. *Handb. Biol. Arbmeth.* 11, 5: 293—480.
DU RIETZ, G. E., – 1930b – Classification and nomenclature of vegetation. *Svensk Bot. Tidskr.* 24: 489—503.
DU RIETZ, G. E., – 1936 – Classification and nomenclature of vegetation units 1930—1935. *Svensk Bot. Tidskr.* 30: 580—589.

DU RIETZ, G. E., T. C. E. FRIES, H. OSWALD & T. A. TENGWALL, – 1920 –
Gesetze der Konstitution natürlicher Pflanzengesellschaften. *Vetensk. prakt.
Unders. Lappl. Flora och Fauna* 7: 1—47.
DUVIGNEAUD, P., – 1946 – La variabilité des associations végétales. *Bull. Soc.
Roy. bot. belg.* 78: 107—134.
DUVIGNEAUD, P., – 1949 – Les savannes du Bas-Congo. Essai de phytosociologie
topographique. *Lejeunia Mem.* 10, 192 pp.
EGLER, F. E., – 1954 – Philosophical and practical considerations of the Braun-
Blanquet system of plant sociology. *Castanea* 19: 45—60.
ELLENBERG, H., – 1939 – Über Zusammensetzung, Standort und Stoffproduktion
bodenfeuchter Eichen- und Buchen-Mischwaldgesellschaften Nordwest-
deutschlands. *Mitt. flor.-soz. Arbeitsgem. Niedersachsen* 5: 1—135.
ELLENBERG, H., – 1950 – Landwirtschaftliche Pflanzensoziologie. I. Unkraut-
gemeinschaften als Zeiger für Klima und Boden. Stuttgart-Ludwigsburg. 141 pp.
ELLENBERG, H., – 1952 – Landwirtschaftliche Pflanzensoziologie. II. Wiesen
und Weiden und ihre standörtliche Bewertung. Stuttgart. 143 pp.
ELLENBERG, H., – 1954a – Landwirtschaftliche Pflanzensoziologie. III. Natur-
gemässe Anbauplanung, Melioration und Landespflege. Stuttgart. 109 pp.
ELLENBERG, H., – 1954b – Zur Entwicklung der Vegetationssystematik in
Mitteleuropa. *Angew. Pfl. Soziol.*, Wien, Festschr. Aichinger 1: 134—143.
ELLENBERG, H., – 1956 – Grundlagen der Vegetationsgliederung. 1. Teil: Auf-
gaben und Methoden der Vegetationskunde. In, 'Einführung in die Phytolo-
gie' IV-1, ed. H. WALTER, Stuttgart, 136 pp.
ELLENBERG, H., – 1963 – Vegetation Mitteleuropas mit den Alpen. In, 'Ein-
führung in die Phytologie' IV-2, ed. H. WALTER, Stuttgart, 943 pp.
ELLENBERG, H., – 1968 – Sichtlochkarten zur Ordnung, Klassifikation und
Analyse pflanzensoziologischer Waldaufnahmen. In, 'Pflanzensoziologische
Systematik', ed. R. TÜXEN, *Ber. Symp. int. Ver. Vegetationskunde*, Stolzenau 1964:
163—175. Junk, The Hague.
ELLENBERG, H. & G. CRISTOFOLINI, – 1964 – Sichtlochkarten als Hilfsmittel
zur Ordnung und Auswertung von Vegetationsaufnahmen. *Ber. geobot. Inst.
ETH Rübel* 35: 124—134,
EMBERGER, L. (ed.), – 1968 – Code pour le relevé méthodique de la végétation
et du milieu. Edit. CNRS, Paris. 292 pp.
EMBERGER, L. et al., – 1957 – Description et mode d'emploi d'une fiche de relevé
pour l'inventaire de la végétation. *Bull. Serv. Carte Phytogéogr.* B-2: 7—23.
ESKUCHE, U., – 1968 – Fisionomía y sociologia de los bosques de Nothofagus
dombeyi en la región de Nahuel Huapi. *Vegetatio* 16: 192—204.
ESKUCHE, U., – 1969 – Berberitzengebüsche und Nothofagus antarctica-Wälder
in Nordwest-Patagonien. *Vegetatio* 19: 264—285.
ETTER, H., – 1949 – De l'analyse statistique des tableaux de végétation. *Vegetatio*
1: 147—154.
EVANS, F. C. & E. DAHL, – 1955 – The vegetational structure of an abandoned
field in Southeastern Michigan and its relation to environmental factors.
Ecology 36: 685—706.
FALINSKI, J., – 1960 – Zastosowanie taksonomii wroclawskiej do fitosocjologii.
(Anwendung der sogenannten 'Breslauer Taxonomie' in der Pflanzensoziolo-
gie.). *Acta Soc. bot. pol.* 29: 333—361.
FALINSKI, J., – 1964 – O róznych sposobach rozumiena pojęcia typu w fitoso-
cjolii. Dyskusje fitosocjologiczne (1). (Sur les différentes façons d'envisager
la notion du type en phytosociology. Discussion phytosociologique (1). *Ekol.
Polska Ser.* B 10: 297—306.
FLAHAULT, CH., – 1893 – Les zones botaniques dans le Bas-Languedoc et les
pays voisins. *Soc. bot. France, Bull.* 40 (Sess. extraord.): 35—62.

FLAHAULT, CH., – 1898 – Introduction, pp. 11—48, in: GAUTIER, G., Catalogue raisonné de la flore des Pyrénées orientales. Perpignan, 550 pp.
FLAHAULT, CH., – 1901 – A project for phytogeographic nomenclature. *Bull. Torrey bot. Club* 28: 391—409.
FLAHAULT, CH. & C. SCHRÖTER, – 1910 – Rapport sur la nomenclature phytogéographique. (Phytogeographische Nomenklatur). *Actes III Congr. int. bot. Bruxelles* 1: 131—164.
FRESCO, L. F. M., – 1969 – Q-type factor analysis as a method in synecological research. *Acta bot. neerl.* 18: 477—482.
FRESCO, L. F. M., – 1971 – Compound analysis: a preliminary report on a new numerical approach in phytosociology. *Acta bot. neerl.* 20: 589—599.
FRESCO, L. F. M., – 1972 – A direct quantitative analysis of vegetational boundaries and gradients. In, 'Grundfragen und Methoden in der Pflanzensoziologie', ed. E. VAN DER MAAREL & R. TÜXEN, *Ber. Symp. int. Ver. Vegetationskunde*, Rinteln 1970: 99—111. Junk, The Hague.
FREY, A., – 1928 – Anwendung graphischer Methoden in der Pflanzensoziologie. *Handb. biol. ArbMeth.* 11, 5: 203—232.
FRIEDERICHS, K., – 1927 – Grundsätzliches über die Lebenseinheiten höherer Ordnung und den ökologischen Einheitsfaktor. *Naturwissenschaften* 15: 153—157, 182—186.
FRIEDERICHS, K., – 1958 – A definition of ecology and some thoughts about basic concepts. *Ecology* 39: 154—159.
FRYDMAN, J. & R. H. WHITTAKER, – 1968 – Forest associations of southeast Lublin Province, Poland. *Ecology* 49: 896—908.
FUKAREK, F., – 1964 – Pflanzensoziologie. Berlin. 160 pp.
FUKAREK, F., M. JASNOWSKI & R. NEUHÄUSL, – 1964 – Termini phytosociologici. Jena. 74 pp.
GAMS, H., – 1918 – Prinzipienfragen der Vegetationsforschung. *Vierteljahrsch. Naturf. Gesellsch. Zürich* 63: 293—493.
GAMS, H., – 1941 – Über neuen Beiträge zur Vegetationssystematik unter besonderer Berücksichtigung des floristischen Systems von Braun-Blanquet. *Bot. Arch.* 42: 201—238.
GAMS, H., – 1954 – Vegetationssystematik als Endziel oder Verständigungsmittel? *Veröff. geobot. Inst. Rübel* 29: 39-40,
GAMS, H., – 1961 – Erfassung und Darstellung mehr-dimensionaler Verwandtschaftsbeziehungen von Sippen und Lebensgemeinschaften. *Ber. geobot. Inst. Rübel* 32: 96—115.
GAMS, H., – 1967 – Anordnung (Ordination), Aufschlüsselung (Klavikation) und Systematik (Klassifikation) von Biozönosen und anderen Naturerscheinungen. *Aquilo, Ser. Bot.* 6: 9—17.
GAMS, H. – 1972 – Die floren- and vegetationsgeschichtliche Erforschung der Alpen. *Ber. dt. bot. Ges.* 85: 7—10.
GÉHU, J. M., – 1969 – Les associations végétales des dunes mobiles et des bordures de plages de la côte atlantique française. *Vegetatio* 23: 122—166.
GILLI, A., – 1969 – Afghanische Pflanzengesellschaften. *Vegetatio* 16: 307—375.
GILLI, A., – 1971 – Afghanische Pflanzengesellschaften. II. Die mesophilen und hygrophilen Pflanzengesellschaften im sommertrockenen Gebiet. *Vegetatio* 23: 199—234.
GITTINS, R., – 1965 – Multivariate approaches to a limestone grassland community. III. A comparative study of ordination and association analysis. *J. Ecol.* 53: 411—425.
GLAHN, H. VON, – 1968 – Der Begriff des Vegetationstyps im Rahmen eines allgemeinen naturwissenschaftlichen Typenbegriffes. In, 'Pflanzensoziologische

Systematik', ed. R. TÜXEN, *Ber. Symp. int. Ver. Vegetationskunde*, Stolzenau 1964: 1—13. Junk, The Hague.
GLEASON, H., – 1926 – The individualistic concept of the plant association. *Bull. Torrey bot. Club* 53: 7—26.
GODRON, M., – 1966 – Application de la théorie de l'information à l'étude de l'homogénéité et de la structure de la végétation. *Oecol. Planta* 1: 187—197.
GOFF, F. G. & G. COTTAM, – 1967 – Gradient analysis: the use of species and synthetic indices. *Ecology* 48: 793—806.
GOODALL, D. W., – 1952 – Quantitative aspects of plant distribution. *Biol. Rev.* 27: 194—245.
GOODALL, D. W., – 1953a – Objective methods for the classification of vegetation. I. The use of positive interspecific correlation. *Austr. J. Bot.* 1: 39—63.
GOODALL, D. W., – 1953b – Objective methods for the classification of vegetation. II. Fidelity and indicator value. *Austr. J. Bot.* 1: 434—456.
GOODALL, D. W., – 1954 – Objective methods for the classification of vegetation. III. An essay int he use of factor analysis. *Austr. J. Bot.* 2: 304—324.
GOODALL, D. W., – 1961 – Objective methods for the classification of vegetation. IV. Pattern and Minimal Area. *Austr. J. Bot.* 9: 162—196.
GOODALL, D. W., – 1963 – The continuum and the individualistic association. *Vegetatio* 11: 297—316.
GOODALL, D. W., – 1969 – A procedure for recognition of uncommon species combinations in sets of vegetation samples. *Vegetatio* 18: 19—35.
GOUNOT, M., – 1957 – Utilisation des fiches perforés pour la comparaison des relevés. *Bull. Serv. Carte Phytogéogr.* B,2: 37—50.
GOUNOT, M., – 1969 – Méthodes d'étude quantitative de la végétation. Masson, Paris. 314 pp.
GRADMANN, R., – 1909/40 – Über Begriffsbildung in der Lehre von den Pflanzenformationen. *Bot. Jb.* 43, Beibl. 99: 91—103. Reprinted 1940.
GRANDTNER, M. M., – 1966 – La végétation forestière du Québec méridional. Presses Univ. Laval, Quebec. XXV + 216 pp.
GREIG-SMITH, P., – 1964 - Quantitative Plant Ecology, 2nd ed. Butterworths, London. xii + 256 pp.
GRISEBACH, A., – 1838 – Über den Einfluss des Climas auf die Begränzung der natürlichen Floren. *Linnaea* 12: 159—200.
GRISEBACH, A., – 1872 – Die Vegetation der Erde nach ihrer klimatischen Anordnung. Leipzig, 2 vols., 603 + 709 p. 2nd ed. 1884, 567 + 693 p.
GROENEWOUD, H. VAN, – 1965 – Ordination and classification of Swiss and Canadian forests by various biometric and other methods. *Ber. geobot. Inst. Rübel* Zürich 35: 28—102.
GUILLERM, J. L., – 1971 – Calcul de l'information fournie par un profil écologique et valeur indicatrice des espèces. *Oecol. Plant.* 6: 209—225.
GUINOCHET, M., – 1955 – Logique et dynamique du peuplement végétal. Coll. Ecol. Sci., Paris. 143 pp.
GUINOCHET, M. & P. CASAL, – 1957 – Sur l'analyse différentielle de Czekanowski et son application à la phytosociologie. *Bull. Serv. Carte phytogéogr.* B,2: 25—33.
HADAČ, E., – 1962 – Übersicht der höheren Vegetationseinheiten des Tatragebirges. *Vegetatio* 11: 46—54.
HADAČ, E., – 1967 – On the highest units in the system of plant communities. *Folia geobot. phytotax.* 4: 429—432.
HARTMANN, H., – 1968 – Über die Vegetation des Karakorum. I. Gesteinsfluren, subalpine Strauchbestände und Steppengesellschaften im Zentral-Karakorum. *Vegetatio* 15: 297—387.

HARTMANN, H., – 1972 – Über die Vegetation des Karakorum. II. Rasen- und Strauchgesellschaften im Bereich der alpinen und der höheren subalpinen Stufe des Zentral-Karakorum. *Vegetatio* 24: 91—157.

HARTOG, C. DEN, – 1959 – The epilithic algal communities occurring along the coast of the Netherlands. *Wentia* I: 1—241.

HARTOG, C. DEN & S. SEGAL, – 1964 – A new classification of the water plant communities. *Acta bot. neerl.* 13: 367—393.

HEER, O., – 1835 – Die Vegetationsverhältnisse des südöstlichen Teiles des Kantons Glarus; ein Versuch die pflanzengeographischen Erscheinungen der Alpen aus klimatischen und Bodenverhältnissen abzuleiten. Mitt. theor. Erdkunde Zürich.

HEGG, O., – 1965 – Untersuchungen zur Pflanzensoziologie und Ökologie im Naturschutzgebiet Hohgant (Berner Voralpen), mit einem Beitrag zur Methodik der floristisch-statistischen Erfassung pflanzensoziologischer Zusammenhänge. *Beitr. geobot. Landesaufn. Schweiz* 46: 1—188.

HEIMANS, J., – 1939 – Plantengeographische elementen in de Nederlandse flora. *Ned. kruidk. Arch.* 49: 416—436.

HOFMANN, G. & H. PASSARGE, – 1964 – Über Homogenität und Affinität in der Vegetationskunde. *Arch. Forstwes.* 13: 1119—1138.

HOLUB, J., S. HEJNÝ, J. MORAVEC & R. NEUHÄUSL, – 1967 – Übersicht der höheren Vegetationseinheiten der Tschechoslowakei. *Rozpr. čs. Akad. Věd, Ser. math. nat.* Praha 77(3): 1—75.

HOPKINS, B., – 1955 – The species-area relations of plant communities. *J. Ecol.* 43: 409—426.

HOPKINS B., – 1957 – The concept of minimal area. *J. Ecol.* 45: 441—449.

HULT, R., – 1881 – Försök till analytik behandling af växtformationerna. (Attempt towards an analytical treatment of plant formations.) *Meddn. Soc. Faun. Flor. fenn.* 8. Helsinki 155 pp.

HUMBOLDT, A. VON, – 1805 – Essai sur la géographie des plantes. Paris, 155 pp.

IVIMEY-COOK, R. B. & M. C. F. PROCTOR, – 1966 – The application of association-analysis to phytosociology. *J. Ecol.* 54: 179—192.

JAKUCS, P., – 1961 – Die phytozönologischen Verhältnisse der Flaumeichen-Buschwälder Südostmitteleuropas. Budapest, 313 pp.

JAKUCS, P., – 1970 – Bemerkungen zur Saum-Mantel-Frage. *Vegetatio* 21: 29—47.

JAKUCS, P., – 1972 – Dynamische Verbindung der Wälder und Rasen. Akademiai Kiadó, Budapest. 228 pp.

JANSSEN, C. R., – 1967 – A floristic study of forests and bog vegetation, northwestern Minnesota. *Ecology* 48: 751—755.

JANSSEN, J. G. M., – 1972 – Detection of some micropatterns of winter annuals in pioneer communities of dry sandy soils. *Acta bot. neerl.* 21: 603—610.

JANSSEN, J. G. M. & E. VAN DER MAAREL, – 1972 – A computer program for structuring a phytosociological table on the basis of relevé similarity (in prep.).

KERNER VON MARILAUN, A, – 1863 – Das Pflanzenleben der Donauländer. Innsbruck, 348 pp.

KLEMENT, O., – 1955 – Prodromus der mitteleuropäischen Flechtengesellschaften. *Repert. Spec. nov. Regni veg.*, Beih. 135: 5—194.

KNAPP, R., – 1942 – Zur Systematik der Wälder, Zwergstrauchheiden und Trockenrasen des eurosiberischen Vegetationkreises. Beil. 12, Rundbrief Zentralstelle Veget. kartierung. 183 pp.

KNAPP, R., – 1948 – Einführung in die Pflanzensoziologie. 1: Arbeitsmethoden der Pflanzensoziologie und Eigenschaften der Pflanzengesellschaften. Stuttgart - Ludwigsburg. 100 pp.

KNAPP, R., – 1957 – Über die Gliederung der Vegetation von Nordamerika. Höhere Vegetationseinheiten. *Geobot. Mitt. Köln* 4: 1—63.

KNAPP, R., – 1958 – Einführung in die Pflanzensoziologie. 1: Arbeitsmethoden der Pflanzensoziologie und Eigenschaften der Pflanzengesellschaften. 2.Ed. Stuttgart. 112 pp.

KNAPP, R., – 1965 – Die Vegetation von Nord- und Mittelamerika und der Hawaii-Inseln. Fischer, Stuttgart xl + 373 pp.

KNAPP, R., – 1966a – Höhere Vegetationseinheiten von West-Afrika unter besonderer Berücksichtigung von Nigeria und Kamerun. *Geobot. Mitt.* 34: 1—16.

KNAPP, R., – 1966b – Höhere Vegetationseinheiten von S.-Patagonien und Feuerland. *Geobot. Mitt.* 35: 1—4.

KNAPP, R., – 1971 – Einführung in die Pflanzensoziologie. Eugen Ulmer, Stuttgart. 388 pp.

KNIGHT, D. H., – 1965 – A gradient analysis of Wisconsin prairie vegetation on the basis of plant structure and plant function. *Ecology* 46: 744—747.

KOCH, W., – 1925 – Die Vegetationseinheiten der Linthebene unter Berücksichtigung der Verhältnisse in der N.O. Schweiz. *Jb. St. Gall. Naturw. Ges.* .61(2): 1—146.

KOHLER, A., – 1968 – Beiträge zur Kenntnis der ephemeren Vegetation am Südrand der Atacama-Wüste. *Ber. dt. bot. Ges.* 80: 563—572.

KOHLER, A., – 1970 – Geobotanische Untersuchungen an Küstendünen Chiles zwischen 27 und 42 Grad südl. Breite. *Bot. Jb.* 90: 55—200.

KORNAS, J., – 1965 – Phytosociological observations on plant communities of the Duke Forest near Durham, N. Carolina, U.S.A. *Fragm. flor. geobot.* 11: 307—338.

KORTEKAAS, W. & E. VAN DER MAAREL, – 1972 – A numerical classification of Spartinetum vegetations. Preliminary paper int. colloquium Rinteln 1972, 9 pp. + tables.

KRAJINA, V. J., – 1969 – Ecology of forest trees in British Columbia. *Ecol. W.N. America* 2: 1—146.

KRAUSE, W., – 1952 – Das Mosaik der Pflanzengesellschaften und seine Bedeutung für die Vegetationskunde. *Planta* 41: 240—289.

KÜCHLER, A. W., – 1967 – Vegetation Mapping. Ronald Press, New York vi + 472 pp. 30 tab. 21 fig.

LACOSTE, A. & M. ROUX, – 1971 – L'analyse multidimensionnelle en phytosociologie et en écologie. Application à des données de l'étage subalpin des Alpes maritimes. 1. L'analyse des données floristiques. *Oecol. Plant.* 6: 353—369.

LAMBERT, J. M., M. B. DALE, – 1964 – The use of statistics in phytosociology. *Adv. ecol. Res.* 2: 59—99.

LEBRUN, J., – 1960 – Études sur la flore et la végétation des champs de Lave au Nord du Lac Kivu (Congo belge). Inst. Parcs nat. Congo Belge Bruxelles pp. 1—352.

LEBRUN, J. & G. GILBERT, – 1954 – Une classification écologique des forêts du Congo. *Publ. Inst. Nat. Étude Agron. Congo Belge, Ser. Sci.* 63: 1—89.

LECOQ, H., – 1844 – Traité des plantes fourragères, ou flore des prairies naturelles et artificielles de la France. Paris, 620 pp.

LECOQ, H., – 1855 – Études sur la géographie botanique de l'Europe et en particulier sur la végétation du plateau central de la France. Vol. 4. Paris, 563 pp.

LEEUWEN, C. G. VAN, – 1965 – Het verband tussen natuurlijke en anthropogene landschapsvormen, bezien vanuit de betrekkingen in grensmilieu's. (Engl. summary) *Gorteria* 2: 93—105.

LEEUWEN, C. G. VAN, – 1966 – A relation theoretical approach to pattern and process in vegetation. *Wentia* 15: 25—46.

LEEUWEN, C. G. VAN, – 1970 – Raum-zeitliche Beziehungen in der Vegetation. In, 'Gesellschaftsmorphologie', ed. R. TÜXEN, *Ber. Symp. int. Ver. Vegetationskunde*, Rinteln 1966: 63—68. Junk, The Hague.
LIETH, H. & G. W. MOORE, – 1971 – Computerized clustering of species in phytosociological tables and its utilization for field work. In: G. P. PATIL, E. C. PIELOU & W. E. WATERS (ed): Spatial patterns and statistical distributions. Statistical Ecology Vol. 1 Penn State Univ. Press p. 403—422.
LIPPMAA, T., – 1935 – La méthode des associations unistrates et le système écologique des associations. *Acta Inst. Horti bot. tartu.* 4 (1—2, art. 3): 1—7.
LOHMEYER, W. & W. TRAUTMANN, – 1970 – Zur Kenntnis der Vegetation der Kanarischen Insel La Palma. *Schr. Reihe Vegetationskde* 5: 209—236.
LONDO, G., – 1971 – Patroon en proces in duinvalleivegetaties langs een gegraven meer in de Kennemerduinen. *Thesis Nijmegen*, 279 pp.
LOOMAN, J., – 1969 – The fescue grasslands of western Canada. *Vegetatio* 19: 128—145.
LOOMAN, J. & J. B. CAMPBELL, – 1960 – Adaptation of Sorensen's K (1948) for estimating unit affinities in prairie vegetation. *Ecology* 41: 409—416.
LORENZ, J. R., – 1858 – Allgemeine Resultate aus der pflanzengeographischen und genetischen Untersuchung der Moore im präalpinen Hügellande Salzburgs. *Flora* 41: 209—221, 225—237, 241—253, 273—286, 289—302, 344—355, 360—376.
LOUCKS, O. L., – 1962 – Ordinating forest communities by means of environmental scalars and phytosociological indices. *Ecol. Monogr.* 32: 137—166.
MAAREL, E. VAN DER, – 1960 – Rapport inzake de vegetatie van het duingebied van de Stichting 'Het Zuid-Hollands Landschap' bij Oostvoorne. Delft, pp. 1-57.
MAAREL, E. VAN DER, – 1965 – Beziehungen zwischen Pflanzengesellschaften und Molluskenfaunen. In, 'Biosoziologie.', ed. R. TÜXEN, *Ber. Symp. int. Ver. Vegetationskunde*, Stolzenau 1960: 184—197. Junk, The Hague.
MAAREL, E. VAN DER, – 1966a – Dutch studies on coastal sand dune vegetation, especially in the Delta region. *Wentia* 15: 47—82.
MAAREL, E. VAN DER, – 1966b – On vegetational structures, relations and systems, with special reference to the dune grasslands of Voorne, The Netherlands. *Thesis Utrecht* 170 pp.
MAAREL, E. VAN DER, – 1969 – On the use of ordination models in phytosociology. *Vegetatio* 19: 21—46.
MAAREL, E. VAN DER, – 1971 – Basic problems and methods in Phytosociology. *Vegetatio* 22: 275—283.
MAAREL, E. VAN DER, – 1972a – Ordination of plant communities on the basis of their plant genus, family and order relationships. In, 'Basic problems and methods in phytosociology', ed. E. VAN DER MAAREL & R. TÜXEN, *Ber. Symp. int. Ver. Vegetationskunde*, Rinteln 1970: 183—190. Junk, The Hague.
MAAREL, E. VAN DER, – 1972b – On the transformation of cover-abundance values in phytosociology. Report Bot. Lab. Nijmegen.
MAAREL, E. VAN DER & J. LEERTOUWER, – 1967 – Variation in vegetation and species diversity along a local environmental gradient. *Acta bot. neerl.* 16: 211—221.
MAAREL, E. VAN DER & R. TÜXEN (eds), – 1972 – Grundfragen und Methoden in der Pflanzensoziologie (Basic problems and methods in phytosociology). *Ber. Symp. int. Ver. Vegetationskunde*, Rinteln 1970: xix + 523 pp. Junk, Den Haag.
MAAREL, E. VAN DER & V. WESTHOFF, – 1964 – The vegetation of the dunes near Oostvoorne, Netherlands (with one vegetation map). *Wentia* 12: 1—61.
MAAREL, E. VAN DER, & V. WESTHOFF & C. G. VAN LEEUWEN, – 1964 – European approaches to the variation in vegetation. Paper 10th. Int. bot. Congress Edinburgh.

MAC INTOSH, R. P., – 1967 – The continuum concept of vegetation. *Bot. Rev.* 33: 131—187.

MARGL, H., – 1967 – Ein Gerät zum raschen Ordnen einer Tabelle. Forstl. BVA Wien Informationsdienst 109, 2 pp.

MATUSZKIEWICZ, W., – 1948 – Roślinnośc lasów okolik Lwowa (Engl. summ.). *Ann. Univ. Mariae Curie-Skłodowska Sect.* C 3: 119—193.

MATUSZKIEWICZ, W., – 1962 – Zur Systematik der natürlichen Kiefernwälder des mittel-und osteuropäischen Flachlandes. *Mitt. flor.-soz. Arbeitsgem.* Stolzenau *NF* 9: 145—186.

MATUSZKIEWICZ, W. & M. BOROWIK, – 1957 – Materialy do fitosociologicznej systematyki losow legowych w Polsce. Zur Systematik der Auenwälder in Polen. *Acta Soc. bot. polon.* 26: 719—756.

MATUSZKIEWICZ, W. & T. TRACZYK, – 1958 – Zur Systematik der Bruchwaldgesellschaften (Alnetalia glutinosa) in Polen. *Acta Soc. bot. polon.* 27: 21—44.

MEIER, H. & J. BRAUN-BLANQUET, - 1934 - Prodrome des Groupements Végétaux. Fasc. 2 Classe des Asplenietales rupestres, groupements rupicoles. Montpellier, 47 pp.

MELTZER, J. & V. WESTHOFF, – 1942 – Inleiding tot de Plantensociologie. 's Graveland. 326 pp.

MEIJER DREES, E., – 1949 – Combined taxation and presence in analysing and comparing association tables. *Vegetatio* 2: 43—46.

MEIJER DREES, E., – 1951 – Capita selecta from modern plant sociology and a design for rules of phytosociological nomenclature. *Rapp. Bosb. proefst. Bogor* 52: 1-68.

MIYAWAKI, A., – 1960 – Pflanzensoziologische Untersuchungen über die Reisfeld-Vegetation auf den Japanischen Inseln mit vergleichender Betrachtung Mitteleuropas. *Vegetatio* 9: 345—402.

MOLINIER, R. fils, – 1960 – Études des biocénoses marines du Cap Corse. *Vegetatio* 9: 121—192, 217—312.

MOORE, G. W., W. S. BENNINGHOFF & P. S. DWYER, – 1967 – A computer method for the arrangement of phytosociological tables. Proc. Ass. Computing Mach. 1967, Washington: 297—299.

MOORE, J. J., – 1962 – The Braun-Blanquet system: a reassessment. *J. Ecol.* 50: 761—769.

MOORE, J. J., – 1971 – Phyto—A suite of programs in Fortran IV for the manipulation of phytosociological tables according to the principles of Braun-Blanquet. Mscr. Dept. Botany Univ. Coll. Dublin, 11 pp.

MOORE, J. J., – 1972 – An outline of computer-based methods for the analysis of phytosociological data. In, 'Grundfragen und Methoden in der Pflanzensoziologie', ed. E. VAN DER MAAREL & R. TÜXEN, *Ber. Symp. int. Ver. Vegetationskunde*, Rinteln 1970: 29—38. Junk, The Hague.

MOORE, J. J., P. FITZSIMONS, E. LAMBE & J. WHITE, – 1970 – A comparison and evaluation of some phytosociological techniques. *Vegetatio* 20: 1—20.

MOORE, J. J. & A. O'SULLIVAN, – 1970 – A comparison between the results of the Braun-Blanquet method and those of 'cluster-analysis'. In, 'Gesellschaftsmorphologie', ed. R. TÜXEN, *Ber. Symp. int. Ver. Vegetationskunde*, Rinteln 1966: 26—29. Junk, The Hague.

MORAVEC, J. – 1968 – Zu den Problemen der pflanzensoziologischen Nomenklatur. In, 'Pflanzensoziologische Systematik', ed. R. TÜXEN, *Ber. Symp. int. Ver. Vegetationskunde*, Stolzenau 1964. 142—151. Junk, The Hague.

MORAVEC, J., – 1969 – Die Anwendung der Typenmethode in der Phytosoziologischen Nomenklatur. *Folia geobot. phytotax.* 4: 23—31.

MORAVEC, J. – 1971 – A simple method for estimating homotoneity of sets of phytosociological relevés. *Folia geobot. phytotax.* 6: 147—170.
MÖRZER BRUYNS, M. F., – 1950 – On biotic communities. *Comm. S.I.G.M.A.* 96: 1—59, Montpellier.
MOSS, C. E., – 1971 – The fundamental units of vegetation. *New Phytol.* 9: 18—53.
MÜLLER, P. J., M. J. A. WERGER, B. J. COETZEE, D. EDWARDS, N. G. JARMAN, – 1972 – An apparatus for facilitating the manual tabulation of phytosociological data. *Bothalia* 10, (in the press).
MÜLLER, TH., – 1968 – Die Gliederung von Pflanzengesellschaften in Rassen und Formen als Beitrag zur Landschaftsökologie, dargestellt am Beispiel von wärmeliebenden Eichen-Hainbuchen-wäldern in Südwestdeutschland. In, 'Pflanzensoziologie und Landschaftsökologie', ed. R. TÜXEN, *Ber. Symp. int. Ver. Vegetationskunde*, Stolzenau 1963: 60—64. Junk, The Hague.
MÜLLER, TH., – 1970 – Mosaikkomplexe und Fragmentkomplexe. In R. TÜXEN (ed): Gesellschaftsmorphologie. *Ber. Symp. int. Ver. Vegetationskunde*, Rinteln 1966: 69—75, Junk, The Hague.
NEUHÄUSL R., – 1968 – Draft proposals for nomenclature principles in floristic phytosociology. *Folia geobot. phytotax.* 3: 47—55.
NEUHÄUSL, R. & Z. NEUHÄUSLOVÁ-NOVOTNÁ, – 1972 – Eine einfache Orientierungsmethode zur Beurteilung des Assoziations-ranges. In, 'Basic problems and methods in phytosociology', ed. E. VAN DER MAAREL & R. TÜXEN, *Ber. Symp. int. Ver. Vegetationskunde*, Rinteln 1970: 211—223. Junk, The Hague.
NORDHAGEN, R., – 1937 – Versuch einer neuen Einteilung der subalpinen – alpinen Vegetation Norwegens. *Bergens Mus. Årbok* Naturv.rekke 1936: 1—88.
NORDHAGEN, R., – 1943 – Sikilsdalen og Norges fjellbeiter. En plantesosiologisk monografi. *Bergens Mus. Skr.* 22: 1—607.
OBERDORFER, E., – 1957 – Süddeutsche Pflanzengesellschaften. Jena, XXXVIII + 564 pp.
OBERDORFER, E., – 1960 – Pflanzensoziologische Studien in Chile, ein Vergleich mit Europa. Flor. Veg. Mundi 2. 208 pp. Weinheim/Bergst.
OBERDORFER, E., – 1968 – Assoziation, Gebietsassoziation, Geographische Rasse. In, 'Pflanzensoziologisch Systematik', ed. R. TÜXEN, *Ber. Symp. int. Ver. Vegetationskunde*, Stolzenau 1964: 124—131. Junk, The Hague.
OBERDORFER, E., – 1970 – Pflanzensoziologische Strukturprobleme am Beispiel Kanarischer Pflanzengesellschaften. In, 'Gesellschaftsmorphologie.', ed. R. TÜXEN, *Ber. Symp. int. Ver. Vegetationskunde*, Rinteln 1966: 273—278. Junk, The Hague.
OBERDORFER, E. et al., – 1967 – Systematische Übersicht der westdeutschen Phanerogamen- und Gefässkryptogamen-Gesellschaften. Ein Diskussionsentwurf. *Schr. Reihe Vegetationskde* 2: 7—62.
ODUM, E. P., – 1969 – The strategy of ecosystem development. *Science* 164: 262—270.
OHBA, T., – 1972 – Übersicht über die Salzwiesengesellschaften Japans. In, 'Grundfragen und Methoden in der Pflanzensoziologie', ed. E. VAN DER MAAREL & R. TÜXEN, *Ber. Symp. int. Ver. Vegetationskunde*, Rinteln 1970: 413—418. Junk, The Hague.
ORLOCI, L., – 1966 – Geometric models in ecology. I The theory and application of some ordination methods. *J. Ecol.* 54: 193—215.
PACZOSKI, J. K., – 1930 – Życie gromadne róslin. Social life of plants. *Bibljot. Bot. (Krakow)* 2: 1—40.
PASSARGE, H., – 1966 – Die Formationen als höchste Einheiten der soziologischen Vegetationssystematik. *Repert. spec. nov. Regni veg.* 73: 226—235.

PASSARGE, H., – 1968 – Neue Vorschläge zur Systematik nord-mitteleuropäischer Waldgesellschaften. *Reprium nov. spec. Regni veg.* 77: 75—103,
PASSARGE, H. & G. HOFMANN, – 1968 – Pflanzengesellschaften des nordostdeutschen Flachlandes II. Pflanzensoziologie Bd XVI. Jena Fischer 288 pp.
PAVILLARD, J., – 1901 – Eléments de Biologie végétale. Paris-Montpellier 589 pp.
PAVILLARD, J., – 1912 – Essai sur la nomenclature phytogéographique. *Bull. soc. langued. Géogr.* 35: 165—176.
PAVILLARD, J., – 1919 – Remarques sur la nomenclature phytogéographique. Montpellier 27 pp.
PAVILLARD, J., – 1920 – Espèces et associations. Essai phytosociologique. Montpellier, 34 pp.
PAVILLARD, J., – 1921 – L'Association végétale, unité phytosociologique. Montpellier, 11 pp.
PAVILLARD, J., – 1935a – Eléments de sociologie végétale (Phytosociologie). *Actual. scient. ind.* 251: 3—102, Paris.
PAVILLARD, J., – 1935b – The present status of the plant-association. *Bot. Rev.* 1: 210—232.
PFEIFFER, H. H., – 1957 – Betrachtungen zum Homogenitätsproblem in der Pflanzensoziologie. *Mitt. flor. soz. Arbeitsgem.* Stolzenau, NF 6/7: 103—117.
PIGNATTI, S., – 1964 – Ein neues Verfahren zur Bearbeitung von Assoziationstabellen. *Acta bot. croat. Vol. extraord.* 89—93.
PIGNATTI, S., – 1968a – Die Verwertung der sogenannten Gesamtarten für die floristische Systematik. In, 'Pflanzensoziologische Systematik', ed. R. TÜXEN, *Ber. Symp. int. Ver. Vegetationskunde*, Stolzenau 1964: 71—73. Junk, The Hague.
PIGNATTI, S., – 1968b – Die Inflation der höheren pflanzensoziologischen Einheiten. In, 'Pflanzensoziologische Systematik', ed. R. TÜXEN, *Ber. Symp. int. Ver. Vegetationskunde*, Stolzenau 1964: 85—97. Junk, The Hague.
POORE, M. E. D., – 1955a – The use of phytosociological methods in ecological investigations. I. The Braun-Blanquet system. *J. Ecol.* 43: 226—244.
POORE, M. E. D., – 1955b – The use of phytosociological methods in ecological investigations. II. Practical issues involved in an attempt to apply the Braun-Blanquet system. *J. Ecol.* 43: 245—269.
POORE, M. E. D., – 1955c – The use of phytosociological methods in ecological investigations. III. Practical applications. *J. Ecol.* 43: 606—651.
POORE, M. E. D., – 1956 – The use of phytosociological methods in ecological investigations. IV. General discussion of phytosociological problems. *J. Ecol.* 44: 28—50.
POORE, M. E. D., – 1962 – The method of successive approximation in descriptive ecology. *Adv. ecol. Res.* 1: 35—68.
POST, H. VON, – 1862 – Försök till en systematisk uppställning af vextställena i mellersta Sverige. Bonnier, Stockholm. 42 p.
PRAEGER, R. LLOYD, – 1934 – The botanist in Ireland. Dublin.
PRESTON, F. W., – 1962 – The canonical distribution of commonness and rarity. *Ecology* 43: 185—215, 410—432.
QUÉZEL, P., – 1965 – La végétation su Sahara, du Tchad à la Mauretanie. Geobotanica Selecta II, xii + 333 pp., Fischer, Stuttgart.
QUÉZEL, P. & P. VERDIER, – 1953 – Les méthodes de la phytosociologie sontelles applicables à l'étude des groupements animaux? *Vegetatio* 4: 165—181.
RAABE, E. W., – 1952 – Über den 'Affinitätswert' in der Pflanzensoziologie. *Vegetatio* 4: 53—68.
RABELER, W., – 1937 – Die planmässige Untersuchung der Soziologie, Ökologie und Geographie der heimischen Tiere, besonders der land- und forstwirtschaftlich wichtigen Arten. :*JBer. naturhist. Ges. Hannover* 81—87: 236—247.

RABELER, W., – 1960 – Biozönotik auf Grundlage der Pflanzengesellschaften. *Mitt. flor.-soz. Arbeitsgem.*, Stolzenau, *NF* 8: 311—332.
RABELER, W., – 1962 – Die Tiergesellschaften von Laubwäldern (Querco-Fagetea) im oberen und mittleren Wesergebiet. *Mitt. flor.-soz. Arbeitsgem.*, Stolzenau, *NF* 9: 200—229.
RABELER, W., – 1965 – Die Pflanzengesellschaften als Grundlagen für die landbiozönotischen Forschung. In R. TÜXEN (ed): Biosoziologie. *Ber. Symp. int, Ver. Vegetationskunde*, Stolzenau 1960: 43—49. Junk, The Hague.
RAMENSKY, L. G., – 1930 – Zur Methodik der vergleichenden Bearbeitung und Ordnung von Pflanzenlisten und anderen Objekten, die durch mehrere, verschiedenartig wirkende Faktoren bestimmt werden. (transl. from the -Russian paper in Trudy soc. geobot.-lugov, 15—20 Jan. 1929 p. 11—36) *Beitr. Biol. Pfl.* 18: 269—304. Breslau.
RAUNKIAER, C., – 1934 – The life forms of plants and statistical plant geography. Oxford. xvi + 632 pp.
RAUSCHERT, S., – 1963 – Beitrag zur Vereinheitlichung der soziologischen Nomenklatur. *Mitt. flor.-soz. Arbeitsgem.*, Stolzenau, *NF* 10: 232—249.
RAUSCHERT, S., – 1969 – Über einige Probleme der Vegetationsanalyse und Vegetationssystematik. *Arch. Naturschutz Landschaftsforsch.* 9: 153—174.
ROGERS, D. J., – 1970 – A preliminary ordination study of forest vegetation in the Kirchleerau area of the Swiss midlands. *Ber. geobot. Inst. Rübel* 40: 28—78.
ROMANE, F., – 1972 – Un exemple d'analyse factorielle des correspondances en écologie végétale. In, 'Grundfragen und Methoden in der Pflanzensoziologie', ed. E. VAN DER MAAREL & R. TÜXEN, *Ber. Symp. int. Ver. Vegetationskunde*, Rinteln 1970: 151—162. Junk, The Hague.
RÜBEL, E., – 1912 – Pflanzengeographische Monographie des Berninagebietes. *Bot. Jb.* 47: 1—616.
SCAMONI, A. & H. PASSARGE, – 1959 – Gedanken zu einer natürlichen Ordnung der Waldgesellschaften. *Arch. Forstwes.* 8: 386—426.
SCAMONI, A. & H. PASSARGE, – 1963 – Einführung in die praktische Vegetationskunde. 2. Ed. Jena xi + 236 pp.
SCAMONI, A., H. PASSARGE & G. HOFMANN, – 1965 – Grundlagen zu einer objektiven Systematik der Pflanzengesellschaften. *Repert. Spec. nov. Regni veg.*, Beih. 142: 117—132.
SCHMID, E., – 1923 – Vegetationsstudiën in den Urner Reusstälern. Thesis, Zürich.
SCHMID, P. & N. KÜHN, – 1970 – Automatische Ordination von Vegetationsaufnahmen in pflanzensoziologischen Tabellen. *Naturwissenschaften* 57: 462.
SCHMITHÜSEN, J., – 1960 – Die Nadelhölzer in den Waldgesellschaften der südlichen Anden. *Vegetatio* 9: 313—327.
SCHMITHÜSEN, J., – 1961 – Allgemeine Vegetationsgeographie. In: OBST, ed.: Lehrbuch der Allgemeinen Geographie 4, Berlin. (3 ed. xxiii + 463 pp. 1968).
SCHNELL, R., – 1952 – Contribution à une étude phytosociologique et phytogéographique de l'Afrique occidentale: les groupements et les unités géobotaniques de la Région Guinéenne. *Mem. Inst. Franc. Afrique Noir Mélanges bot.* 18.
SCHOUW, J. F., – 1823 – Grundzüge einer allgemeinen Pflanzengeographie. Reimer, Berlin. 524 pp.
SCHRÖTER, C., – 1894 – Notes sur quelques associations de plantes rencontrées pendant les excursions dans la Valais. *Soc. bot. France, Bull.* 41 (sess. extraord.): 322—325.
SCHRÖTER, C. & O. KIRCHNER, – 1902 – Die Vegetation des Bodensees. *Schr. Ver. Gesch. Bodensees Lindau* 9 (2): 1—86.

SCHUBERT, R., – 1960 – Die Zwergstrauchreichen azidiphilen Pflanzengesellschaften Mitteldeutschlands. Pflanzensoziologie Band 11, Fischer, Jena. 235 pp.
SCHWICKERATH, M., – 1931 – Die Gruppenabundanz (Gruppenmächtigkeit); ein Beitrag zur Begriffsbildung der Pflanzensoziologie. *Bot. Jb.* 64: 1—16.
SCHWICKERATH, M., – 1940 – Ausgleich – und Richtungsprinzip als Grundlage der Pflanzengesellschaftslehre. *Repert. Spec. nov. Regni veg.*, Beih. 121: 53-67.
SCHWICKERATH, M., – 1942 – Bedeutung und Gliederung des Differentialartenbegriffs in der Pflanzengesellschaftslehre. *Beih. bot. Zentralbl.* 61B: 351—383.
SCHWICKERATH, M., – 1954a – Lokale Charakterarten - geographische Differentialarten. *Veröff. geobot. Inst. Rübel* 29: 96—104.
SCHWICKERATH, M., – 1954b – Die Landschaft und ihre Wandlung auf geobotanischer und geographischer Grundlage entwickelt und erläutert im Bereich der Messtischblattes Stolberg. Aachen.
SCHWICKERATH, M., – 1963 – Assoziationsdiagramme und ihre Bedeutung für die Vegetationskartierung. In R. TÜXEN (ed): Ber. Int. Symposium für Vegetationskartierung 1959 p. 11—35. Cramer, Weinheim.
SEGAL, S. & V. WESTHOFF, – 1959 – Die Vegetationskundliche Stellung von Carex buxbaumii Wahlenb. in Europa, besonders in den Niederlanden. *Acta bot. neerl.* 8: 304—329.
SEIBERT, P., – 1968 – Gesellschaftsring und Gesellschaftskomplex in der Randschaftsgliederung. In, 'Pflanzensoziologie und Landschaftsökologie', ed. R. TÜXEN, *Ber. Symp. int. Ver. Vegetationskunde*, Stolzenau 1963: 48—60. Junk, The Hague.
SHIMWELL, D. W., – 1972 – Description and Classification of Vegetation. Sidgwick & Jackson, London xiv + 322 pp.
SISSINGH, G., – 1950 – Onkruid-associaties in Nederland. Thesis Wageningen Versl. Landbouwk. Ondcrz. 56, 14: 1—24.
SOKAL, R. R. & P. H. A. SNEATH, – 1963 – Principles of Numerical Taxonomy. San Francisco xvi + 359 pp.
SØRENSEN, TH. A., – 1948 – A method of establishing groups of equal amplitude in plant sociology based on similarity of species content. *Biol. Skr. K. danske Vidensk. Selsk.* 5 (4): 1—34.
SPATZ, G., – 1969 – Elektronische Datenverarbeitung bei pflanzensoziologischer Tabellenarbeit. *Naturwissenschaften* 56: 470—471.
SPATZ, G., – 1972 – Eine Möglichkeit zum Einsatz der elektronischen Datenverarbeitung bei der pflanzensoziologischen Tabellenarbeit. In, 'Grundfragen und Methoden in der Pflanzensoziologie' ed. E. VAN DER MAAREL & R. TÜXEN, *Ber. Symp. int. Ver. Vegetationskunde*, Rinteln 1970: 251—258. Junk, The Hague.
STEBLER, F. G. & C. SCHRÖTER, – 1893 – Beiträge zur Kenntnis der Matten und Weiden der Schweiz. 10. Versuch einer Übersicht über die Wiesentypen der Schweiz. *Landw. Jahrb. Schweiz* 6: 95—212, Bern.
STOCKINGER, J. J. & W. F. HOLZNER, – 1972 – Rationelle Methode zur Auswertung pflanzensoziologischer Aufnahmen mittels Elektronenrechner. In, 'Grundfragen und Methoden in der Pflanzensoziologie', ed. E. VAN DER MAAREL & R. TÜXEN, *Ber. Symp. int. Ver. Vegetationskunde*, Rinteln 1970: 239—248. Junk, The Hague.
SUKATSCHEW, W. N. (SUKACHEV, V. N.) – 1929 – Über einige Grundbegriffe in der Phytosoziologie. *Ber. dt. bot. Ges.* 47: 296—312.
SUKATSCHEW, W. N. (SUKACHEV, V. N.), – 1954 – Die Grundlagen der Waldtypen. *Angew. Pfl. Soziol.* Wien. Festschr. Aichinger 2: 956—964.

SUKACHEV, V. N., – 1960 – The correlation between the concept 'forest ecosystem' and 'forest biogeocenose' and their importance for the classification of forests. *Silva fenn.* 105: 94—97.

SUZUKIO-TOKIO, – 1954 – L'Alliance du Shiion sieboldi. *Vegetatio* 5/6: 361—372.

SZAFER, W. & B. PAWŁOWSKI, – 1927 – Die Pflanzenassoziationen der Tatra-Gebirges. A. Bemerkungen über die angewandte Arbeitsmethodik (zu den Teilen III, IV, und V). *Bull. Int. Acad. pol. Sci. Lett.*, Cl. Sci. Nat. Math., sér B, 1926 (12) Suppl: 1—12.

TAKHTAJAN, A., – 1969 – Flowering Plants, Origin and Dispersal. Transl. by C. JEFFREY. Edinburgh Oliver & Boyd. x + 310 pp.

TANSLEY, A. G., – 1920 – The classification of vegetation and the concept of development. *J. Ecol.* 8: 118—149.

TANSLEY, A. G., – 1922 – The new Zürich-Montpellier school. *J. Ecol.* 10: 241—243.

TANSLEY, A. G., – 1935 – The use and abuse of vegetational concepts and terms. *Ecology* 16: 284—307.

THIENEMANN, A., – 1956 – Leben und Umwelt. Vom Gesamthaushalt der Natur. Hamburg, 153 pp.

THURMANN, J., – 1849 – Essai de phytostatique appliqué à la chaîne du Jura et aux contrées voisines. Bern.

TRENTEPOHL, M. W., – 1968 – Ein mechanisch-elektromagnetisches Gerät zur Schnellbearbeitung pflanzensoziologischer Tabellen. Ref. Symposium Tatsachen und Probleme der Grenzen in der Vegetation Rinteln 1968.

TUOMIKOSKI, R., – 1942 – Untersuchungen über die Untervegetation der Bruchmoore in Ostfinnland. I. Zur Methodik der pflanzensoziologischen Systematik. *Ann. bot. Soc. zool. bot. fenn. Vanamo* 17 (1) vi + 203 pp.

TÜXEN, R., – 1933 – Klimaxprobleme des nw. europäischen Festlandes. *Ned. kruidk. Arch.* 43: 293—309.

TÜXEN, R., – 1937 – Die Pflanzengesellschaften Nordwestdeutschlands. *Mitt. Flor.-soz. Arbeitsgem. Niedersachsen* 3: 1—170.

TÜXEN, R., – 1950 – Grundriss einer Systematik der nitrophilen Unkrautgesellschaften. *Mitt. flor.-soz. Arbeitsgem.*, Stolzenau, *N.F.* 2: 94—175.

TÜXEN, R., – 1952 – Hecken und Gebüsche. *Mitt. geogr. Ges. Hamburg* 50: 85—117.

TÜXEN, R., – 1955 – Das System der nordwestdeutschen Pflanzengesellschaften. *Mitt. flor.-soz. Arbeitsgem.*, Stolzenau, *N.F.* 5: 155—176.

TÜXEN, R., – 1956 – Die heutige potentielle natürliche Vegetation als Gegenstand der Vegetationskartierung. *Angew. PflSoziol.*, Stolzenau 13: 5—42.

TÜXEN, R., – 1957 – Entwurf einer Definition der Pflanzengesellschaft (Lebensgemeinschaft). *Mitt. flor.-soz. Arbeitsgem.*, Stolzenau, *N.F.* 6/7: 112—113.

TÜXEN, R., – 1962 – Zur systematischen Stellung von Spezialisten-Gesellschaften. *Mitt. flor.-soz Arbeitsgem.*, Stolzenau, *N.F.* 9: 57—59.

TÜXEN, R. (ed.), – 1963 – Bericht über das Internationale Symposion für Vegetationskartierung Stolzenau 1959. Cramer, Weinheim. vii + 500 pp.

TÜXEN, R. (ed), – 1965a – Biosoziologie. *Ber. Symp. int. Ver. Vegetationskunde*, Stolzenau 1960. xvi + 350 pp. Junk, The Hague.

TÜXEN, R., – 1965b – Wesenszüge der Biozönose: Gesetze für das Zusammenleben von Pflanzen und Tieren. In, 'Biosoziologie', ed. R. TÜXEN, *Ber. Symp. int. Ver. Vegetationskunde*, Stolzenau 1960: 10—13. Junk, The Hague.

TÜXEN, R., – 1966a – Über nitrophile Elymus-Gesellschaften an nordeuropäischen, nordjapanischen und nordamerikanischen Küsten. *Ann. bot. fenn.* 3: 358—367.

TÜXEN, R., (ed.), – 1966b – Anthropogene Vegetation. *Ber. Symp. int. Ver. Vegetationskunde*, Stolzenau 1961. xvi + 398 pp. Junk, The Hague.

TÜXEN, R., – 1967a – Ausdauernde nitrophile Saumgesellschaften Mitteleuropas. Contributii Bot. Cluj p. 431—453.

TÜXEN, R., – 1967b – Pflanzensoziologischen Beobachtungen an südwestnorwegischen Küsten-Dünengebieten. *Aquilo Ser. Bot.* 6: 241—272.

TÜXEN, R., (ed), – 1967c – Pflanzensoziologie und Palynologie. *Ber. Symp. int. Ver. Vegetationskunde,* Stolzenau 1962, xvii + 275 pp.

TÜXEN, R. (ed), – 1968a — Pflanzensoziologie und Landschaftsökologie. *Ber. Symp. int. Ver. Vegetationskunde,* Stolzenau 1963. xvii+426 pp. Junk, The Hague.

TÜXEN, R. (ed), – 1968b – Pflanzensoziologische Systematik. *Ber. Symp. Int. Ver. Vegetationskunde,* Stolzenau 1964. xii+348 pp. Junk, The Hague.

TÜXEN, R., – 1969a – Stand und Ziele geobotanischer Forschung in Europa. *Ber. geobot. Inst. Rübel* 39: 13—26.

TÜXEN, R. (ed), – 1969b – Experimentelle Pflanzensoziologie. *Ber. Symp. Int. Ver. Vegetationskunde,* Rinteln 1965, xvii + 256 pp. Junk, The Hague.

TÜXEN, R., – 1970a – Entwicklung, Stand und Ziele der pflanzensoziologischen Systematik (Syntaxonomie). *Ber. dt. bot. Ges.* 83: 633—639.

TÜXEN, R., – 1970b – Pflanzensoziologie als synthetische Wissenschaft. *Misc. Papers Landbouwhogeschool Wageningen* 5: 141—159.

TÜXEN, R., – 1970c – Einige Bestandes- und Typenmerkmale in der Struktur der Pflanzengesellschaften. In, 'Gesellschaftsmorphologie', ed. R. TÜXEN, *Ber. Symp. int. Ver. Vegetationskunde,* Rinteln 1966. p. 76—98. Junk, The Hague.

TÜXEN, R., – 1971 – Besprechung von H. Passarge & G. Hofmann 1968 - Pflanzengesellschaften des nordostdeutschen Flachlandes II. *Vegetatio* 23: 382—383.

TÜXEN, R., – 1972 – Kritische Bemerkungen zur Interpretation pflanzensoziologischer Tabellen. In, 'Grundfragen und Methoden in der Pflanzensoziologie', ed. E VAN DER MAAREL & R. TÜXEN, *Ber. Symp. int. Ver. Vegetationskunde,* Rinteln 1970: 168—173. Junk, The Hague.

TÜXEN, R. & W. H. DIEMONT, – 1937 – Klimaxgruppe und Klimaxschwarm. Ein Beitrag zur Klimaxtheorie. *JBer. naturhist. Ges. Hannover* 88,89: 73—87.

TÜXEN, R. & H. ELLENBERG, – 1937 – Der Systematische und der ökologische Gruppenwert. Ein Beitrag zur Begriffsbildung und Methodik der Pflanzensoziologie. *Mitt. flor.-soz. Arbeitsgem. Niedersachsen* 3: 171—184.

TÜXEN, R., A. MIYAWAKI & K. FUJIWARA, – 1972 – Eine erweiterte Gliederung der Oxycocco-Sphagnetea. In, 'Grundfragen und Methoden der Pflanzensoziologie', ed. E. VAN DER MAAREL & R. TÜXEN, *Ber. Symp. int. Ver. Vegetationskunde,* Rinteln 1970: 500—509. Junk, The Hague.

TÜXEN, R. & E. PREISING, – 1951 – Erfahrungsgrundlagen für die pflanzensoziologische Kartierung des westdeutschen Grünlandes. *Angew. Pfl.Soziol.* Stolzenau 4: 1—28.

VARESCHI, V. – 1931 – Die Gehölztypen des obersten Isartales. *Ber. naturw. -med. Ver. Innsbruck* 42: 79—184.

VESTAL, A. G., – 1949 – Minimum areas for different vegetations. *Univ. Illinois biol. Monogr.* 20 (3):1—129.

VRIES, D. M. DE, J. P. BARETTA & G. HAMMING, – 1954 – Constellation of frequent herbage plants, based on their correlation in occurrence. *Vegetatio* 5/6: 106—111.

WAGNER, H., – 1954 – Gedanken zur Berücksichtigung der mehr-dimensionalen Beziehungen der Pflanzengesellschaften in der Vegetationssystematik. *Rapp. Comm. 8th Congr. Int. Bot., Paris,* Sect. 7—8: 9—11.

WAGNER, H., – 1968 – Prinzipienfragen der Vegetationssystematik. In. 'Pflanzensoziologische Systematik', ed. R. TÜXEN, *Ber. Symp. int. Ver. Vegetationskunde* Stolzenau 1964: 15—20. Junk, The Hague.

WAGNER, H., – 1972 – Zur Methodik der Erstellung und Auswertung von Vegetaitonstabellen. In, 'Grundfragen und Methoden in der Pflanzensoziologie', ed. E. VAN DER MAAREL & R. TÜXEN, Ber. Symp. int. Ver. Vegetationskunde Rinteln 1970: 225—233. Junk, The Hague.

WALTER, H. & H. STRAKA, – 1970 – Arealkunde. Floristisch-historische Geobotanik. 2. Aufl. Ulmer, Stuttgart 478 pp.

WARMING, E., – 1895 – Plantesamfund. Kjøbenhavn. 335 pp.

WEAVER, J. E., & F. E. CLEMENTS, – 1938 – Plant Ecology. 2.ed. New York-London. xxii + 601 pp.

WENDELBERGER, G., – 1951 – Das vegetationskundliche System Erwin Aichingers und seine Stellung im pflanzensoziologischen Lehrgebäude Braun-Blanquets. Angew. Pfl. Soziol. Wien. 1: 69—92.

WERGER, M. J. A., – 1973 – Phytosociology of the upper Orange River Valley, South Africa. A syntaxonomical and synecological study. Thesis Nijmegen. Pretoria, 222 pp.

WERGER, M. J. A., F. J. KRUGER & H. C. TAYLOR, – 1972 – A phytosociological study of the Cape Fijnbos and other vegetation at Jonkershoek, Stellenbosch. Bothalia 10 (4): 1—19, German version Vegetatio 24: 71—89 (1972).

WESTHOFF, V., – 1947 – The vegetation of dunes and salt marshes on the Dutch islands of Terschelling, Vlieland and Texel. Thesis Utrecht, 131 pp.

WESTHOFF, V., – 1949 – De plantengezelschappen van Botshol. In: Landschap, flora en vegetatie van de Botshol nabij Abcoude. Baambrugge p. 43—102.

WESTHOFF, V., – 1950 – Het associatiebegrip in geografisch verband. Ned. kruidk. Arch. 57: 98—100.

WESTHOFF, V., – 1951 – An analysis of some concepts and terms in vegetation study or phytocenology. Synthese 8: 194—206.

WESTHOFF, V., – 1954 – Some remarks on synecology. Vegetatio 5—6: 120—128.

WESTHOFF, V., – 1959 – The vegetation of Scottish pine woodlands and Dutch artificial coastal pine forests; with some remarks on the ecology of Listera cordata. Acta bot. neerl. 8: 422—448.

WESTHOFF, V., – 1965 – Plantengemeenschappen. In: 'Het leven der planten'. 2e druk Zeist-Arnhem p. 288—349.

WESTHOFF, V., – 1967 – Problems and use of structure in the classification of vegetation. The diagnostic evaluation of structure in the Braun-Blanquet system. Acta bot. neerl. 15: 495—511.

WESTHOFF, V., – 1970 – Vegetation study as a branch of biological science. Misc. Papers Landbouwhogeschool Wageningen 5: 11—30.

WESTHOFF, V., – 1971a – The dynamic structure of plant communities in relation to the objectives of conservation. In E. DUFFEY & A. S. WATT (ed) The scientific management of animal and plant communities for conservation. Oxford. p. 3—14.

WESTHOFF, V., – 1971b – Choice and management of nature reserves in the Netherlands. Bull. Jard. Bot. Nat. Belgique 41 (1): 231—245.

WESTHOFF, V., J. J. BARKMAN, H. DOING & C. G. VAN LEEUWEN, – 1959 – Enige opmerkingen over de terminologie in de vegetatiekunde. Jaarb. Kon. ned. bot. Ver. 1958: 44—46.

WESTHOFF, V. & A. J. DEN HELD, – 1969 – Plantengemeenschappen in Nederland. Zutphen Thieme, 324 pp.

WESTHOFF, V. & C. G. VAN LEEUWEN, – 1966 – Ökologische und systematische Beziehungen zwischen natürlicher und anthropogener Vegetation. In, 'Anthropogene Vegetation', ed. R. TÜXEN, Ber. Symp. int. Ver. Vegetationskunde, Stolzenau 1961: 156—172. Junk, The Hague.

WESTHOFF, V., H. PASSCHIER & G. SISSINGH, – 1946 – Overzicht der plantengemeenschappen in Nederland. 2e dr. Breughel, Amsterdam 118 pp.

WESTHOFF, V. & J. N. WESTHOFF-DE JONCHEERE, – 1942 – Verspreiding en nestoecologie van de mieren in de Nederlandse bossen. *Tijdschr. Plantenziekten* 48: 138—212.
WHITTAKER, R. H., – 1951 – A criticism of the plant association and climatic climax concepts. *NW. Sci.* 25: 17—31.
WHITTAKER, R. H., – 1953 – A consideration of climax theory; the climax as a population and pattern. *Ecol. Monogr.* 23: 41—78.
WHITTAKER, R. H., – 1954a – Plant populations and the basis of plant indication. *Angew. Pfl. Soziol.* Wien, Festschr. Aichinger 1, 183—206.
WHITTAKER, R. H., – 1954b – The ecology of serpentine soils. IV. The vegetational response to serpentine soils. *Ecology* 35: 275—288.
WHITTAKER, R. H., – 1956 – Vegetation of the Great Smoky Mountains. *Ecol. Monogr.* 26: 1—80.
WHITTAKER, R. H., – 1960 – Vegetation of the Siskiyou Mountains, Oregon and California. *Ecol. Monog.* 30: 279—338.
WHITTAKER, R. H., – 1962 – Classification of natural communities. *Bot. Rev.* 28: 1—239.
WHITTAKER, R. H., – 1967 – Gradient analyses of vegetation. *Biol. Rev. London* 42: 207—264.
WHITTAKER, R. H., – 1970 – The population structure of vegetation. In R. Tüxen (ed) Gesellschaftsmorphologie. *Ber. Symp. int. Ver. Vegetationskunde*, Rinteln 1966: 39—59. Junk, The Hague.
WHITTAKER, R. H., – 1972 – Convergences of ordination and classification. In, 'Basic problems and methods in phytosociology', ed. E. VAN DER MAAREL & R. TÜXEN, *Ber. Symp. int. Ver. Vegetationskunde* Rinteln 1970: 39—57. Junk, The Hague.
WILLIAMS, C. B., – 1964 – Patterns in the Balance of Nature and Related Problems in Quantitative Ecology. London-New York Academic Press. vii + 324 pp.
WILLIAMS, W. T. & J. M. LAMBERT, – 1959 – Multivariate methods in plant ecology. I Association analysis in plant communities. *J. Ecol.* 47: 83—101.
WILLIAMS, W. T. & J. M. LAMBERT, – 1961 – Multivariate analysis in plant ecology. III Inverse association analysis. *J. Ecol.* 49: 717—729.
WILLIAMS, W. T., J. M. LAMBERT & G. N. LANCE, – 1966 – Multivariate Methods in Plant Ecology. V Similarity Analyses and information-analysis. *J. Ecol.* 54: 427—445.
WILMANNS, O., – 1959 – Ein Gerät zur Mechanisierung von Tabellenarbeit. *Ber. dt. bot. Ges.* 72: 419—420.
WILMANNS, O., – 1970 – Kryptogamen - Gesellschaften oder Kryptogamen - Synusien. In 'Gesellschaftsmorphologie', ed. R. TÜXEN, *Ber. Symp. int. Ver. Vegetationskunde* Rinteln 1966: 1—6. Junk, The Hague.
WISHART, D., – 1969 – Clustan Ia. User manual. St. Andrews Computing Centre.
ZINDEREN BAKKER JR., E. M. VAN, – 1971 – Ecological investigations on ravine forests of the Eastern Orange Free State. Thesis Bloemfontein, 123 pp.
ZOLLER, H., – 1954 – Die Typen der Bromus erectus-Wiesen des Schweizer Juras. Ihre Abhängigkeit von den Standortsbedingungen und wirtschaftlichen Einflüssen und ihre Beziehungen zur ursprünglichen Vegetation. *Beitr. geobot. Landesauf. Schweiz* 33: 1—309.
ZONNEVELD, I. S., – 1960 – De Brabantse Biesbosch. Een studie van bodem en vegetatie van een zoetwatergetijdendelta. With English summary, Thesis, Wageningen, part A, p. 1—210 - part B, p. 1—396.

REFERENCES ADDED TO THE SECOND EDITION

BARKMAN, J. J., J. MORAVEC & S. RAUSCHERT, – 1976 – Code of phytosociological nomenclature. (Engl., Germ., French). *Vegetatio* 32: 131-185.
BEEFTINK, W. G. & J.-M. GÉHU, – 1973 – Spartinetea maritimae *Prodrome des groupements végétaux d'Europe*. Vol. 1. Cramer, Lehre. 43 pp.
BOTTLIKOVÁ, A. et al., – 1975 – Quelques résultats obtenus par l'analyse factorielle et les profils écologiques sur des observations phytoécologiques receuillies dans la vallée de Liptov (Tchécoslovaquie). *Vegetatio* 31: 79–91.
BOUXIN, G., – 1975 – Ordination and classification in the savanna vegetation of the Akagera Park. (Rwanda, Central Africa). *Vegetatio* 29: 155—167.
BRAAKHEKKE, W. G. & E. I. BRAAKHEKKE-ILSINK, – 1976 – Nitrophile Saumgesellschaften im Südosten der Niederlande. (Engl. summ.) *Vegetatio* 32: 55-60.
COETZEE, B. J. & M. J. A. WERGER, – 1975 – On association-analysis and the classification of plant communities. *Vegetatio* 30: 201—206.
DALE, M. B. & H. T. CLIFFORD, – 1976 – On the effectiveness of higher taxonomic ranks for vegetation analysis. *Austr. J. Ecol.* 1: 37—62.
DALE, M. B. & L. QUADRACCIA, – 1973 – Computer assisted tabular sorting of phytosociological data. (Ital. summ.) *Vegetatio* 28: 57-73.
DALE, M. B. & L. J. WEBB, – 1975 – Numerical methods and the establishment of associations. (Germ. summ.) *Vegetatio* 30: 77–87.
ELLENBERG, H., – 1974 – Zeigerwerte der Gefässpflanzen Mitteleuropas. *Scripta Geobot.* 9: 1—97.
FEOLI, E., – 1973 – Un esempio di ordinamento di tipi fitosociologici mediante l'analisi delle componenti principali. *Not. Fitosoc.* 7: 21—27.
GÉHU, J.-M., – 1974 – Sur l'emploi de la méthode phytosociologique sigmatiste dans l'analyse, la définition et la cartographie des paysages. *Compt. Rend. Acad. Sci. Paris* 279: 1167—1170.
GÉHU, J.-M. (ed.), – 1975 – La végétation des dunes maritimes. *Coll. Phytosoc.* 1. Cramer, Vaduz. 283 pp.
GILS, H. VAN, E. KEYSERS & W. LAUNSPACH, – 1975 – Saumgesellschaften im klimazonalen Bereich des Ostryo-Carpinion orientalis. *Vegetatio* 31: 47—64.
GOUNOT, M. & M. CALLÉJA, – 1962 – Coefficient de communauté, homogenéité et aire minimale. *Bull. Serv. Carte phytogéogr.*, Paris, B, 7: 181—210.
GUINOCHET, M., – 1973 – Phytosociologie. Masson et Cie., Paris, VI+228 pp.
HILL, M. O., – 1973 – Reciprocal averaging: an eigenvector method of ordination. *J. Ecol.* 61: 237—249.
HILL, M. O., – 1974 – Correspondence analysis: a neglected multivariate method. *J.R. statist. Soc., Ser. C (Appl. Statist.)* 23: 340–354.
HILL, M. O., R. G. H. BUNCE & M. W. SHAW, – 1975 – Indicator species analysis: a devisive polythetic method of classification, and its application to a survey of native pinewoods in Scotland. *J. Ecol.* 63: 597—613.
JANSSEN, J. G. M., – 1975 – A simple clustering procedure for preliminary classification of very large sets of phytosociological relevés. (Germ. summ.) *Vegetatio* 30: 67–71.
KOPECKÝ, K., – 1974 – Die anthropogene nitrophile Saumvegetation des Gebirgés Orlické Lory (Adlergebirge) und seines Vorlandes. *Rozpr. Čs. Akad. Věd.*, Ser. math.-nat., Praha. 84(1): 1–173.
KOPECKÝ, K. & S. HEJNÝ, – 1973 – Neue syntaxonomische Auffassung der Gesellschaften ein- bis zweijähriger Pflanzen der Galio-Urticetea in Böhmen. *Folia Geobot. Phytotax.* 8: 49—66.
KOPECKÝ, K. & S. HEJNÝ, – 1974 – A new approach to the classification of anthropogenic plant communities. *Vegetatio* 29: 17–20.
KORTEKAAS, W. M. & E. VAN DER MAAREL, – 1973 – A numerical classification

of Spartinetum vegetations. II Comparison of the computer-based numerical system with the system published in the Prodrome des Groupements Végétaux. Paper Prague Conference Working Group for Data-Processing in Phytosociology.

LACOSTE, A., – 1975 – La végétation de l'étage subalpin du bassin supérieur de la Tinée (Alpes-Maritimes). *Phytocoenologia* 3: 83—345.

LACOSTE, A., – 1976 – Relations floristiques entre les groupements prairiaux du Triseto-Polygonion et les Megaphorbiaies (Adenostylion) dans les Alpes occidentales. *Vegetatio* 31: 161—176.

MAAREL, E. VAN DER, – 1974 – The Working Group for Data-Processing of the International Society for Plant Geography and Ecology in 1972-1973. *Vegetatio* 29: 63—67.

MAAREL, E. VAN DER, L. ORLÓCI & S. PIGNATTI, – 1976 – Data-processing in phytosociology, retrospect and anticipation. (Germ. summ.) *Vegetatio* 32:65-72.

MAAREL, E. VAN DER & A. H. P. STUMPEL, – 1975 – Landschaftsökologische Kartierung und Bewertung in den Niederlanden. In P. Müller (ed.), Verhandlungen der Geschellschaft für Ökologie, Erlangen 1974, p. 231—240, Junk, Den Haag.

MORAVEC, J., – 1973 – The determination of the minimal-area of phytocoenoses. *Folia Geobot. Phytotax.* 8: 23—47.

MORAVEC, J., – 1973b – Some notes on estimation of the basic homogeneity-coefficient of sets of phytosociological relevés. *Folia Geobot. Phytotax.* 8: 429—434.

MUELLER-DOMBOIS, D. & H. ELLENBERG, – 1974 – Aims and methods of vegetation ecology. John Wiley & Sons, New York. XX + 547 pp.

ORLÓCI, L., – 1975 – Multivariate analysis in vegetation research. Junk, The Hague. VIII+276 pp.

PIGNATTI, S., – 1975 – Pflanzensoziologie am Scheideweg. *Vegetatio* 30: 149—152.

PIGNATTI, E. & S. PIGNATTI, – 1975 – Syntaxonomy of the Sesleria varia-grasslands of the calcareous Alps. (Germ. summ.) *Vegetatio* 28: 5-14.

SPATZ, G. & J. SIEGMUND, – 1973 – Eine Methode zur tabellarischen Ordination, Klassifikation und ökologischen Auswertung von pflanzensoziologischen Bestandsaufnahmen durch den Computer. (Engl. summ.) *Vegetatio* 28: 1—17.

STANEK, W., – 1973 – A comparison of Braun-Blanquet's method with sum-of-squares agglomeration for vegetation classification. (French summ.) *Vegetatio* 27: 323-338.

STOCKINGER, F. & W. HOLZNER, – 1973 – Prinzipien einer Implementierung des pflanzensoziologischen Zerlegungsproblems. *Mitt. Bot. Linz* 4: 87—106.

TÜXEN, R., – 1973 – Vorschlag zur Aufnahme von Gesellschaftskomplexen in potentiellen natürlichen Vegetationsgebieten. *Acta Bot., Acad. Sci. Hung.* 19: 379—384.

WERGER, M. J. A., – 1972 – Species-area relationship and plot size: with some examples from South-African vegetation. *Bothalia* 10: 583—594.

WERGER, M. J. A., – 1974 – The place of the Zürich-Montpellier method in vegetation science. *Folio Geobot. Phytotax.* 9: 99—109.

INDEX

A

Abstract units 3-4, 127-129, 294-295
Abundance 309-310
Additions to Second Edition 374, 398
Agglomerative approaches 262, 269-274, 361, 364
Agropyro repentis-Aegopodietum 368
ALEKHIN, V. V. 9, 173, 174, 177, 186
ALEKSANDROVA, V. D., article 167
Algorithm 268, 273
Alliance 18, 215, 291, 332-334, 340-344, 376
Alpha diversity 360
American Tradition 10, 67-69
Analogy, organismic 67-68, 293
Analytical phase 301-313, 372, 375
Animal communities
- classification 11, 348
Applied phytosociology 351
Aquatic communities 14, 16, 122, 136-142, 308
Area
- minimum 143, 146, 152, 209-210, 258, 305-307, 375
- sample 146, 209-210, 218-220, 258, 305-307, 375
- species curves 210-211, 306, 318
- representative 210
Areal types of species 13-14, 230-231
Artemisietum 329-333
Artificial vs. natural units 4-6, 299-301
Association (see under community-types)
Association analysis 10, 261, 263-271
Association-complex 207, 224, 349
Association-group 123, 178, 222, 291
Association-individual 294
Association of species 358-359
Association, Uppsala theory 207-217, 234
Associes 16, 68
Associon 122-123, 132, 215
Australian ecology 10, 46, 48
Authors' names 342-343
Axes, interpretation 368 (see also *Ordination*)

B

BARKMAN, J. J. 18, 113, 118-119, 125, 132-135, 145-149, 152, 296, 310, 317-319, 327, 337, 339, 349
- article, 111

Basal communities 376
BEARD, J. S. 14, 49-50, 56-58
- article 33
Beech forests (hêtraies) 339
Bell-shaped curves (see Gaussian curves)
Beta diversity 319
Biochore 41
Bioclimatic diagram 43
Biocoenose 13, 120, 138-139, 142, 293
Biogeocoenose 9, 12, 19, 100, 174, 293
Biome and biome-type 11, 36, 38
Biotic community 348
Biotic provinces 14
Biotope 37, 84-85, 103, 113, 120, 353-354, 368
Biotope-types 12
BÖCHER, T. W. 8, 229-230, 354
Bog vegetation 85-87, 213, 224-227
Botanical Congresses 215-216, 291, 295, 339
Boundaries 5, 72, 89-90, 117, 153, 208-209, 232, 302-305
BRAUN-BLANQUET school 8, 18, 72, 76, 102, 117-122, 137-138, 144, 215, 289-378
BRAY & CURTIS ordination 366
British Tradition 9, 16, 48-60
Bundesanstalt (Zentralstelle) für Vegetationskartierung 292, 352, 356

C

CAJANDER, A. K. 8-9, 14, 17, 83-105, 228, 339
Cards for relevé data 356-357
Caricetum 342-343
Carpinion 364, 369
Centre d'Etudes Phytosociologiques et Ecologiques 354, 356
Centroid 269, 368
Chaining 272
Character-species 18, 211, 215, 222, 290-291, 324-328, 333, 376
Character-species, geographic relations 326-328, 376
Character-taxon 297
Characteristic species combination 291, 295, 330
Chi square tests 261, 264, 362
Chronocoenose, chronocoenon 120, 128-129
Circle of vegetation 186, 334, 350
Circularity 315
Class-concept 3-4, 250
Class and class group 18, 224, 332-

4, 340–344
:ula 124, 129–130
ication
:oaches 3–20
ethod 3–4
s of 3–6
ate 12, 43–45
parative 6
ria 3, 11–18, 76
minance 15, 67–78, 180–185, 2–214, 222, 231, 347, 376
ronmental 11–12, 75, 207
uation 275–276
istic 18, 58–59, 76, 289–344
irchical 18, 73, 77, 122–124,)–131, 171, 178, 259, 263, 5–276, 289–344
ral 4–6, 259, 299–301
erical 10, 17, 175, 180–181,)–283, 361–364, 377
iognomic 7–11, 35–61, 73, 352
ciples 3–6, 249–251
al and synusial 8–10, 16–17, –92, 113–160, 177–178, 348–

essional 8, 15–16, 68, 74
s (see Community-types)
ordination 60, 276–277, 281–, 365, 371
NTS, F. E. 9–10, 16, 51, 67–115, 293
e, classification of 12, 43–44
c gradients 43–45, 49–50, 351
9, 16, 55–56, 67, 74–75, 96, –358
-complex 75, 84, 350
ing of samples and species 258–, 268–270
ient of community 269, 359,

ell 178
line 98
(community-type) 113, 121, –129, 294
opulation 118, 120, 129
uant 178
ve types 74
ed estimation scale 219, 309–

nity
act and concrete 3–6, 127–, 293–295
daries 5, 72, 89–90, 153, 177, –209, 232, 302–305
lex 75, 98–99, 179, 349–350,

epts 293
nuity 5, 9, 49, 170, 179, 209, 300, 305, 365
– patterns (see Patterns)
– principle of continuity 5
– ring 349
– tables 4, 280, 320–332
Community-types (For subordinate communities, see Synusiae, Merocoenose) 3–6
– alliance 18, 215, 224, 291, 332–334, 340–344
– association 18, 215–216, 291, 295, 329–344
– association, American of CLEMENTS 10, 68, 75
– association, British, dominance 10, 15, 54, 58–59
– association, naming 183, 228, 339–343
– association, Russian 9, 180–183, 190
– association, Uppsala (sociation) 206–212
– association, Uppsala theory 207–217, 234
– association-complex 207, 224, 349
– association-group 123, 178, 222, 291
– associes 16, 68
– basal and derivate 376
– biocoenose 13, 120, 138–139, 142, 293
– biome and biome-type 11, 36, 38
– biotic province 14
– biotope-type 12
– chronocoenon 128–129
– circle of vegetation 189, 334
– class (of BRAUN-BLANQUET) 18, 224, 332–333, 340–344
– class group 334
– coenon 113, 121, 128–129, 294
– collective type 74–75
– community ring 349
– concept 3–4
– consociation 54, 123, 217-218, 224
– consortium 143, 337
– definitions 3–4
– division 335
– dominance-type 15, 59, 67–68
– facies 18, 299, 338–339
– federation (see also under Synusiae) 215, 226
– forest site-type 10, 17, 83–105
– form 337
– formation 11, 35–36, 54–61, 68, 73, 207, 215–216, 229, 353
– formation (Russian) 178, 184–185, 188, 190
– formation-series 10, 14, 49–51, 56–57, 60

401

- formation-type 11, 38–42, 55, 60, 73
- habitat-type 12
- merocoenon 122, 128–129, 294
- microcoenon 128–129
- microlandscape-type 12
- nature-complex 8, 12
- noda 10, 18, 294
- order (of BRAUN-BLANQUET) 18, 218, 292, 332–333, 340–344
- ordination of 376
- panformation 215–216, 224
- phratria 175, 187–189
- phytocoenon 294–295
- series (NILSSON) 218, 225–226
- sigmassociation 377
- site-type 10, 17, 83–105
- sociation 7, 9–10, 16, 92, 100, 132, 213–216, 222–223, 229, 233–234, 337
- societies (see Synusiae)
- sociotype (soziotypus) 220
- stratocoenon 128–129, 294
- subassociation 18, 297, 344–346
- subformation 73–74, 215–216
- subtypes 75, 86, 92
- syntaxon 18, 113, 296, 329–388
- taxocoenon 128–129
- twin-formation 117
- variant (see also under Synusiae) 18, 180, 338, 344
- vegetation-development-type 16, 98
- vegetation-girdle 13
- vegetation type (Russian) 171, 185–186, 188
- vicariant 129–130, 331–332, 335–336, 351
- zone 2, 4

Companions (compagnes) 292, 298
Comparative classification 6
Comparative (polar) ordinations 261, 366
Competition 88, 95
Complexes 179, 215, 224, 349–350, 377
Complex-gradient 5
Complex population continuum 300
Composite sample 366
Compositional gradient analysis 366
Compositional trends 366–370
Compound analysis 358
Concrete communities 3–4, 127–129, 293–295
Consociation 54, 123, 218–219, 222
Consocion 132, 214–215, 337
Consortium 143, 337
Constancy 219, 316–317
- rule of 210–212
Constant companions 292–298

Constant species 207–212, 222–223, 298, 318
Contagion 310
Contingency (2 x 2) tables 264
Continuity of communities 9, 49, 170, 179, 209, 300, 305, 365
Convergence of formations 38–42, 189
Correlation of species distributions 17, 264–265, 357–358
Cover-abundance scale 219, 309–310, 319
Coverage 219, 220, 309
Criteria of classification 3, 11–18, 76
Criticisms
- of BRAUN-BLANQUET approach 301, 315, 330, 365, 371
- of Uppsala association theory 208–217
CURTIS, J. T. 11, 299

D

DAGNELIE, P. 319, 324, 358, 366
DAHL, E. 18, 218, 231–232, 302, 317–318, 360–361
Danish phytogeography 8, 229–231, 354
DANSEREAU, p. 41–43, 46, 48, 308
Data cards for relevés 356–357
Deductive approach 205
Dendrite 366
Dendrogram 259–261, 273–274, 364
Density 93
Derivate communities 376
Deserts 39–45
Deviant index 266
Diagnostic species (see also Character-species, Differential species), 54, 59, 89, 215, 222, 230, 289, 296–298, 372
Dicrano-Juniperetum 146–150
Differential-species 18, 215, 222, 297–298
Differentiating species 298, 321
Discontinuities 232, 252–253, 279–280, 300, 305
Distance, Euclidean 267
Distributional similarity of species (see also correlation) 358
Division 335
Divisive approaches 262–269, 272
Diversity (see Species-diversity)
DOMIN scale 309
Dominance classification 5, 67–78, 180–185, 212–214, 222, 231, 347
Dominance-types 5, 59, 67–68
Dominant species 69–72, 85–92, 122–

23, 129, 180–185, 212–214, 376
RIETZ, G. E. 17, 118–124, 206–226, 306, 308, 316
iomorph (growth-form) 187, 189
line 49, 97
ogical groups 113, 358
ogical series 4, 7, 75, 96–103, 71–172
ystem 268–269
ARDS and CAVALLI-SFORZA 56–267
tion gradient 70
nation of outliers 265
ENBERG, H. 18, 127, 136, 299, 00, 312, 319–323, 327, 329, 30, 352, 355, 366, 377
onmental classification 11–12, 75, 07
hic communities 118–121
ytes 16, 37, 102, 121, 132–136, 49
ian phytocoenology 6, 8, 87–91, 31, 228–229
dean distance 267
ation of classifications 277–278
sional definition 4, 249, 253

s 18, 290, 338–339
on 399, 341, 369
ion 122–123, 215
ation 124, 129–130, 178, 215, 24
226–227
ity (see Soil Fertility) 311–312
ty 291, 296, 324–328, 358–9
sh phytosociology 7, 83–105, 228
tic analysis 308–313
tic classification 18, 58–59, 289–4
tic composition 18, 289
tic-sociological approach 292
coenotype 187
ts
ssification 39–45, 83–105, 169–0, 181, 324, 334
munities 48, 53, 227
nations 39–45
ient analysis 50, 70, 96–97
agement 93–95
terns of types 88, 91, 99, 103, 1–172, 336
-types 10, 17, 83–105
cession and retrogresssion 84–86 337

Formation (see under Community-types)
Formion 122, 178, 215
Frequency 219, 316
Frequency-dominants 229
FREY, T. E. A., article 81
FRIES, T. C. E. 7, 206
Fungus communities 142–145, 148

G

GAMS, H. 7, 16, 115–117, 130, 228, 312, 365–366
Gaussian curve 70–71
General character-species 328
Geographic relations 13–14, 35–46, 229, 325–326, 326–328, 330–332, 354
GLEASON, H. A. 5, 10, 150, 293
GOODALL, D. W. 17, 262, 264, 270, 359
– article 247
Gradient analysis
– physiognomic 43–46, 48–51, 56–57, 60
Grassland communities 10, 39–46, 184–185, 307, 370
Group analysis 265
Groups of species (see also Ecological group) 5, 113, 182, 362, 368
Growth forms
– classification by 11
– classification of 36–38, 46–47, 221
– distribution of 5
Growth rates of trees 87, 93–95
Guild 120, 155

H

Habitat-types 12
HALL heterogeneity method 271–272
HANSEN, H. MØLHOM 230–231, 354
Haptophytes 137–140
Heath communities 10, 39–45, 87–89
Heterogeneity method (of HALL) 271–272
Hierarchies 18, 73, 77, 122–124, 129–131, 171, 178, 188, 215, 229, 291, 296, 344
History of classification 6–11, 35–36, 114–124, 290-292, 375
HOLDRIDGE, L. R. 44–46
Holocoene 293
Homogeneity 17, 143, 218, 255, 302, 317–318, 360

403

Holocoene 293
Homogeneity 17, 143, 218, 255, 302, 317–318, 360
Homotoneity 218, 317–318, 360
Horizontal classification 336
HULT, R. 7, 114, 205, 290, 308
HULT-SERNANDER coverage scale 219
Hydrophytes 137–141

I

Icelandic phytosociology 8, 230–231, 345
Importance values 69, 308–310, 319
Indicator species 54, 59, 230, 353–355
Indices
– sample similarity 359–365
– species association 264–265, 358–359
– species correlation 358
Individualistic concept 5, 293, 300
Inductive approach 205
Inflation of syntaxa 334
Information analysis 267–268, 272–273
Intensional definition 249
Intercommunity patterns 43–45, 88, 91, 97, 172, 227, 366–370
Interspecific correlation (see Species association and Species correlation)
Isocenes 368
Isopleths, isodems 367–370

J

Juniper scrub, synusiae of 146–150

K

KATZ (KATS), N. J. 9, 115, 173, 180, 212
Kennart 296
KNAPP, R. 318, 320–321, 328, 331, 336–337, 342
KÜCHLER, A. W. 46

L

Landscape-types 12
LAVRENKO, E. M. 174, 176, 184–187
Layers (see also strata) 307–308
Leading-types 98
Leaf size classes 36, 47, 55
Leitpflanzen 291
Leningrad school 9, 170, 174

Lichens 96, 132–136, 140, 349
Life-forms (RAUNKIAER) 36, 125, 229, 368
Limes convergens and divergens 303–305
Limestone 92, 97, 226, 326
LINNÉ, C. von 204
LIPPMAA, T. 7, 16, 123, 131, 177, 211, 228
Local character-species 328

M

MAAREL, E. van der 127–128, 290, 306, 319, 348, 360, 366–370
– article 287
MACNAUGHTON-SMITH, P. 267
MALMER, N., article 201
Mapping vegetation 41, 46–48, 185, 352
Matrices
– arrangement 221, 258, 280
– correlation 2
– primary (data) 261–262, 313–316
– Q and R 262
– sample similarity 258, 361, 369
Matrix and plexus arrangement techniques 258–259, 361
MEIJER DREES, E. 295, 305, 327–328, 331–332, 336, 338–339, 359
Merocoenose, merocoenon 120, 128–129, 294
Microbiotopes or microhabitats 126, 148, 154–155
Microcoenon 128–129
Microcoenose (microcommunity), 118–120, 126, 128–129, 146–149, 176, 348–349
Microlandscape-types 22
Minimum area 143, 146, 152, 209–210, 228, 305–307, 375
Mires 226–228
Moisture gradient 70, 370
Monothetic methods 262–263, 268, 283
Moscow school 9, 173, 174
Mosaic chart (see also intercommunity pattern) 366
Mosaic community 176–177, 337–338, 351
Multidimensional approaches 180–181, 225, 365, 368

N

Names for syntaxa 339–343
Natural classification 4–8, 229–301, 259

ral units 4-6
re-complexes 8, 12
e 118, 155, 300
ophilous edge communities 367-68
a 10, 18, 294
al analysis 361
enclature 339-343, 376
DHAGEN, R. 211, 216, 220, 231
hern Tradition 7, 102, 203-233
vegian mountain vegetation 231-32
hypothesis 254, 263, 265, 267
erical classification 10, 17, 176, 80-181, 249-283, 361-364, 377
erical syntaxonomy 363-365

communities 227, 305, 307, 334-35, 338, 341, 344-345
RDORFER, E. 295, 299, 303, 30-332, 338, 341
mization 280
r (of BRAUN-BLANQUET) 18, 16, 292, 332-333, 340-344
nation
mmunity-types 368, 378
mparative (polar) 261, 366-370
ncept 365
ysiognomic 49-61, 60
lar (comparative) 261, 366-370
inciple components analysis 10, 368
uctural 369
ighted averages, 355
nation vs. classification 60, 276-77, 281-282, 365, 371
lus 124, 129-130
nismal concept 67, 301
OCI agglomerative technique 270-72

ormation 215-216, 224
ormion 122, 178, 218
limax 350
lel type-series 98
rns of communities 43-45, 88, 1, 97, 172, 227, 366-370
dicity 312
s of BRAUN-BLANQUET approach 301-343, 373, 375
ria 175, 187-189
ognomic classification 7-11, 35-61, 73, 352-354

Physiognomic convergence 38-42
Physiognomic ordination 49-51, 60
Physiognomic Tradition 7, 11, 35-61, 290
Phytocoenon 294-295
Phytocoenose (plant community) 113, 120, 171, 175-177, 294-295
Phytosociology 176
Pine forests 93-94, 97, 183-186, 212
Placor 171, 186, 189
Plankton 138-139
Pleustophytes 137-141
Plexus techniques 88, 91, 171-172, 181
Polar (comparative) ordination 261, 366-370
Polythetic methods 262, 283
Population continuum 300
Population distribution 70-71
Population structure of vegetation 300
POST, H. von 204, 290
Potential natural vegetation 350
Presence 316-317, 321-322
Presence and absence data 357
Presence class 316-317
Primary table (see Matrix, primary) 315-316
Principal components analysis 10, 368
Principle of community continuity 5, 300
Principle of species individuality 5, 293
Principles of classification 3-6, 249-251
Probabilistic classification 272, 280
Profile diagram 52-53, 55
Progression, sociological 291, 298-299
Protocol of a relevé 313

Q

Quadrat samples 218-220, 229
Querco-Betuletum 338, 341
Querco-Fagetea 300, 334-335, 341, 344-345

R

Rainforest 39-45, 48, 53, 55
RAMENSKY, L. G. 5, 9, 172, 179, 209-211, 299
RAUNKIAER, C. 8, 229-231, 317
RAUNKIAER life-forms 36, 123, 229, 368
Rearrangement of tables 221, 320-324, 361-363, 377
Reconnaissance 301
Regional associations 331-332

405

Regional character-taxa 326–328
Relevé 146, 217, 301–302, 312–313, 356–357, 373
Representative area 210
Rhizophytes 137, 141
Rich vs. poor vegetation types 226–227, 347
RUBEL formation-types 40–42
Russian Tradition 9, 100, 169–191

S

Saltmarsh communities 332, 351, 357, 363–364
Samples
– homogeneity 17, 143, 218, 255, 302, 317–318, 360
– size, area 146, 218, 219, 258, 305–307, 375
– composite 366
– location 219
– quadrat 218–220, 229
– selection 218, 301–302
– sets 218
– similarity 18, 359–365
– small square 218–221
Savanna 39–45
Scales
– combined estimation 219, 309–310
– coverage 219, 309–310
– ecological 355
– periodicity 312
– sociability 310–311
– vitality and fertility 311–312
SCHIMPER and FABER formation-types 39–43
SCHMID, E. 13, 292
Schools of ecology and phytosociology 6–20, 76
SCHWICKERATH, M. 298, 331, 337, 349, 366
Seasonal aspects 119–122, 126
Series, ecological 14, 17, 75, 96–103, 171–172
Series (NILSSON) 215, 225–226
SERNANDER, R. 225
S.I.G.M.A. 292, 356
Sigmassociation 377
Sigmatism 292
Significance tests 264–267
Similarity measurements 18, 231–232, 267, 269, 359–365
Site-types 10, 17, 83–105
Site-type classes 86–87
Small square analysis 218–220
SOCHAVA, V. B. 175, 177, 186–189
Sociability 310–311
Sociation 7, 9–10, 16, 92, 100, 132, 213–216, 222–223, 229, 233–234, 337
Society 113–114, 120, 129–130, 157
– classification of, see Synusiae
Sociological group 113, 353–354
Sociological progression 291, 298–299, 369
Socion 122–123, 129–130, 132, 215
Sociotype (soziotypus) 224
Soil fertility 96–97, 226–227, 311–312
Soil moisture 96–97, 370
Southern Tradition 8
Spartina communities 351, 357, 364–365
Species areas 13–14, 230–231, 328
Species-area curves 210–211, 306, 318
Species association 264–265, 358–359
Species correlation 7, 357–358
Species, diagnostic (see Diagnostic species, Character-species, Differential-species)
Species-diversity 96
– alpha 360
– beta 319
Species groups (see also ecological group) 5, 113, 182, 362, 368, 377
Species individuality 5, 293, 300
Species numbers 96
Species weights 319
Spectra 355, 368
Spruce forests 96–97, 171, 277
Standardization of classification 6
Standardization of data 356–357
Steppe 39, 42, 169, 184, 187
Stopping rule 259–262, 265
Storage of relevés 355–357
Strata (see also Synusiae) 307–308
Stratal independence 85, 100, 127, 153, 308
Stratocoenose, stratocoenon 118–120, 128–129, 294
Structural description 44–46, 307–308
Structural-functional ordination 369
Subassociation 18, 297, 336–338
Subformation 73–74, 215–216
Subjectivity in sample choice 301
Subordinate communities (see Synusiae, Society, Merocoenose, Microcoenose)
Subtypes 75, 86, 92
Succession 55–56, 67, 74–75, 84
Successional complex 75, 350
Suffixes 131–132, 183, 339–343
SUKACHEV, V. N. 9, 170–171, 174–176, 180, 293
Swedish phytosociology 7, 17, 100–101, 203–208, 235
Swiss-French School 8, 292

tic table 314, 322, 329, 373
;on 18, 113, 295, 329–334
;on table 329, 373
;onomical phase 329–344, 373,
5
;onomy 296, 329–344
etical phase 313–328, 375
.ae
:oach 9, 16, 113–159, 177–178,
), 223, 228–229, 348–349
cion 122–123, 132, 215
atic 136–142
;icula 124, 129–130
:ept 129–130, 157
;ocion 132, 214–215
ria 125–128
hytic 132–136, 148
rion 122–123, 215
ration (see also under Commu-
/-types) 124, 129–130, 178
ity or specificity of 134, 149–
, 153–158
ion 122, 178, 215
es 117, 131
juniper scrub 146–150
fication of 150–157
lus 124, 129–130
ormion 122, 178, 215
s 121–124, 129–130, 157
n 122–123, 129–130, 132, 215
nd fungal 142–145, 148
ns 7, 16, 85, 101, 128–132,
–150, 178, 211
131–141, 178, 215
nt (see also under Community-
es) 129–130
lar 130–132
ant 129–130
R and PAWLOWSKI fidelity
325

(see also Matrices)
nunity 221, 228, 280, 313–
, 361–362
ngency 264
EY, A. G. 9, 51, 55, 293
enose, taxocoenon 115, 120,
–129, 348
my 296
ial-character-taxa 327
of association (Uppsala school)
–217
phic moisture gradient 70, 370
ns 6–11
rmation 309, 319

Transgressive 309
Transgressive character-taxon 333
TRASS, H. H. 177–178, 182
– article 201
Trennart 298
Tundra 39–45, 176, 230
TÜXEN, R. 8, 120, 143, 149, 292–
293, 306, 318, 333–334, 350,
352, 377
Twin-formations 115
Type-series 98
Typification 38, 60, 73, 85, 300–
301

U

Uniformity 302, 360–361
Union 7, 16, 85, 101, 128–132, 147–
150, 211
Units of vegetation
– See also Community-types
– natural vs. artificial 4–6, 207–217
– Uppsala theory 207–217
Uppsala school (see also DU RIETZ)
7, 17, 204, 206–217, 233

V

Variant 18, 129–130, 180, 338, 344
VASILEVICH, V. I. 178–181
Växtbiologiska Institution 206
Vegetation, definition of 294
Vegetation-development-type 16, 100
Vegetation-girdle 13
Vegetation type (Russian) 171, 185–
188, 190
Vertical classification 336
Vicariant site-types 97
Vicariant units 129–130, 331–332,
335–336, 351
Vitality 311–312

W

WARMING, E. 35, 229, 290
Weighted averages 355
Weighting of factors 256–257
Weighting of species 319
WESTHOFF, V. 121, 140, 294–297,
303, 306, 311–313, 337–339, 353–
355
– article 287
WHITTAKER, R. H., articles 1, 65
Wisconsin comparative ordination 366–
370
Wisconsin school (see also CURTIS,
BRAY & CURTIS) 11

407

Woodlands 39–45
Working Group for Data Processing 356–357, 377

Z

Zentralstelle (Bundesanstalt) für Vegetationskartierung 18, 292, 352, 356
Zonal soils and vegetation 173, 189
Zones 12, 14
Zoocoenoses 120
Zürich-Montpellier school (see also BRAUN-BLANQUET) 8, 390–392

DATE DUE